RADIO
PROPAGATION
HANDBOOK

Dedication

To Pilar and Acendi

No. 1146
$17.95

RADIO PROPAGATION HANDBOOK
BY PETER N. SAVESKIE

TAB BOOKS Inc.
BLUE RIDGE SUMMIT, PA. 17214

FIRST EDITION

FIRST PRINTING—JUNE 1980

Copyright © 1980 by TAB BOOKS Inc.

Printed in the United States of America

Reproduction or publication of the content in any manner, without express permission of the publisher, is prohibited. No liability is assumed with respect to the use of the information herein.

Library of Congress Cataloging in Publication Data

Saveskie, Peter N.
  Radio propagation handbook.

  Includes index.
    1. Ionospheric radio wave propagation.
  2. Radio wave propagation. I. Title.
  QC973.4I6S28   621.384I'I   79-25205
  ISBN 0-8306-9949-X
  ISBN 0-8306-1146-0 pbk.

Preface

Radio propagation, being in its "full glory" an extremely capricious and profound topic, the author over his telecommunications career has often dreamed of a Radio Propagation Handbook something better and more utile than the ordinarily available simplism but, at the same time, on "this side" of the highly scholastic, abstruse and intractable treatises on the subject. It became increasingly obvious that the only manner in which he would ever come into possession of such a work is to write it himself. *Radio Propagation Handbook*, dealing with terrestrial radio propagation in the LF through EHF gamut accentuates method and procedure for maximal use in minimal time while at the same time lending instruction and piquancy of interest by intertwining explanatory and clarifying information. May this book's service to you be commensurate with the author's sense of fulfilment accruing from its writing.

Peter N. Saveskie

Acknowledgement

I would like to thank the following organizations for assistance in providing material for this book:

Bell Laboratories
British Broadcasting Corporation
European Broadcasting Union
International Radio Consultive Committee (CCIR)
Microflect Co., Inc.
National Telecommunications and Information Agency (NTIA)
Radio Research Laboratories, Japan
TAI Incorporated
U.S. Department of Commerce (Chapter 3 maps, charts and tables)

Contents

1 Ground Wave Propagation 9
Surface Wave—Ground Wave Problems

2 Ionospheric Waveguide Mode Propagation 21
Characteristics—Field Strength Examples

3 High Frequency Ionospheric Propagation 25
HF Calculation Procedures—HF Propagation Prediction by Computer

4 Ionospheric Scatter Propagation 144
Rayleigh Fading—Forward-Scatter Problem

5 Microwave and VHF/UHF Propagation 148
Microwave Path Selection (Office Phase)—Microwave Path Selection (Field Phase)

6 Diffraction Propagation 190
General Rules—Operations for Diffraction Problems—Diffraction Problems

7 Tropospheric Forward Scatter Propagation 223
Forward Scatter Path Calculations—The Procedure

8 Millimeter Wave Propagation 251
Refractivity and K-Factor—Windows—Fresnel Clearance—Terrestrial Reflections—Fading—Rain—Polarization—Fog—Snow and Hail—Ducting (Trapping)—Diversity—Millimeter-Wave Path Design

Appendices 263

1. Earth Constants 264
2. dBμV/m to dBm Conversions 266
3. CCIR Noise Maps 269
4. Standard Man-Made Noise and
   Galactic Noise in 1.0-Hz Bandwidth 319
5. dB Powers—Addition and Subtraction Curves 320
6. List of Decibel Terms 323
7. Free Space 326
8. Line-of-Sight 329
9. Structure of Our Ionosphere 331
10. Great Circle Radio Path Calculations 332
11. Lightning 335
12. Diversity Reception and Combining 339
13. Rayleigh Fading 342
14. White Noise 343
15. Viva HF 346
16. Standard Time Zones of the World
    and Their Relationship to UTC or GMT 350
17. Secret Spectrum 352
18. Parabolic World 355
19. Refractivity and K-Factor Data 357
20. Radio Fresnel Zone Radii 420
21. Rain Attenuation 431
22. Fading Distributions 463
23. Outage 466
24. Digital Microwave 468
25. Fade Margin Curves 470
26. Special Protractor to Measure Angles
    on K-Factor Profiles 473
27. System Performance 475
28. Radiogeology 478
29. Antenna Height 480
30. Nature's Combiner 483
31. Aperature-to-Medium Coupling Loss 486
32. Effective Distance 488
33. Neither Snow nor Rain 492

Index 495

Chapter 1
Ground Wave Propagation

At the outset, we should put our frames of reference in order and define the sometimes nebulously understood "ground wave". This ground wave term is not-infrequently interchanged with the "surface wave." The complete ground wave, as shown in Fig. 1-1, is classically comprised of three sub-components:

1. The direct wave through the atmosphere between transmitting and receiving antennas.
2. The terrestrially-reflected wave.
3. The surface wave. The combination of the direct wave and terrestrially-reflected wave constitute what is known as the space wave portion of the total ground-wave.

As might be imagined, however, the real-world situation is complicated by such factors as local terrain features, weather details, and the fact that distinction between radio-frequency energy received via the surface wave and space-wave components of the ground-wave is, unfortunately, not always discernible. Furthermore, to point up the capricious nature of radio propagation, the limit of surface-wave service (as in standard AM broadcasting) may often be set by surface-wave/sky-wave interference which causes fading.

SURFACE WAVE

Nevertheless, good use may be made of this surface wave, necessitating a closer investigation. The surface wave results from the tendency of the wave to propagate along the earth/atmosphere boundary as a result of earth currents induced by the transmitted signal's magnetic field. More precisely, transmitted radio energy impinging upon, and reacting with the earth induces electric charges upon, and corresponding radio-frequency currents into the earth, both governed respectively by the earth's permittivity and conductivity. This action foments surface-wave radio-frequency energy progression (propagation) parallel to the earth, in turn again inducing earth charges and currents.

The wave front of our surface wave is composed, as is any radio-wave front, of mutually orthogonal (perpendicular) lines of magnetic and electric force, one half the total radio-frequency energy residing in each (see Fig.

1-2). The surface-wave front propagates along the earth/atmosphere boundary in a layer whose vertical dimension is approximately one wavelength over land (one wavelength, for example, at 10 kHz = 30,000 meters, while at 10 MHz it is 30 meters) and several wavelengths over water.

This traveling wave front is "tilted forward" at the top, while "dragging along" the earth at the bottom, all the while dissipating radio-frequency energy (attenuation) in our resistive/capacitive earth. That is, since the earth is imperfect, neither a perfect conductor or perfect dielectric, energy is dissipated at the wave front/earth interface (and below with lower and more earth-penetrating frequencies). This lost radio-frequency energy is extracted from the wavefront's upper areas, resulting in a wavefront-to-earth energy flow, constituting a net absorptive attenuation. The attenuation is, of course, in addition to the radio-frequency energy loss resulting from wave-front expansion with distance. The loss is greater over so-called "poor" earth (e.g. dry and rocky) than so-called "good" earth (e.g. rich damp pastoral soil) and least over sea water, our closest practical approximation to an ideal and perfect electrical earth.

While the space wave may contribute substantially to the received radio-frequency field at WRH (within radio horizon) distances, the surface wave is instrumental in propagating radio-frequency energy considerably beyond the horizon. The surface wave is relatively stable (free of fading) when free from other radio-propagation modes (e.g. ionospheric "reflection"). The surface wave is, of necessity, vertically polarized with respect to the earth's surface (antenna's electric field perpendicular to the earth's surface) since if horizontal polarization were employed at low antenna heights (in terms of wavelength), the electric component of the transmitted field would be short circuited by virtue of its position parallel to the earth. Another way of saying this is that the electric field (of the total transmitted electromagnetic field) induced in the earth would cancel out with the electric field in the antenna and thus be virtually non-existent. And, since the magnetic component of a total radio-frequency electromagnetic field cannot exist without the electric component, the entire wave front would virtually collapse and be rendered too weak to be useful.

The ground (surface) wave intensity is related to radiated power, propagation path distance, frequency of operation, and earth constants (conductivity and permittivity). (See Appendix 1, entitled "Earth Constants" for listing and discussion of this subject.) Though the surface wave becomes increasingly unimportant toward the VHF band (30 to 300 MHz) and above, it does figure importantly at lower frequencies. As a general rule, at the present state of the art, its use has most usually been restricted to frequencies below 10 MHz.

The ground (surface) wave curves shown in Figs. 1-3, 1-4, and 1-5 depict the available radio-frequency field strengths in terms of μV/m (microvolts per meter) and corresponding dBμV/m (decibels referenced to 1.0 microvolt per meter), as a function of operating frequency, propagation distance, and earth constants. They also assume 1.0 kilowatt of radio-frequency power is being radiated from a short vertical (equal to, or less than, 0.125 wavelength) antenna installed over a good ground radial system (i.e. a quasi-perfect ground). The curves in Appendix 2, entitled "dBμV/m

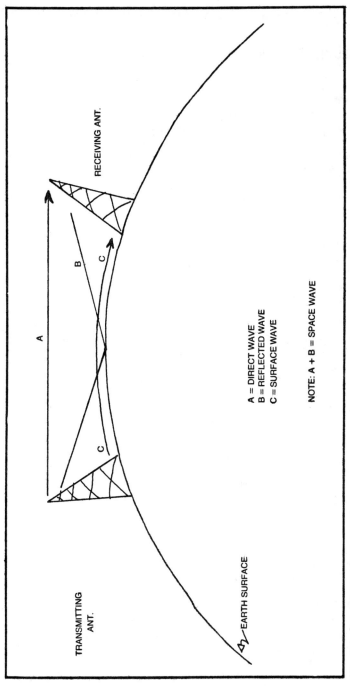

Fig. 1-1. Ground wave components.

to dBm Conversions," may be used as desired to convert from one unit to the other.

GROUND WAVE PROBLEMS

As illustration, let's try several ground (surface) wave examples.

Example 1

We are in the process of designing a standard AM broadcast station. The operating frequency is to be 700 kHz; the desired ground (surface) wave signal is 1.0 mV/m (millivolts per meter) in a rural area 240 kilometers, from the transmitting antenna. The terrain between transmitting and receiving antennas, on our particular radial of concern, is rich agricultural land with low hills. Assumed transmitter power is 1.0 kilowatt into a short vertical antenna (0.125 wavelengths long) installed over a well designed ground. Now, if our required received field strength is not attainable, what can be done?

Solution

Consulting Appendix 1, we find that our rich agricultural land with low hills is characterized as: conductivity = 0.01 mhos per meter, permittivity = 15.0, and that "good" earth has a conductivity of 0.001 to 0.02 mhos per meter with a permittivity of 4.0 to 30.0. Generally, we can then conclude that Fig. 1-4 is the most applicable. In some problems, after choosing the best conductivity and permittivity values for your particular terrain, interpolation between the curves may then be required.

From Fig. 1-4 we find that at our distance of 240.0 kilometers and frequency of 700.0 kHz, the field strength is only 100.0 μV/m, or in other terms, + 40.0 dB above 1.0 μV/m (+ 40.0 dBμV/m). Now, since we require a field strength of 1000.0 μV/m, + 60.0 dBμV/m, we must find some method of making up the 20-dB deficit. What can be done about this? Well, we might increase our transmitter power by 20.0 dB, from 1.0 kilowatt, which is + 30.0 dBw (decibels referred to one watt) or + 60.0 dBm (decibels referred to 1.0 milliwatt), to 100.0 kilowatts (+ 50.0 dBw or + 80.0 dBm). Or, we might increase our antenna gain and transmitter power by 10.0 dB each—or some other combination between the two which results in a power gain of 20 dB. It's a matter of basic economy or other convenience as to how this is ultimately accomplished. As another possible solution, we might consider reducing the distance required for our field strength of 1000.0 μV/m to the shorter distance of 85.0 kilometers Fig. 1-4 shows is obtainable. All these trade-offs are subject to available finances, radio regulations, available real estate (for larger and higher-gain antennas), zoning laws, and other considerations.

Example 2

Let's say that you're faced with a "mixed-path" problem, a path in which the surface wave first traverses a span of terrain having one set of conductivity/permittivity constants and then a terrain segment of another set of such constants. This type of surface-wave propagation problem is not solved by interpolation between two given curves, of Figs. 1-3, 1-4 and 1.5

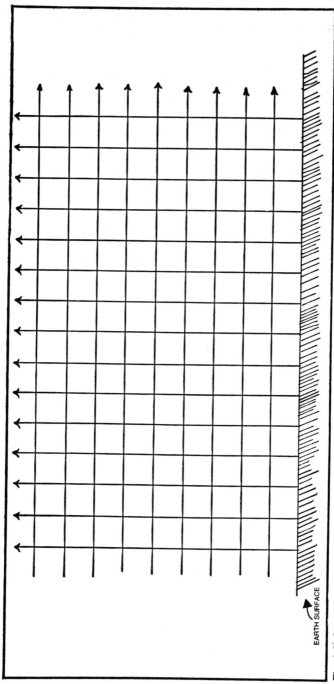

Fig. 1-2. Ground wave wavefront. The vertical force lines are electric, while the horizontal force lines are magnetic. The wavefront is propagating perpendicular to the page.

but must be worked out by path segments, both in the forward and reciprocal directions with the answers averaged. A 1.0-kilowatt transmitter, at a frequency of 1.0 MHz, is fed into a vertical antenna 14.75 dBi (dB gain over isotropic) gain. The wave travels (propagates) over a mixed path, first over 90.0 kilometers of sea water and then 85.0 kilometers of sea ice for a total path length of 175.0 kilometers. What is the field strength at the receiving location?

Solution

From Appendix 1, the conductivity and permittivity characteristics of sea water match with Fig. 1-3 and those of sea ice match with Fig. 1-5. To solve the problem, proceed as follows; from Fig. 1-3, the surface-wave field strength at the sea-water/ sea-ice boundary, after propagating over 90.0 kilometers of sea water is close to + 71.0 dBμV/m. From Fig. 1-5, the field strength at the same boundary after propagating as if this segment of the propagation path were sea ice instead of sea water, is about 29.0 dBμV/m. Therefore the "gain" due to this propagation path segment being over sea water is 71.0 dBμV/m − 29.0 dBμV/m = 42.0 dB.

Now, if the entire path had been over sea ice, Fig. 1-5 shows that the field strength would have been + 15 dBμV/m. Applying our "gain" from the sea water segment of the path, we obtain 15 dBμV/m + 42.0 dB = 57.0 dBμV/m. Record this value.

At this point, we work out the path as above, but in the reciprocal direction. From Fig. 1-5, the field strength at the sea-ice/sea-water boundary, after traveling over 85.0 kilometers of sea ice, is close to + 31.0 dBμV/m. The field strength over the same stretch if it had been over sea water, would be, from Fig. 1-3, about +71.0 dBμV/m. The "loss" then, due to the 85.0 kilometers of sea-ice segment over an equivalent sea-water segment, calculates as +71.0 dBμV/m − 31.0 dBμV/m = 40.0 dB. Had the entire path been over sea water, the resultant surface-wave field strength would be approximately +63.0 dBμ V/m. But, applying the "loss" derived for the sea-ice path portion, we obtain +63.0 dBμ V/m − 40.0 dB +23.0 dBμV/m. Record this value also.

Now, taking our two surface-wave field-strength values and averaging them, we obtain (using the method of BBC England):

$$\frac{+57.0 \text{ dB}\mu\text{V/M} + 23.0 \text{ dB}\mu\text{V/m}}{2}$$

$$= + 40.0 \text{ dB}\mu\text{V/m}$$

At this point we must recall that our antenna has a gain of 14.75 dBi. This means that it has a gain of 14.75 dB over an isotropic antenna (one of unity or 0-dB gain). A short antenna approximately 0.125 wavelengths long has a gain of 1.76 dBi, and if installed vertically over a good system of copper radials the gain increases to 4.75 dBi.

It is interesting to observe, at this point, that the 1.76-dBi (sometimes given as 1.75-dBi) gain is derived by the antenna being short as opposed to an isotropic antenna. An additional 3.0 dB is added because the vertical antenna's power is radiated into a hemisphere (formed by the ground and the atmosphere) instead of spherically. Another way of stating this is that this antenna will lay down, with 1.0 kilowatt of radio-frequency power fed to

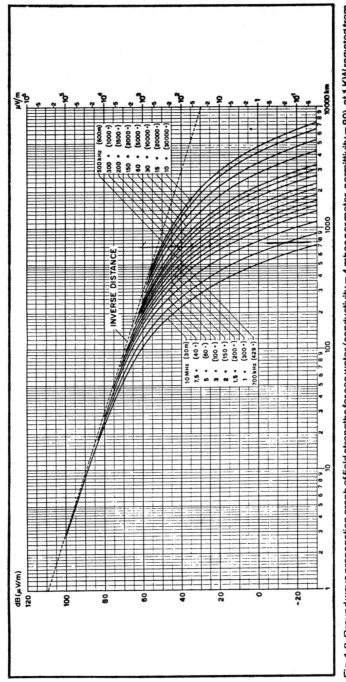

Fig. 1-3. Ground wave propagation graph of field strengths for sea water (conductivity = 4 mhos per meter, permittivity = 80) at 1 KW radiated from a short vertical antenna (courtesy CCIR).

it, an unattenuated field of 300.0 mV/m at a distance of 1.0 kilometer. The unattenuated field strength at the 1.0-kilometer range is found from $300.0 \times \sqrt{P}$ (in rms millivolts per meter) where P is the power fed to antenna in kilowatts.

The term unattenuated field strength means that in this standard ideal antenna, against which practical antennas are assigned figures of merit or gain, there is no radio-frequency energy dissipation into the earth over the 1.0 kilometer distance, only loss due to progressive wave front expansion similar to that in "free space" loss—this standard ideal antenna assumes a perfect earth. Actually, this standard ideal antenna is as fictitious as an isotropic antenna (the isotropic antenna hypothetically radiating or receiving equally in all directions). It does, however, like the isotropic antenna, afford us a standard against which to compare real antennas of the type discussed. As it turns out, our short grounded vertical antenna is the standard in surface wave type of propagation work.

Returning to our problem, recall that the antenna's gain is 14.75 dBi. This represents a 10.0-dB gain over our grounded vertical standard (14.75 dBi − 4.75 dBi = 10.0 dB). Accordingly, we simply add 10.0 dB to our + 40.0 dBμV/m figure to obtain + 40.0 dBμV/m + 10.0 dB = + 50.0 dBμV/m as the surface wave field strength available at the receiver location after propagation over our mixed sea-water/sea-ice path.

We are able to solve any number of related problems by similar means. Let's say, for example, that we want to find the basic transmission loss (loss between two isotropic antennas) on this particular path. First we convert our real transmitting antenna into a hypothetical isotropic source by reducing its gain to 0.0 dBi. This also diminishes our calculated field strength at the receiving location by the same amount, + 50.0 dBμV/m − 14.75 dB = +35.25 dBμV/m. The transmitted radio-frequency power into our hypothetical isotropic source is one kilowatt or + 60.0 dBm. The power delivered by an isotropic receiving antenna from the + 35.25 dBμV/m field is, from Appendix 2B, 41.5 dBm. With + 60.0 dBm of radio-frequency power transmitted, and −41.5 dBm of radio-frequency power received by a hypothetical isotropic antenna, the basic transmission loss simply becomes the difference, + 60.0 dBm−(−41.5 dBm) = 101.5 dB.

Example 3

A certain transmitter lays down a measured (via field-strength meter) surface-wave signal level of + 50.0 dBμV/m at a noisy (man-made noise) receiving location in Washington, D.C. at 0300 hours during the winter. The frequency is 3.0 MHz and our receiving antenna has a gain of 3.85 dBd (gain over a half wave dipole). The receiver's intelligence (post-detection) bandwidth is 2.0 kHz. Solve for the S/N (signal-to-noise) ratio.

Solution

First for convenience, we convert our + 50.0 dBμV/m field-strength signal to a corresponding level received by an isotropic antenna. From Appendix 2B, this would be − 36.0 dBm or −6.0 dBW. We made this conversion now because our noise will later be calculated in dBm.

Remember the graphs of Appendix 2 show dBm values as the power delivered at the output terminals of an isotropic antenna; therefore, we

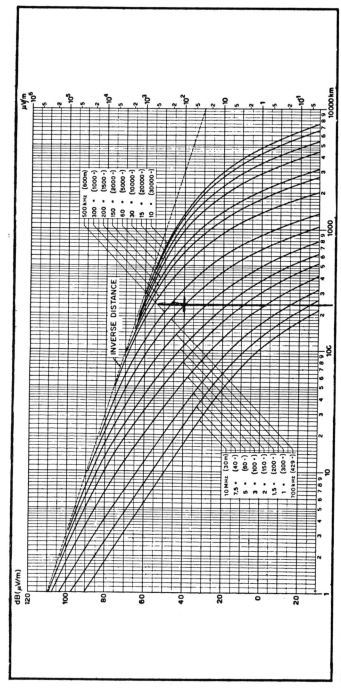

Fig. 1-4. Ground wave propagation graph of field strengths for good earth (conductivity = 0.01 mhos per meter, permittivity = 4) at 1KW radiated from a short vertical antenna (courtesy CCIR).

must add our antenna gain. The antenna in this problem has a gain of 3.85 dBd, and since a half-wave dipole has a gain of 2.15 dB isotropic our antenna's total gain over isotropic is + 2.15 dB + 3.85 dB = 6.0 dB. Therefore, the receiver input signal is (ignoring receiver antenna feedline loss) − 36.0 dBm + 6.0 dB = − 30.0 dBm (or − 60.0 dBw). Record this value.

Now that we have our received signal value into the receiver, we must calculate the noise into the receiver to obtain our S/N ratio. (Note, the noise component of the S/N ratio at 3.0 to 30.0 MHz is generated external to the receiver and is composed of atmospheric noise, man-made noise, and galactic (cosmic) noise. These noises will be individually determined and the values summed). Consulting Appendix 3 we find that information for 0300 hours local time in winter is provided by graphs A3-1A, A3-1B, and A3-1C in the appendix.

From graph A3-1A the atmospheric radio noise (F_{am}) for Washington, D.C. is 70.0 dB above kT_ob at a frequency of 1.0 MHz in a bandwidth of 1.0 Hz. With this 70.0 dB value, we proceed to graph A3-1B to adjust this atmospheric noise for our actual operating frequency (3.0 MHz). To do this, simply enter the graph at 3.0 MHz, move vertically until you intersect the 70.0 dB, and then left to read 57.0 dB.

At this point in our calculations atmospheric noise at 3.0 MHz, in a 1.0 Hz bandwidth, is 57.0 dB above kT_ob. Since kT_ob = 174 dBm we calculate the atmospheric noise as − 174.0 dBm + 57.0 dB = − 117 dBm. Now, we must correct this − 117.0-dBm atmospheric noise value for the uncertainty of our F_{am} because of data spread during measurement and smoothing of the graphs. To do this, simply enter graph A3-1C at 3.0 MHz and move vertically until intersection with curve σF_{am}, moving left to read + 3.8 dB. Correcting our − 117.0 dBm figure, we obtain a final atmospheric noise level of − 117.0 dBm + 3.8 dB = − 113.2 dBm.

Recall at this point that the graph A3-1B shows the man-made noise at a quiet receiving location, and that our receiving site was noisy. Consequently, we use Appendix 4 to determine this man-made noise. If you have, or are able to obtain, a measured man-made noise value, this should be used, but if not, use the standard CCIR values shown in Appendix 4. In our example, we have simply assumed an arbitrary man-made noise level of − 136.5 dBw in a bandwidth of 1.0 Hz. We next account for the galactic (cosmic) noise component of the total noise. To do this, enter graph A3-1B again at 3.0 MHz, using the dashed line to obtain + 40.0 dB above kT_ob. With kT_ob being − 174.0 dBm, we calculate our cosmic noise in a 1.0 Hz bandwidth as − 174.0 dBm + 40.0 dB = − 134.0 dBm. Tabulating the various signal and noise levels provides:

Signal Level	=	− 30.0 dBm
Atmospheric Noise Level	=	− 113.2 dBm
Man-Made Noise Level	=	− 106.5 dBm
Cosmic Noise Level	=	− 134.0 dBm

where all noise contributors are considered in a 1.0-Hz bandwidth

We next correct the three noise values defined in a 1.0-Hz bandwidth to the noise values existing in the receiver's 2.0-kHz bandwidth (or 2,000.0

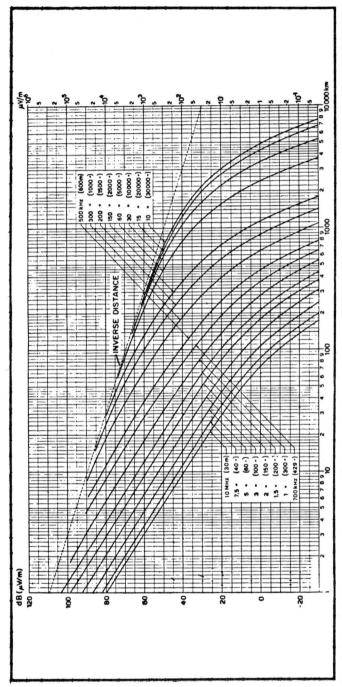

Fig. 1-5. Ground wave propagation graph of field strengths for poor earth (conductivity = 0.001 mhos per meter, permittivity = 4) at 1 KW radiated from a short vertical antenna (courtesy CCIR).

Hz bandwidth). This correction factor is calculated as follows:

$$dB = 10 \log_{10} \frac{2,000.0}{1.0} = 33.0$$

In other words, the noise in a 2.0-kHz bandwidth is considered to be 33.0 dB higher than that in a 1.0-Hz bandwidth. We now add 33.0 dB to each of the three 1.0-Hz bandwidth noises. Our Levels thus become:

Signal Level (no charge) = − 36.0 dBm
Atmospheric Noise Level = − 80.2 dBm
Man Made Noise Level = − 73.5 dBm
Cosmic Noise Level = − 101.0 dBm

Next, we sum up the three noise levels into a single resultant quantity for use in our S/N (signal-to-noise) ratio calculation. The difficult way to add (or subtract) power levels expressed in dB terms is to convert these dB values to power values (watts), add (or subtract) them as desired, and then reconvert to dB terms (dBm, dBW, dBk, etc.). The convenient way of adding or subtracting dB values is by the use of Appendix 5 curves. Simply follow the directions in the appendix. To add three dB values, first add any two, and then with this resultant sum the third one. For addition (or subtraction) of more dB quantities, simply continue the process until the last dB value has been handled. Thus, for our example we have:

− 80.2 dBm + (− 73.5 dBm) + (− 101.0 dBm) = − 72.65 dBm

Remember, you cannot directly add or subtract dB values to obtain sums or differences of power level, 10.0 dBm + 10.0 dBm = 13.0 dBm and not 20.0 dBm. Direct addition and subtraction of dB values may be done only to represent gains and losses respectively, i.e. multiplication and division functions.

Returning to the problem, we may now solve for S/N ratio. This is simply the difference between the signal and noise dBm values entering our receiver.

S/N (dB) = −30.0 dBm − (− 72.65 dBm) = 42.65 dB

Chapter 2
Ionospheric Waveguide Mode Propagation

Below 30.0 kHz, our ionosphere's D-layer (see Appendix 9 for the ionosphere's configuration) represents a relatively well-defined "reflecting" layer. This fact, coupled with low earth absorption or dissipation at these extremely long wavelengths (those between about 8.0 and 30.0 kHz), good terrestrial-reflectivity characteristics, and the D-layer/earth-separation distances, combine to provide for an ionosphere/earth mode of radio propagation. Terrain roughness (hills, mountains, etc.) at these wavelengths is generally less important since the dimensions of the irregularities, compared with the wavelength, tend to be small. Distances of approximately 1000.0 to 8000.0 kilometers may be spanned by this earth/ionosphere wave guide transmission mode. Grounded, vertically polarized antennas are used in terrestrial circuits in preference to horizontally polarized antennas since at the small physical distances (in terms of wavelength) above ground the horizontal electric component would be short circuited in the earth, i.e. the signal from the horizontal antenna and that induced in the earth would cancel.

CHARACTERISTICS

Some of the characteristics of this frequency range and transmission mode are as follow. The useful frequency range is generally limited to about 10.0 kHz to 30.0 kHz, due in no small part to the fact that it is difficult to erect antennas at these tremendously long wavelengths (wavelengths at 3.0 and 30.0 kHz respectively are 100.0 and 10.0 kilometers). High atmospheric noise levels prevail at these lower frequencies (see Appendix 3). Propagation is characterized by relatively low attenuation, with signals quite stable in the time domain. Bandwidth is seriously limited (in the order of 20.0 to 150.0 Hz) due to the high-Q antennas by lumped reactances. Large transmitter powers are generally required for adequate S/N (signal-to-noise) ratios due mainly to the prevalence of high atmospheric noise levels. The low-frequency spectrum (30.0 to 300.0 kHz) is characterized by higher attenuation than that at VLF, lower atmospheric noise levels, and more stable propagation time delays than VLF. The transmission/reception range is approximately 1000.0 to 5,000.0 kilometers, with bandwidths somewhat greater than those at VLF.

The signal strength at distances of approximately 1,000 kilometers to 8,000 kilometers may be calculated by employing the following formula:

$$20 \log_{10} E_1 = 20 \log_{10} E_0 - (Ad \times 10^{-3}) - (10 \log_{10} fd) - 8.0$$

where E_1 = Distant field strength in uV/m, and the term $20 \log_{10} E_1$ is the same quantity in terms of $dB\mu V/m$.

E_0 = The signal field strength laid down by our chosen transmitter/transmitting antenna combination at a distance of 1.0 kilometer.

A = Ionosphere/Earth "wave guide" attenuation in terms of dB/1,000.0 kilometers (decibels per thousand kilometers of path distance).

d = Propagation path distance in kilometers (great circle path distance which may be known or calculated or from appropriate maps or charts or calculated...or calculated, as shown in Appendix 10, or from appropriate...as shown in Appendix 10).

f = operating frequency in kHz.

FIELD STRENGTH EXAMPLES

Let us take an example.

Example

Calculate the field strength from a great circle distance of 6,000.0 kilometers between local time 0800-1200 during the Autumn season in Tampa, Florida. The frequency is 15.0 kHz with 9.0 kilowatts of radio-frequency power fed to a vertically polarized antenna with 0.0-dB gain over a standard short vertical antenna. The Tampa receiver has a noise bandwidth of 5.0 kHz. In addition to the field strength in Tampa, what S/N (signal-to-noise) ratio would the receiver antenna see if in a quiet rural receiving site?

The Solution

First, we determine E_0, the field strength laid down by our transmitting antenna at a distance of 1.0 kilometer. One kilowatt of radio-frequency power into a standard short vertical antenna will lay down a field strength of 300,000.0 NV/m at a distance of 1.0 kilometer. Our actual antenna will provide the same field strength since the antenna has a gain of 1.0 (0.0 dB) relative to the standard. However, we have a power of 9.0 kilowatts, instead of 1.0 kilowatt, feeding into the actual antenna. Therefore, since the field strength at a distance of 1.0 kilometer is proportional to the square root of the power (P), our 1.0 kilometer signal E_0 would be 300,000.0 × $\sqrt{9.0}$ = 900,000.0 $\mu V/m$.

The next step is to determine the value of A. Consult the graph in Fig. 2-1. Entering this graph without operating frequency of 15.0 kHz, we find that our value of A (attenuation per 1,000.0 kilometers), is approximately 1.7 dB. Now, substituting these values in our formula, we obtain:

$$20 \log_{10} E_1 = 20 \log_{10} 900,000.0 - (1.7 \times 6,000.0 \times 10^{-3}) -$$
$$(10 \log_{10} 15 \times 6000.0) - 8.0$$
$$= 119.08 - 10.2 - 49.54 - 8.0$$
$$= +51.34 \ dB\mu V/m$$

Our signal field strength at the receiving site location in Tampa, Florida is 51.34 $dB\mu V/m$. Thus, we have met the first part of the problem and we

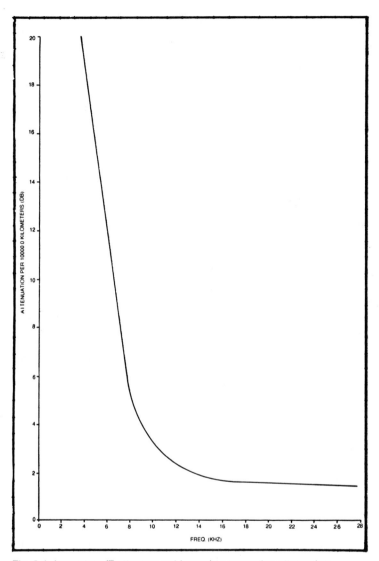

Fig. 2-1. Ionosphere/Earth wave-guide mode propagation attenuation.

should record this value. We also have the signal (S) value of our S/N (signal to noise) ratio.

Next, we need to derive the required S/N ratio at the Tampa Florida receiving site. Recall that at these frequencies, noise external to the receiver (atmospheric, galactic or cosmic, and man-made noise) is the controlling factor. That is, external noise is the combined noise against which the signal intelligence must compete in the S/N ratio. In our present case, it turns out that this noise is mainly atmospheric, generated by a multitude of lightning strokes (see Appendix 11). This noise is then propagated to the receiver location mechanisms similar to that which propagates our signal. To compute our receiver site noise level we can use Appendix 3 (CCIR noise data). Each graph set represents world wide noise distributions presented seasonally and in diurnal local time blocks. We find from graph A3-21A in Appendix 3 approximately 38.0 dB above kT_ob in a bandwidth of 1.0 Hz. Using this 38.0-dB noise value and graph A3-21B we enter at 1.0 MHz, and then proceed to the 38.0 dB level. At this point we proceed along the 38.0 dB level curve to our 15.0 kHz frequency. We can now read our noise value at 15.0 kHz as F_{am} n = 147.0 dB above kT_ob. With $kT_ob = -174.0$ dBm, our noise value thus far is $147.0 - 174.0 = -27$ dBm. Correcting this value for measuring and plotting uncertainties, we use graph A3-21C; entering this graph at our operating frequency of 15 kHz, and proceeding vertically to the σF_{am} curve, we find that our uncertainty correction factor is approximately 4.2 dB. Applying this correction factor to the noise level of -27 dBm, we obtain $-27.0 + 4.2 = -22.8$ dBm. Recalling at this point that the noise data are given in terms of 1.0 Hz bandwidth, we must correct the noise accepted by a receiver having a bandwidth of 5.0 kHz. This decibel bandwidth correction factor is $10 \log_{10} 5,000.0 = 37.0$ dB. Our noise value then is:

$$-22.8 \text{ dBm} + 37.0 \text{ dB} = +14.2 \text{ dBm}.$$

It should be noted at this point that we have considered only atmospheric noise, and have ignored noises of the man-made and galactic variety. The reason for this is that these latter two noise intensities appear on graph A3-21B to be some 45.0 dB below the atmospheric noise. With this 45.0–dB difference, the noise contributed by man-made and galactic noise would be insignificant.

Now, to derive a meaningful S/N ratio, we must have our signal and noise values in the same units. Recall that our signal strength is in terms of $dB\mu V/m$. Since our noise is in terms of dBm, we must either convert the signal to dBm or the noise to $dB\mu V/m$. Let's do the latter since this might best express our desired S/N ratio, i.e. signal versus noise densities incident upon the receiving area. Consulting Appendix 2, we see from graph A3-1A 2A that the noise value of +14.2 dBm converts to +54.5 $dB\mu V/m$.

Since our signal strength is +51.34 $dB\mu V/m$ and our noise value is +64.5 $dB\mu V/m$, our S/N ratio is $+51.34 - 54.5 = -3.16$ dB. Whether this negative S/N ratio can serve a useful purpose or not, would depend upon the application, the particular equipment involved, detection methods, signal-enhancement techniques, and so on. The required S/N ratio can be determined for your particular equipment specifications. Should, for example, a 0.0 dB S/N ratio be required, the 3.16–dB signal deficit could be made up by increasing the transmitter power by this same amount and/or increasing the gain of the transmitting (or receiving) antenna.

Chapter 3
High Frequency
Ionospheric Propagation

High-frequency (3.0 to 30.0 MHz) skywave propagation may be utilized for circuit up to half way around the world, and, in fact, totally around. The traffic capacity of this transmission mode is relatively thin-line, generally on the order of four voice communications channels of ISB (Independent Side Band). Fading, in its worst case, is generally considered to be Rayleigh distributed (see Appendix 13 entitled "Rayleigh Fading"). Transmitting and receiving antennas and power output in this propagation mode are relatively modest. The high-frequency spectrum, scheduled for relegation to a low order of use due to the advent of satellites, has never been more occupied (see Appendix 15 entitled "Viva HF").

High-frequency ionospheric sky wave propagation is possible because radio frequency energy is transmitted from the earth surface (or aircraft) to the ionospheric layers and reflected (actually refracted) back to earth. The various ionized layers (mainly the E and F layers) are important in returning these waves to earth. The ionosphere (mainly the D-layer) also absorbs radio frequency energy from the wave in its transit to and from the E and F layers.

In order to perform the manual HF propagation calculations, it will be necessary to procure the following three volumes of graphs.

1. *Ionospheric Predictions*, Vol. 2, Maximum Usable Frequencies MUF (Zero) F2, MUF (4000) F2, MUF (2000) E for a period of minimum solar activity, $R_{12} = 10$, OT/TRER-13.

2. *Ionospheric Predictions*, Vol. 3, Maximum Usable Frequencies MUF (Zero) F2, MUF (4000) F2, MUF (2000) E for a maximum solar activity period of an average solar cycle, $R_{12} = 110$, OT/TRER-13.

3. *Ionospheric Predictions*, Vol. 4, Maximum Usable Frequencies MUF (Zero) F2, MUF (4000) F2, MUF (2000) E for a maximum solar activity period of an unusually high solar cycle, $R_{12} = 160$, OT/TRER-13.

These volumes are available from: National Technical Information Services (NTIS), Order Processing Branch, 5285 Royal Road, Springfield, Virginia 22161. Prepaid prices are $14.00 for each volume.

HF CALCULATION PROCEDURES

There are two current calculation procedures for an HF (3.0 to 30.0 MHz) ionospheric propagation path. They cover the *short path* (4,000.0 kilometers or less) and *long path* (over 4,000.0 kilometers). Distances are along the great circle path. The basic reason for using two sets of calculations is one of computational convenience. The average height of the F2 ionospheric layer is approximately 320.0 kilometers. This earth/ionosphere relationship permits maximum one-hop (earth-ionosphere-earth) distance of close to 4,000.0 kilometers. Using this "long-distance" F-layer (sometimes termed the most important of the layers), it can be readily appreciated that to cover HF path distance greater than 4,000.0 kilometers would normally require more than one hop (Pedersen ray, chordal hops, and the like excepted).

Let's take first things first, and examine the procedures which will probably be as enlightening, from a practical point of view, as alot of explanatory text. As we proceed in our step-by-step procedures, we'll keep track of the results of each step. The first procedure to be covered will be for great circle paths distances of 4,000.0 kilometers or less. The worksheet will be shown in Fig. 3-1. Don't be dismayed at the seemingly endless number of steps. There is NO shorter manual method as good. After you complete the worksheet by following the steps in the problem, you will find it very educational and quite satisfying.

HF Short Path Manual Procedure

1. Enter the names and/or geographic coordinates of the terminals at the top of the worksheet (see Fig. 3-1). Also enter the geographic latitude of the path midpoint, M. In our short-path example, the end terminals, A and B, are Denver, Colorado and Philadelphia, Pennsylvania, respectively. From a map or other source, obtain their geographic coordinates (an accuracy of 1.0° is sufficient in this type of HF propagation work). Philadelphia's coordinates are 40° N/ 75° W while those of Denver are 39.5° N/ 105° W. Enter this data at the top of the worksheet as shown.

2. Overlay a sheet of transparent paper (onion skin) on the world map of Fig. 3-2. (Note, I prefer a thin piece of transparent plastic approximately ⅛-inch thick, 7.0-inches high, and about 20.0-inches wide.)

3. On the transparent material, trace the equator line and the 0° and 180° meridians. Label the equator, meridians, and A and B terminals. (See Fig. 3-3). (If you choose to use the plastic sheet, it is best to permanently scribe the equator and meridans, and make subsequent markings, unique to a given problem, with soft grease pencil. In this manner, such markings may simply be erased with a cloth at the end of the calculations.)

4. Transfer the transparency to the great-circle chart shown in Fig. 3-4. In this figure the solid lines represent great circles (on a flat surface) and the numbered dot-dash lines indicate distance in thousands of kilometers. The distance between dot-dashed and dotted lines is 500.0 kilometers. Line up the transparency and great circle chart equators. Now, maintaining the equator alignment, slide the transparency horizontally until the HF terminal locations, A and B, lie on the same great-circle (solid) line or both terminals lie the same proportionate distance between two solid line great circles. Trace this HF great-circle path on the transparency. Read the path

TERMINAL A DENVER, COLORADO-39½° N 105° W
TERMINAL B PHILADELPHIA, PENNSYLVANIA-40° N 75° W
PATH DISTANCE (KILOMETERS) 2500
PATH MIDPOINT "M" (LAND OR WATER) LAND
PATH MIDPOINT "M" GEOGRAPHIC LATITUDE 41° N

MONTH AND YEAR JUNE-1976
RSSN (ZSSN) (R12) 10
IF PATH LONGER THAN 2000 KM, PATH HALF DISTANCE KM 1250
PATH MIDPOINT GEOMAGNETIC LATITUDE 50° N

GYRO FREQUENCY (MHZ) 1.51
E-LAYER HEIGHT, KILOMETERS 110.0
ANTENNA RADIATION ANGLE (DEGREES) FOR E-LAYER 6°
SYSTEM LOSS ABOVE QUASI-MINIMUM (dB) 9.0

WORKSHEET FOR H.F. IONOSPHERIC PROPAGATION PATH CALCULATION (SHORT PATH-4,000.0 KILOMETERS OR LESS).

(1)	(2)	(3)	(4)	(5)	(6)	(7)	(8)	(9)	(10)	(11)	(12)	(13)	(14)	(15)	(16)	(17)	(18)	(19)	(20)	(21)	(22)	(23)	(24)	(25)	(26)
UT, GMT, OR ZULU TIME	TERMINAL "A" LOCAL TIME (STANDARD TIME)	TERMINAL "B" LOCAL TIME (STANDARD TIME)	PATH MIDPOINT "M" LOCAL TIME (STANDARD TIME)	PATH MIDPOINT "M" MUF (ZERO) F2 (MHZ)	PATH MIDPOINT "M" MUF (4000) F2 (MHZ)	PATH MIDPOINT "M" MUF (2000) E (MHZ)	F-2 LAYER HEIGHT (KILOMETERS)	SUN'S ZENITH ANGLE (DEGREES)	IONOSPHERIC ABSORPTION INDEX (I)	ANTENNA RADIATION ANGLE (DEGREES) FOR F2 LAYER	PATH F2 MUF (MHZ)	PATH F2 FOT (MHZ)	PATH E MUF/FOT (MHZ)	PATH LUHF (MHZ)	EFFECTIVE FREQUENCY F2 (MHZ) F2	EFFECTIVE FREQUENCY E (MHZ) E	IONOSPHERIC ABSORPTION FOR F2 PROPAGATION (dB)	IONOSPHERIC ABSORPTION FOR E PROPAGATION (dB)	TERRESTRIAL REFLECTION LOSS FOR F2 PROPAGATION (dB)	TERRESTRIAL REFLECTION LOSS FOR E PROPAGATION (dB)	RAY PATH DISTANCE LOSS (F2) dB.	RAY PATH DISTANCE LOSS (E) dB.	QUASI-MINIMUM SYSTEM LOSS F2 (dB)	QUASI-MINIMUM SYSTEM LOSS E (dB)	SYSTEM LOSS, MONTHLY MEDIAN OF HOURLY MEDIANS F2 (dB)
00	17	19	18	5.3	21.8	10.0	300	78	0.28	7.0	17.3	14.7	8.5	7.5	16.21	10.01	3.2	18.0	0.0	2.0	124.0	120.0	127.0	140.0	136.2
02	19	21	20	4.9	19.8	5.5	270	96	0.05	5.0	16.0	13.6	4.6	4.1	15.11	6.11	1.0	8.0	0.0	2.0	124.0	114.0	125.0	124.0	134.0
04	21	23	22	3.8	15.2	3.1	280	*	0.00	6.0	12.2	10.4	2.6	2.3	11.91	4.11	0.0	0.0	0.0	2.0	122.0	109.0	122.0	111.0	131.0
06	23	01	00	3.0	12.0	3.0	285	*	0.00	6.5	9.6	8.2	2.3	2.25	9.71	3.81	0.0	0.0	0.0	2.0	120.0	108.0	120.0	110.0	129.0
08	01	03	02	3.0	10.0	3.5	285	*	0.00	6.5	8.1	6.9	2.9	2.62	8.41	4.41	0.0	0.0	0.0	2.0	119.0	110.0	119.0	112.0	128.0
10	03	05	04	3.0	12.0	6.0	285	94	0.07	6.5	9.6	8.2	4.9	4.50	9.71	6.41	2.3	11.0	0.0	2.0	120.0	115.0	122.3	128.0	131.3
12	05	07	06	3.8	15.2	12.0	285	72	0.36	6.5	12.2	10.4	10.0	9.0	11.91	11.51	7.6	16.6	0.0	2.0	122.0	121.0	129.6	139.6	138.6
14	07	09	08	4.6	18.0	15.0	290	51	0.66	10.0	14.5	12.3	12.0	11.3	13.81	13.51	8.8	22.4	0.0	2.0	123.0	122.0	131.8	146.4	140.8
16	09	11	10	4.9	18.0	16.0	360	28	0.91	12.0	14.5	12.3	13.0	12.0	13.81	14.51	10.9	27.0	0.0	2.5	123.0	123.0	133.9	152.5	142.9
18	11	13	12	4.9	18.0	16.0	400	17	0.99	12.0	14.5	12.3	13.0	12.0	13.81	14.51	12.0	29.0	0.0	2.5	123.0	123.0	135.0	154.5	144.0
20	13	15	14	5.0	18.0	16.0	400	34	0.85	10.0	14.6	12.4	13.0	12.0	13.81	14.51	12.6	25.0	0.0	2.5	123.0	123.0	135.6	150.5	144.6
22	15	17	16	5.0	17.7	15.0	375	55	0.61	10.0	14.4	12.3	12.0	11.3	13.86	13.51	8.0	20.6	0.0	2.0	123.0	122.0	131.0	144.6	140.0
24	17	19	18	5.3	21.8	10.0	300	78	0.28	7.0	17.3	14.7	8.5	7.5	16.21	10.01	3.2	18.0	0.0	2.0	124.0	120.0	127.2	140.0	136.2

Fig. 3-1. Worksheet for HF ionospheric propagation path calculations (short path).

TRANSMITTING ANTENNA GAIN (dB) 10.0
RECEIVING ANTENNA GAIN (dB) 10.0
TRANSMITTER TENTATIVE POWER (KW. AND dBW) 1.0 KW. or +30.0 dBW
DIVERSITY RECEPTION S/N IMPROVEMENT (dB) +3.0 (DUAL DIVERSITY) (MAXIMAL RATIO COMBINING)

TYPE OF SERVICE FACSIMILE (SSB-SUBCARRIER-2800 HZ B.W.)
REQUIRED MEDIAN S/N RATIO (dB) (NOISE IN 1.0 HZ B.W.) 55.0
REQUIRED RELIABILITY 99%

(27)	(28)	(29)	(30)	(31)	(32)	(33)	(34)	(35)	(36)	(37)	(38)	(39)	(40)	(41)	(42)	(43)	(44)	(45)
SYSTEM LOSS, MONTHLY MEDIAN OF HOURLY MEDIANS E (dB)	COLUMN 26 CORRECTED FOR COMBINED TRANSMITTER AND RECEIVER ANTENNA GAINS (dB)	COLUMN 27 CORRECTED FOR COMBINED TRANSMITTER AND RECEIVER ANTENNA GAINS (dB)	TENTATIVE RECEIVER INPUT SIGNAL F2 TERMINAL A OR B (dBW)	TENTATIVE RECEIVER INPUT SIGNAL E TERMINAL A OR B (dBW)	TERMINAL A ATMOSPHERIC NOISE (dBW) FOR PATH F2 FOT (MHZ) FREQUENCY	TERMINAL A ATMOSPHERIC NOISE (dBW) FOR PATH E MUF (FOT) MHZ FREQUENCY	TERMINAL A GALACTIC NOISE (dBW) FOR PATH F2 FOT (MHZ) FREQUENCY	TERMINAL A GALACTIC NOISE (dBW) FOR PATH E MUF (FOT) MHZ FREQUENCY	TERMINAL A MAN-MADE NOISE (dBW) FOR PATH F2 (FOT) MHZ FREQUENCY	TERMINAL A MAN-MADE NOISE (dBW) FOR PATH E MUF (FOT) MHZ FREQUENCY	TERMINAL A TOTAL NOISE (dBW) FOR PATH E MUF (FOT) MHZ FREQUENCY	TERMINAL B ATMOSPHERIC NOISE (dBW) FOR PATH E MUF(FOT) MHz FREQUENCY	TERMINAL B ATMOSPHERIC NOISE (dBW) FOR PATH F2 FOT (MHz) FREQUENCY	TERMINAL B ATMOSPHERIC NOISE (dBW) FOR PATH E MUF (FOT) MHZ FREQUENCY	TERMINAL B GALACTIC NOISE (dBW) FOR PATH F2 FOT (MHZ) FREQUENCY	TERMINAL B GALACTIC NOISE (dBW) FOR PATH E MUF (FOT) MHZ FREQUENCY	TERMINAL B MAN-MADE NOISE (dBW) FOR PATH F2 FOT (MHZ) FREQUENCY	TERMINAL B MAN-MADE NOISE (dBW) FOR PATH E MUF (FOT) MHZ FREQUENCY
149.0	116.2	129.0	−86.2	−99.0	−153.8	−149.7	−179.0	−172.0	−150.2	−136.5	−148.5	−136.3	−155.8	−151.8	−179.0	−173.0	−150.2	−136.5
133.0	114.0	113.0	−84.0	−83.0	−153.0	−143.7	−177.0	−167.0	−150.2	−136.5	−148.2	−134.7	−165.7	−137.8	−166.0	−166.0	−150.2	−136.5
120.0	111.0	100.0	−81.0	−70.0	−153.1	−124.4	−175.0	−161.0	−150.2	−136.5	−148.4	−124.2	−155.1	−131.3	−160.0	−160.0	−150.2	−136.5
119.0	109.0	99.0	−79.0	−69.0	−150.7	−122.3	−172.0	−160.0	−136.5	−136.5	−136.5	−122.1	−145.9	−120.0	−159.0	−159.0	−136.5	−150.2
121.0	108.0	101.0	−78.0	−71.0	−141.0	−126.2	−171.0	−162.0	−136.5	−136.5	−135.2	−125.8	−141.8	−123.2	−162.0	−162.0	−136.5	−150.2
137.0	111.3	117.0	−81.3	−87.0	−146.9	−134.6	−172.0	−169.0	−136.5	−136.5	−136.5	−132.4	−158.0	−153.6	−167.0	−167.0	−136.5	−150.2
148.6	118.6	128.6	−88.6	−98.6	−161.8	−160.8	−175.0	−175.0	−150.2	−136.5	−150.0	−136.5	−162.8	−161.8	−175.0	−175.0	−150.2	−150.2
155.4	120.8	135.4	−90.8	−105.4	−165.6	−166.6	−177.0	−176.0	−150.2	−150.2	−150.0	−150.2	−168.8	−168.8	−176.0	−176.0	−150.2	−150.2
161.6	122.9	141.6	−92.9	−111.6	−167.8	−167.8	−177.0	−177.0	−150.2	−150.2	−150.0	−150.2	−168.8	−168.8	−177.0	−177.0	−150.2	−150.2
163.5	124.0	143.5	−94.0	−113.5	−167.8	−167.8	−177.0	−177.0	−150.2	−150.2	−150.0	−150.2	−164.7	−163.8	−177.0	−177.0	−150.2	−150.2
159.5	124.6	139.5	−94.6	−109.5	−163.8	−164.0	−177.0	−177.0	−150.2	−150.2	−150.0	−150.2	−164.7	−163.8	−177.0	−177.0	−150.2	−150.2
153.6	120.0	133.6	−90.0	−103.6	−163.8	−164.0	−177.0	−177.0	−150.2	−150.2	−150.0	−150.2	−155.6	−156.3	−177.0	−177.0	−150.2	−150.2
149.0	116.2	129.0	−86.2	−99.0	−153.8	−149.7	−179.0	−172.0	−150.2	−136.5	−148.5	−136.3	−155.8	−151.8	−173.0	−173.0	−150.2	−136.5

#	Parameter												
(46)	TERMINAL B TOTAL NOISE (dBW) FOR PATH F2 FOT (MHZ)	-149.0	-150.0	-149.0	-136.0	-135.3	-136.5	-149.2	-150.2	-150.2	-150.0	-149.0	-149.0
(47)	TERMINAL B TOTAL NOISE (dBW) FOR PATH E MUF (FOT) MHZ FREQUENCY.	-136.3	-133.9	-130.0	-119.9	-123.0	-136.5	-149.9	-150.2	-150.2	-150.0	-149.3	-136.3
(48)	TENTATIVE TERMINAL A S/N RATIO (dB) FOR F2 PROPAGATION (MEDIAN, 50%)	62.3	64.2	67.4	57.5	57.2	55.2	61.4	59.2	57.1	56.0	55.4	60.0	62.3
(49)	TENTATIVE TERMINAL A S/N RATIO (dB) FOR F2 PROPAGATION (MEDIAN, 50%)	37.3	51.7	54.2	53.1	54.8	45.5	37.9	44.8	38.6	36.7	40.7	46.6	37.3
(50)	TENTATIVE TERMINAL B S/N RATIO (dB) FOR F2 PROPAGATION (MEDIAN, 50%)	62.8	66.0	68.0	57.0	57.3	55.2	60.6	59.4	57.3	56.0	55.4	59.0	62.8
(51)	TENTATIVE TERMINAL B S/N RATIO (dB) FOR E PROPAGATION (MEDIAN, 50%)	37.3	50.9	60.0	50.9	52.0	49.5	51.3	44.8	38.6	36.5	40.5	45.7	37.3
(52)	COLUMN-48 S/N (dB) VALUES CORRECTED FOR DIVERSITY (F2)	65.3	67.2	70.4	60.5	60.2	58.2	64.4	62.2	60.1	59.0	58.4	63.0	65.3
(53)	COLUMN-49 S/N (dB) VALUES CORRECTED FOR DIVERSITY (E)	40.3	54.7	57.2	56.1	57.8	48.5	40.9	47.8	41.6	39.7	43.7	49.6	40.3
(54)	COLUMN-50 S/N (dB) VALUES CORRECTED FOR DIVERSITY (F2)	65.8	69.0	71.0	60.0	60.3	58.2	63.6	62.4	60.3	59.0	58.4	62.0	65.8
(55)	COLUMN-51 S/N (dB) VALUES CORRECTED FOR DIVERSITY (E)	40.3	53.9	63.0	53.9	55.0	52.5	54.3	47.8	41.6	39.5	43.5	48.7	40.3
(56)	S/N (dB) RATIO REQUIRED FOR 99% RELIABILITY FOR F2 PROPAGATION	75.5	76.0	76.5	77.0	77.5	77.0	76.5	68.0	68.0	68.0	68.0	68.0	75.5
(57)	S/N RATIO (dB) REQUIRED FOR 99% RELIABILITY FOR E PROPAGATION	77.0	78.0	81.0	81.0	81.0	78.0	76.0	68.0	68.0	68.0	68.0	68.0	77.0
(58)	TERMINAL A S/N (dB) RATIO DEFICIT FOR F2 PROPAGATION FOR 99% RELIABILITY.	10.2	8.8	6.1	16.5	17.3	18.8	12.1	5.8	7.9	9.0	9.6	5.0	10.2
(59)	TERMINAL A S/N (dB) RATIO DEFICIT FOR E PROPAGATION FOR 99% RELIABILITY.	36.7	23.3	23.8	24.9	23.2	29.5	35.1	20.2	26.4	28.3	24.3	18.4	36.7
(60)	TERMINAL B S/N (dB) RATIO DEFICIT FOR F2 PROPAGATION FOR 99% RELIABILITY.	9.7	7.0	5.5	17.0	17.2	17.8	12.9	5.6	7.7	9.0	9.6	6.0	9.7
(61)	TERMINAL B S/N (dB) RATIO DEFICIT FOR E PROPAGATION FOR 99% RELIABILITY.	36.7	24.1	18.0	27.1	26.0	25.5	21.7	20.2	26.4	28.5	24.5	19.3	36.7
(62)	NEW TERMINAL A S/N (dB) RATIO DEFICIT FOR F2 PROPAGATION FOR 99% RELIABILITY	0.2	-1.2	-3.9	6.5	7.3	8.8	2.1	-4.2	-2.1	-1.0	-0.4	-5.0	0.2
(63)	NEW TERMINAL B S/N (dB) RATIO DEFICIT FOR F2 PROPAGATION FOR 99% RELIABILITY	-0.3	-3.0	-4.5	7.0	7.2	8.8	2.9	-4.4	-2.3	-1.0	-0.4	-4.0	-0.3
(64)	SERVICE RELIABILITY %, TERMINAL B TO TERMINAL A.	98.9	99+	99+	95.0	94.5	93.5	98.3	99+	99+	99+	99+	99+	98.9
(65)	SERVICE RELIABILITY %, TERMINAL A TO TERMINAL B	99+	99+	99+	95.5	94.5	93.0	98.0	99+	99+	99+	98.5	99+	99+

Fig. 3-1. Worksheet for HF ionospheric propagation path calculations (short path).

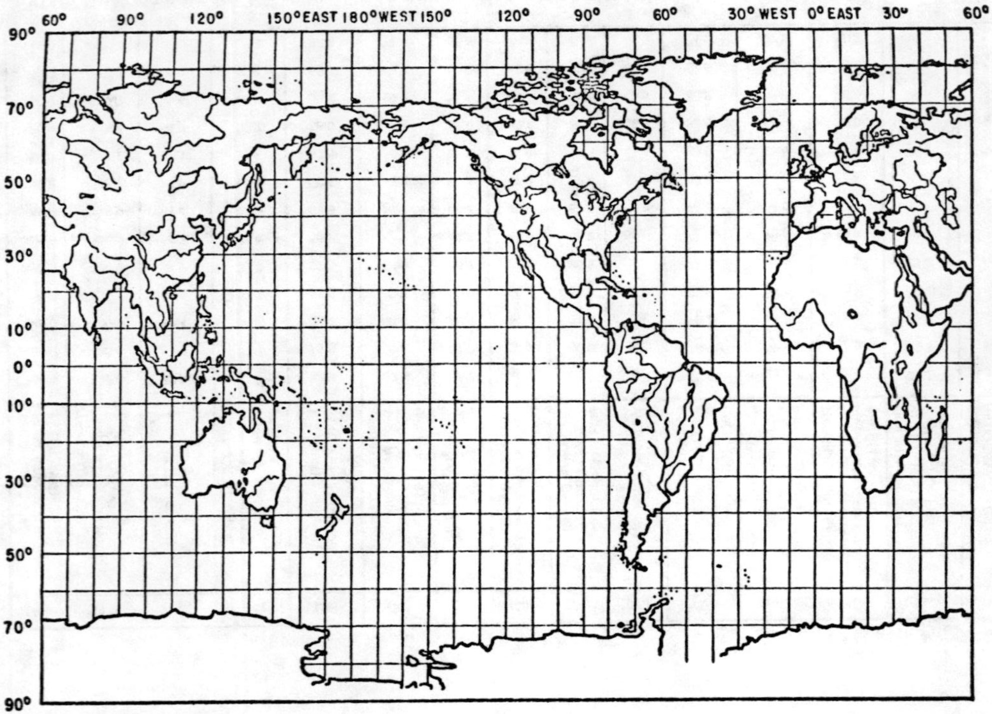

Fig. 3-2. World map for use with the overlays in HF path calculations (courtesy U.S. Department of Commerce).

distance using the 2,500.0 kilometers. Enter this information at the top of the worksheet. Mark the midpath point between A and M. (The maximum distance for one-hop E-layer type of propagation is about 2,000.0 kilometers. Therefore, this Denver/Philadelphia path will require two hops for E-layer propagation, since the great-circle distance is 2,5000.0 kilometers.)

5. Determine from the world map (Fig. 3-2) with the overlaid transparency (Fig. 3-3) the nature (land or water) of the path midpoint, M. In our example, it is land. Indicate this at the top of the worksheet as shown. Also, indicate at the top of the worksheet the geographic latitude of this path midpoint, M.

6. Enter month, year, and RSSN (running smoothed sunspot number) for this calculation at the top of the worksheet. Let's assume for our example that we are calculating for June 1976, and that the RSSN is about 10.0 (typical of the low part of the sunspot cycle). (This entire HF calculation, as they are all, is made for a given month and year in accordance with the changing RSSN throughout the sunspot cycle.) This sunspot number information is available from several sources, such as Dr. Waldmeier at the solar observatory in Switzerland, from George Jacobs' HF propagation

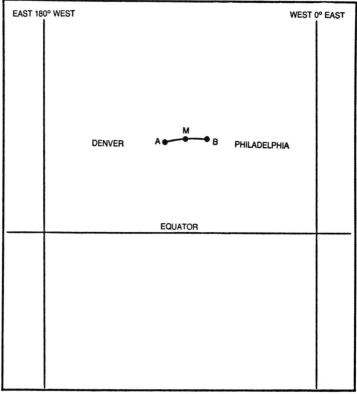

Fig. 3-3. Overlay for use in the HF path calculations.

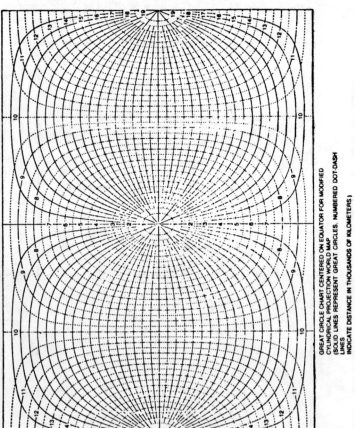

Fig. 3-4. Great circle chart for use with the HF path overlays.

articles in *CQ* magazine, and so. I obtained data through the courtesy of Ms. Carlene Mellicker at NTIA/ITS (National Telecommunications and Information Administration/Institute for Telecommunications Sciences), Boulder, Colorado 80303. This data, as shown in Fig. 3-5A and 3-5B, shows past and predicted RSSN values for the current sunspot cycle, 21.

MONTH	JAN.	FEB.	MAR.	APR.	MAY	JUNE	JULY	AUG.	SEPT.	OCT.	NOV.	DEC.
1976	15.2 (0)	13.2 (0)	12.2 (0)	12.6 (0)	12.5 (0)	12.2 (0)	12.9 (0)	13.9 (0)	14.1 (0)	13.3 (0)	13.3 (0)	14.7 (0)
1977	16.6 (0)	17.9 (0)	19.8 (0)	21.9 (0)	23.7 (0)	25.6 (0)	28.1 (0)	32.3 (0)	36.7 (2)	40.9 (4)	45.2 (6)	49.6 (9)
1978	54.5 (12)	59.9 (15)	65.2 (17)	70.7 (20)	76.8 (23)	82.2 (26)	86.5 (30)	90.3 (33)	93.9 (36)	97.6 (36)	102.2 (36)	108.2 (35)
1979	113.4 (35)	117.7 (36)	121.8 (37)	125.0 (38)	127.8 (39)	131.8 (42)	136.0 (43)	138.9 (44)	140.4 (47)	140.9 (49)	140.8 (52)	139.9 (54)
1980	138.3 (54)	137.0 (54)	136.4 (52)	136.6 (53)	135.7 (54)	132.8 (55)	129.8 (55)	127.3 (55)	126.0 (54)	125.3 (54)	124.1 (55)	123.2 (57)
1981	123.2 (59)	122.6 (59)	120.4 (57)	117.7 (55)	115.8 (54)	114.3 (52)	113.8 (51)	114.0 (49)	113.6 (4i)	112.5 (47)	110.7 (45)	107.9 (43)
1982	104.6 (41)	101.5 (4p)	99.0 (40)	97.3 (38)	95.1 (36)	92.3 (35)	88.4 (34)	84.1 (31)	80.0 (29)	76.4 (26)	73.4 (24)	70.2 (24)
1983	66.6 (23)	64.3 (22)	62.6 (23)	60.8 (23)	58.9 (23)	56.7 (23)	55.1 (24)	53.6 (26)	52.1 (27)	50.8 (29)	49.8 (30)	49.1 (31)
1984	47.9 (31)	46.1 (30)	43.5 (29)	40.3 (29)	37.9 (30)	36.6 (31)	3.54 (31)	33.7 (31)	32.3 (30)	31.3 (29)	30.2 (28)	28.9 (27)
1985	27.9 (27)	26.9 (26)	26.0 (26)	25.5 (26)	24.7 (26)	23.8 (25)	23.0 (23)	22.2 (23)	21.5 (22)	20.7 (23)	19.7 (24)	19.1 (24)
1986	19.0 (24)	18.6 (24)	18.1 (24)	17.1 (23)	16.1 (22)	15.1 (21)	14.3 (19)	13.7 (19)	13.3 (17)	12.9 (16)	12.7 (15)	12.4 (13)
1987	12.4 (12)	12.6 (11)	13.1 (11)	13.9 (12)	14.7 (13)	15.6 (14)	16.4 (15)	17.3 (16)	18.5 (18)	19.9 (20)	21.1 (23)	23.2 (26)

BEGINNING OF CYCLE 21
JUNE 1976

3/7/78
CM OT/ITS

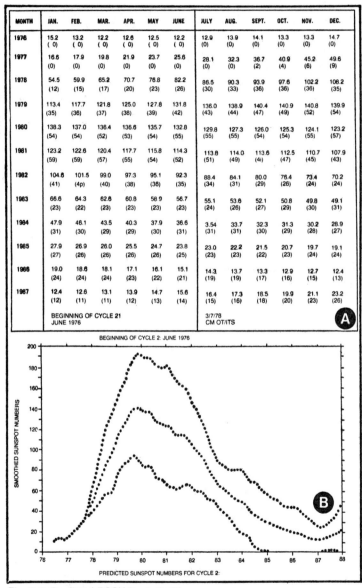

Fig. 3-5. Predicted RSSN for cycle 21.

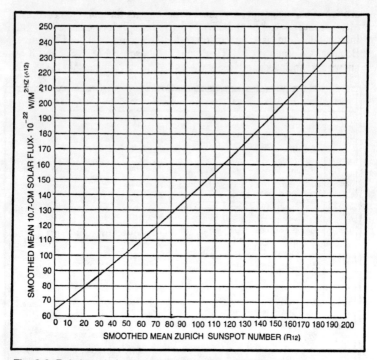

Fig. 3-6. Relationship between smoothed mean Zurich sunspot number and smoothed mean 10.7-cm solar flux.

This 21st cycle is given that number because it is the 21st nominal eleven-year cycle observed by man. The values in parenthesis in Fig. 3-5A are the 90% confidence limits of the predicted sunspot data. Where (0) is indicated, the actual value of the sunspot number has been computed. In Fig. 3-5B, the middle curve is the smoothed RSSN prediction while the upper and lower curves represent the plus and minus confidence limits.

At this point, a few words regarding the history of sunspots and sunspot numbers are necessary. Sunspots indicate the thermo-nuclear activity on the sun (eruptions, flares, etc.). These solar explosions are accompanied by radiant and corpuscular emissions. In brief, it is these radiant/electromagnetic emissions (ultra-violet, x-ray, etc.) which ionizes our upper atmosphere and creates the ionosphere. Herr Rudolpf Wolf devised methods for describing sunspot activity. The Zurich (Switzerland) solar observatory has been recording solar activity since 1749. Sunspot numbers vary considerably from day to day resulting in little correlation between the daily sunspot number count and HF ionospheric propagation conditions. It is not uncommon for the daily sunspot numbers (SSN) during a month to vary between 10.0 and 130.0. For reasons yet unknown, there does not appear to be much correlation between monthly SSN averages of the daily sunspot count and HF ionospheric propagation conditions. Empirical considerations, however, have shown that a quantity known as the twelve-month running average (or RSSN—running smoothed sunspot

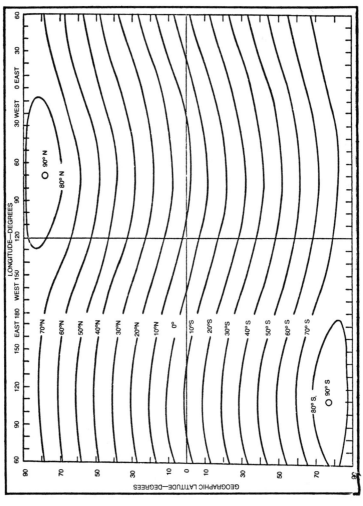

Fig. 3-7. World map of geomagnetic latitudes.

number, or ZSSN—Zurich smoothed sunspot number, or R_{12}—twelve-month running average smoothed sunspot number, it has many names) is a useful index of the upper atmosphere's state of ionization, and therefore, its ability to reflect (refract) HF radio waves. As a matter of information, there is currently underway, however, a swing toward the use of an improved solar activity index which is based upon measurements of the sun's 10.7 centimeter (2.8 MHz) emissions (see Fig. 3-6).

7. Transfer the transparency to the World Map of Geomagnetic Latitudes (Fig. 3-7). With equators and meridians aligned, read the geomagnetic latitude of the path midpoint. Record this value at the top of the worksheet. In our example, it turns out to be 50° North.

8. Opposite the UTC (or GMT or Zulu) times in column 1 of the worksheet, write in, respectively in columns 2, 3, and 4, the corresponding local times of terminals A and B and the path midpoint, M (see Appendix 16 entitled "Standard Time Zones Of The World And Their Relationships To UTC or GMT").

9. Determine the following:

a. Monthly Median MUF (Zero) F2 in MHz.—The monthly median MUF (maximum usable frequency, defined as that frequency at which the ionosphere will return signals to the earth for 50.0% of the days of the month) at zero great circle distance or vertical incidence (90° takeoff or wave angle) from the F2 ionospheric layer. Frequency in MHz.

b. Monthly Median MUF (4000) F2 in MHz.—Same as above, but for 4,000.0 kilometers great-circle distance (0.0° take-off angle with the tangent to earth at the point of the antenna location).

c. Monthly Median MUF (2000) E in MHz.—Same as above, but for the E-layer and maximum E-layer propagation distance of some 2,000.0 kilometers at 0.0° wave angle. (Recall that due to the E-layer's height of approximately 100.0 kilometers, the maximum one-hop great-circle range is about 2,000.0 kilometers).

For this step, we use volume 2, 3, or 4, of OT/TRER-13 as appropriate. Note that if our sunspot number of interest is 10, 110, or 160, we use the data directly from these volumes, but if our sunspot number falls more than slightly outside these numbers, linear interpolation is necessary (the relationship between sunspot number and frequency may be considered linear with only a small error). In our example, since the running sunspot number (R_{12}) = 10, we select volume 2. And, since our month is June, we begin with "JUNE UT 00 MONTHLY MEDIAN MUF (ZERO)F2 MHZ" and "JUNE UT 00 MONTHLY MEDIAN MUF (4000)F2" (see Figs. 3-8A and 3-8B).

Overlay the transparency consecutively on each one (being sure to align the equator and meridian lines at each step) and read the corresponding MUF in MHz at the path midpoint, M. Record these frequency values in the appropriate columns, 5 and 6, opposite each corresponding two-hour period. Repeat this procedure for each time period until 2300 UT. Next, repeat the above basic procedure, but for "JUNE MONTHLY MEDIAN MUF (2000 E) MHZ" (see Fig. 3-8C). Record these values in worksheet column 7 for UT times 00 through 2200 hours.

10. Determine the F2 ionospheric layer height, in kilometers, at the path midpoint. This is accomplished by placing the transparency over the

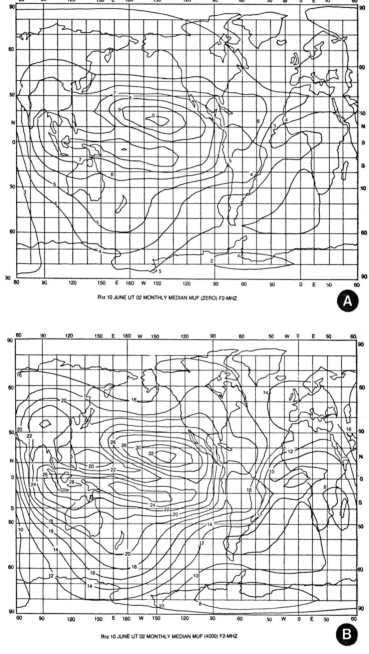

Fig. 3-8. Examples of monthly median MUF maps.

37

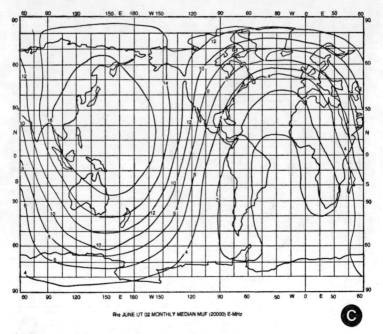

R12 JUNE UT 02 MONTHLY MEDIAN MUF (20000) E-MHz

Fig. 3-8. Examples of monthly median MUF maps.

chart (there is one for each month) entitled, *"Typical Height of the F2 Layer"* (Fig. 3-9). In our example, the month is June so we use Fig. 3-9F. Keeping this chart and the transparency equators aligned, slide the transparency so that its 0° meridian covers, with successive movements, all the local times in the two-hour increments shown in column 1. Remember, since we are using the 0° meridian, our F2-layer heights will come out in terms of UT. Record these F2-layer heights in column 8 on the worksheet opposite the corresponding UT times. (You will note that due to the geographical location of midpoint, M, it is possible to start only at the 06-hours reading; the readings from 00 through 04 hours will be picked up by sliding the transparency to the right, in the usual two-hour increments, past 00 hours.

11. Determine the sun's zenith angle at the path midpoint, M. To accomplish this, place the transparency over the appropriate "Sun's zenith angle" chart. For our example, the month of June, use Fig. 3-10F. Line up the transparency horizontally as in the previous step, and record the results for each two-hour interval. If the solar angle is greater than 105°, simply mark a dash in the appropriate time slot in column 9.

12. In this step we determine the ionospheric absorption index (I) which is a function of the previously determined solar zenith angle and the RSSN (R_{12}), the smoothed sunspot number logged at the top of our worksheet. See Fig. 3-11, entitled *"Nomogram of ionospheric absorption index,"* with its self explanatory example. Now, in our particular example, we enter the nomogram of Fig. 3-11 with the RSSN of

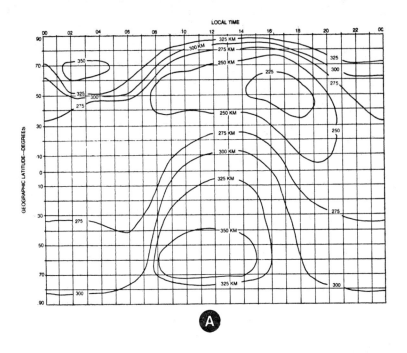

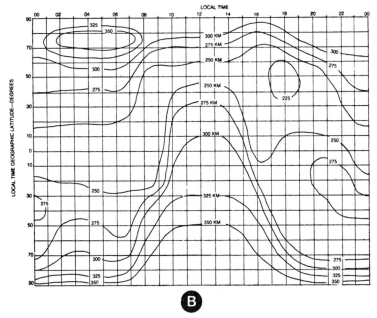

Fig. 3-9. Typical height of the F2 layer for the months of January through December.

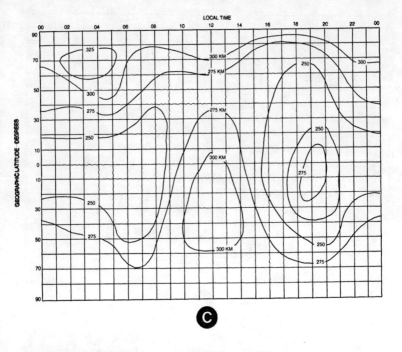

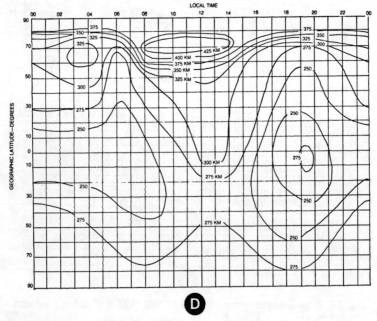

Fig. 3-9. (Continued from page 39).

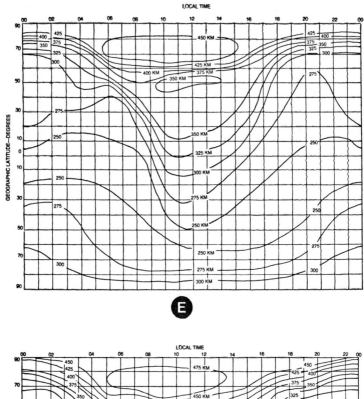

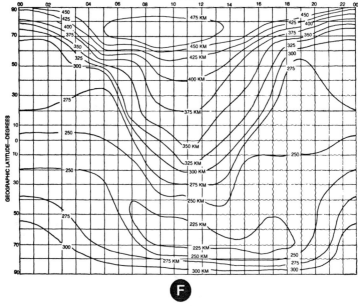

Fig. 3-9. (Continued from page 40).

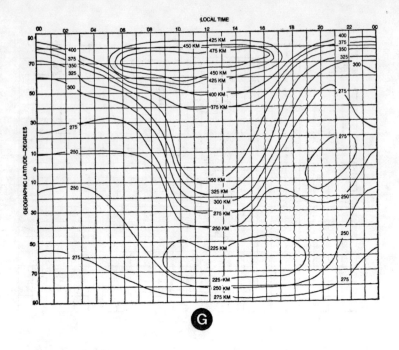

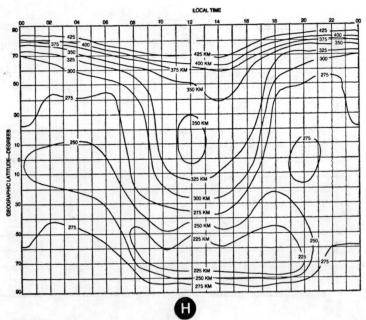

Fig. 3-9. (Continued from page 41).

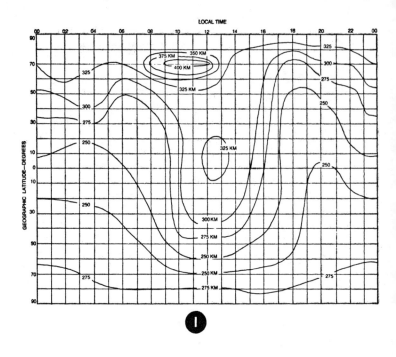

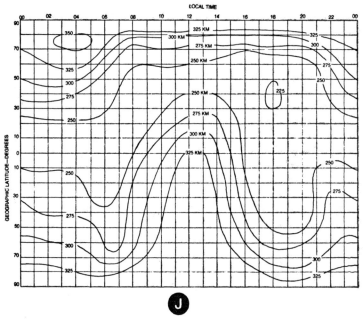

Fig. 3-9. (Continued from page 42).

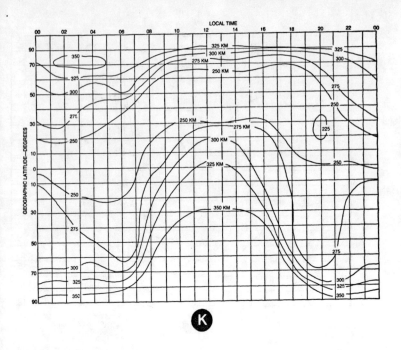

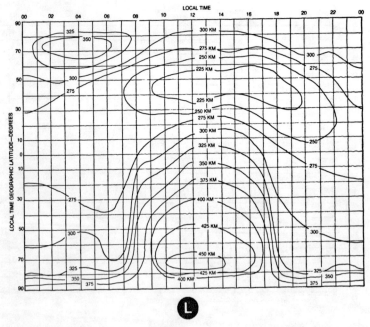

Fig. 3-9. (Continued from page 43).

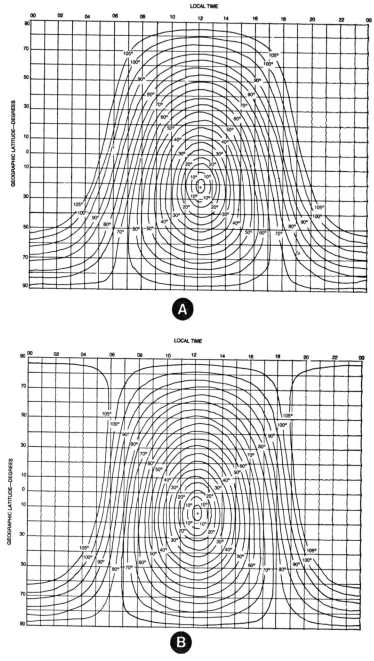

Fig. 3-10. Maps of the sun's zenith angle for January through December.

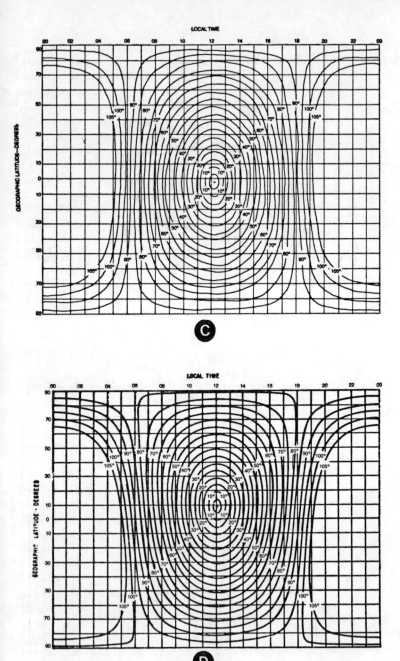

Fig. 3-10. Continued from page 45.

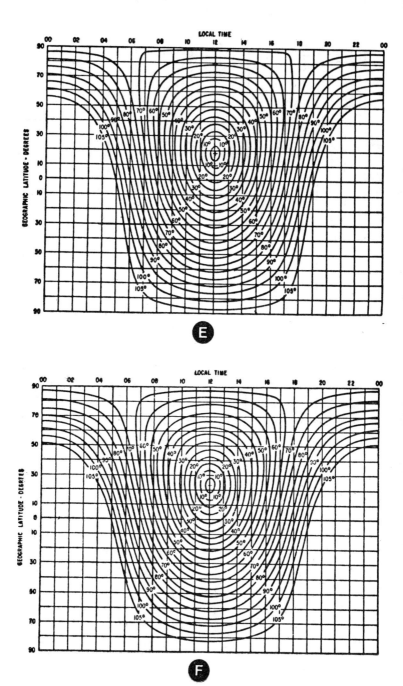

Fig. 3-10. Continued from page 46.

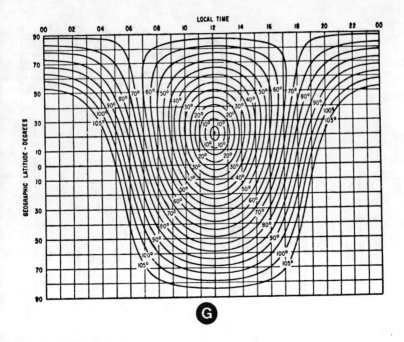

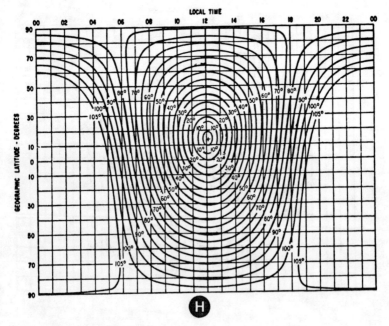

Fig. 3-10. Continued from page 47.

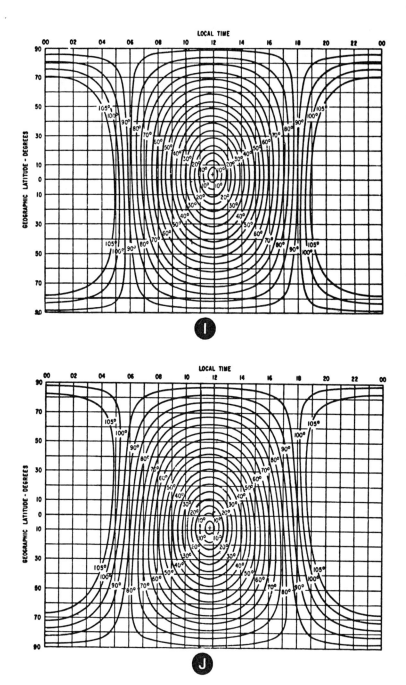

Fig. 3-10. Continued from page 48.

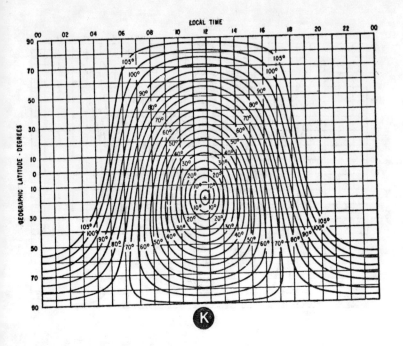

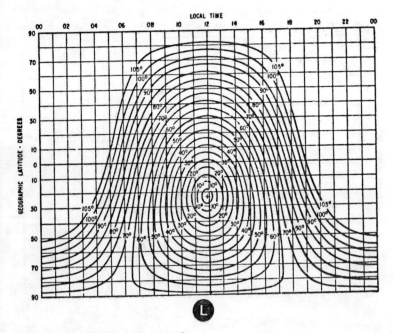

Fig. 3-10. Continued from page 49.

10 and the sun's zenith angles from column 9 to determine, for each corresponding two-hour period, our ionospheric absorption index (I). Record the results in the appropriate time-slots, in column 10.

A portion of the ionosphere (mainly due to the D-layer) is generally "destructive" to HF waves in that it absorbs their radio-frequency energy. The absorption index (I) is proportional to the sunspot number and to the sun's zenith angle. The higher the sunspot activity, and the higher the sun in the sky, the greater the D-layer absorption.

13. Determine the ionosphere's E-region gyro-frequency. This is done simply by placing our transparency over the map in Fig. 3-12 (*World Map of E-Region Gyro Frequency*), and, as before, with the equators and meridians coincident, read the gyro-frequency value at the path midpoint. Log this value at the top of the worksheet in Fig. 3-1. The gyro-frequency,

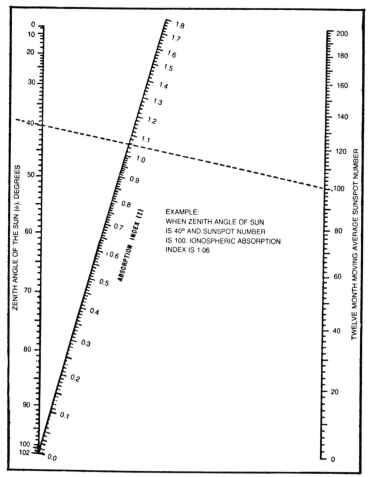

Fig. 3-11. Nomograph of ionospheric absorption index.

or gyromagnetic frequency, is the natural frequency at which charged particles rotate around the lines of force of the earth's magnetic field. In the case of free electrons in our ionosphere, this frequency falls between 0.7 and 1.9 MHz. The gyro-frequency is important in magneto-ionic wave propagation parameters such as absorption.

14. Find the necessary radiation angle (take-off or wave angle), the vertical angle of the antenna's main lobe with respect to a plane tangential to the earth at the antenna's location. This is generally done for each mode of HF propagation, 1 hop F, 2 hop E, in our case. These modes generally suffice in short hop (4000 kilometers or less) HF propagation calculations. However, the entire procedure may be carried out for any of several additional modes as desired (2 hop F, 3 hop E, etc.). The radiation angles for our current problem, are found by the use of the nomogram of Fig. 3-13 entitled, *Radiation Angle (Δ) As A Function Of Great Circle Distance and Ionospheric Layer Height.*

The ionospheric E layer is more stable in layer height, as well as ionization density, and may be assumed to have a constant height of 110 kilometers for practical purposes. The F layer is much more capricious in these respects. Its height, as you will recall, has already been recorded in column 8 for the various 2-hour periods of the day. A note of caution—recall that our path, being over 2,000.0 kilometers in great-circle path length, caused us to use the two-hop propagation mode for the E layer. From a geometrical point of view, our path length is one half for each of the two hops we record at the top of the worksheet. It can be seen that this will affect the radiation-angle value. Our F-layer propagation geometry, of course, is based upon the entire path length of 2500 kilometers. Using Fig. 3-13, record the E-layer radiation angle at the top of the worksheet and also, in column II, the value for the F-layer. We see from the results in our example that an antenna vertical radiation pattern of 9° (down 3.0 dB at the 6° and 12° vertical angle points) would possibly serve nicely. (It is customary in HF ionospheric propagation work to design the antennas at both transmitting and receiving terminals for the same wave angle).

15. Determine the F2 MUF for our path. The F2 MUF is the maximum usable frequency (frequency supported for at least 50.0% of the days of the month) for a given path distance and time of day, by the F2 layer. To derive the MUF F2 for our 2,500-kilometer path, we use worksheet columns 5 and 6, along with the path distance in kilometers (2500 kilometers). Actually, we are applying the cosecant law which states that the frequency "reflected" from the ionosphere is proportional to the wave's angle of incidence upon the ionosphere. Enter Fig. 3-14, entitled *Nomogram To Estimate Minimum Distance or Maximum Frequency For Propagation Via The F2 Layer,* with data for each two-hour time period. Since we have the frequency values for vertical (90°) incidence (zero-path distance), and that for maximum F2 path distance (4000 kilometers), we are, in effect, interpolating for our actual path distance of 2500 kilometers. The shorter the path, the lower the path MUF. The example shown in dotted lines on Fig. 3-14 is for 00 UT, 5.3 MHz MUF (ZERO) F2 and 21.8 MHz MUF (4000) F2, giving a path MUF for our 2500 kilometer path, of 17.3 MHz. As shown in this example, connect the F2 (ZERO) MUF to the F2 (4000) MUF by a straight line. Draw another straight line through our path distance

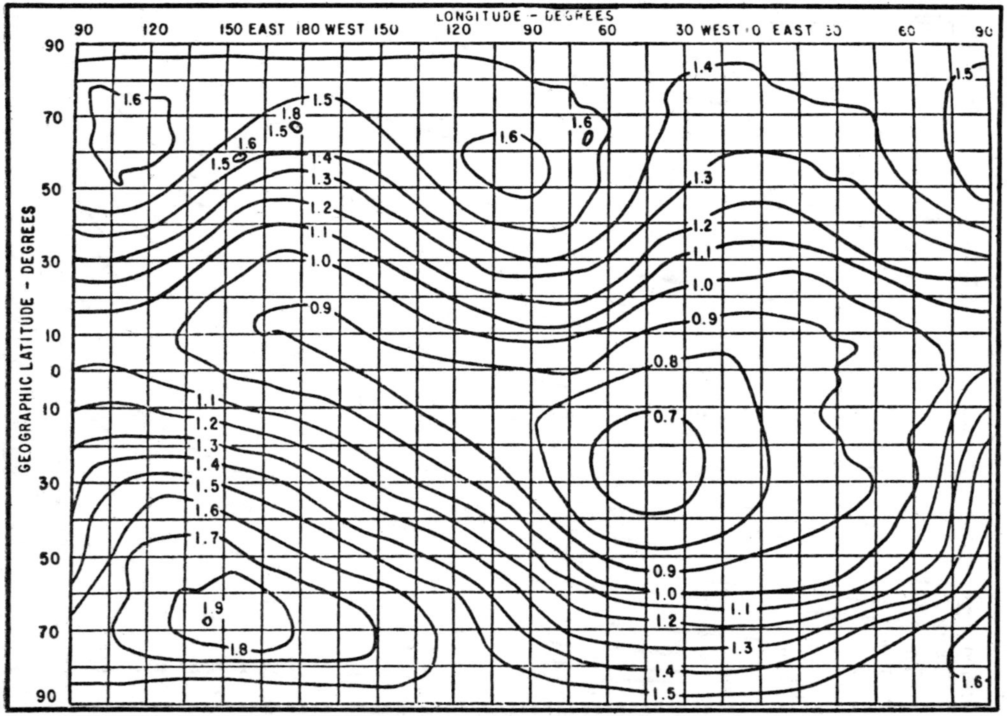

Fig. 3-12. World map of E-region gyro frequency.

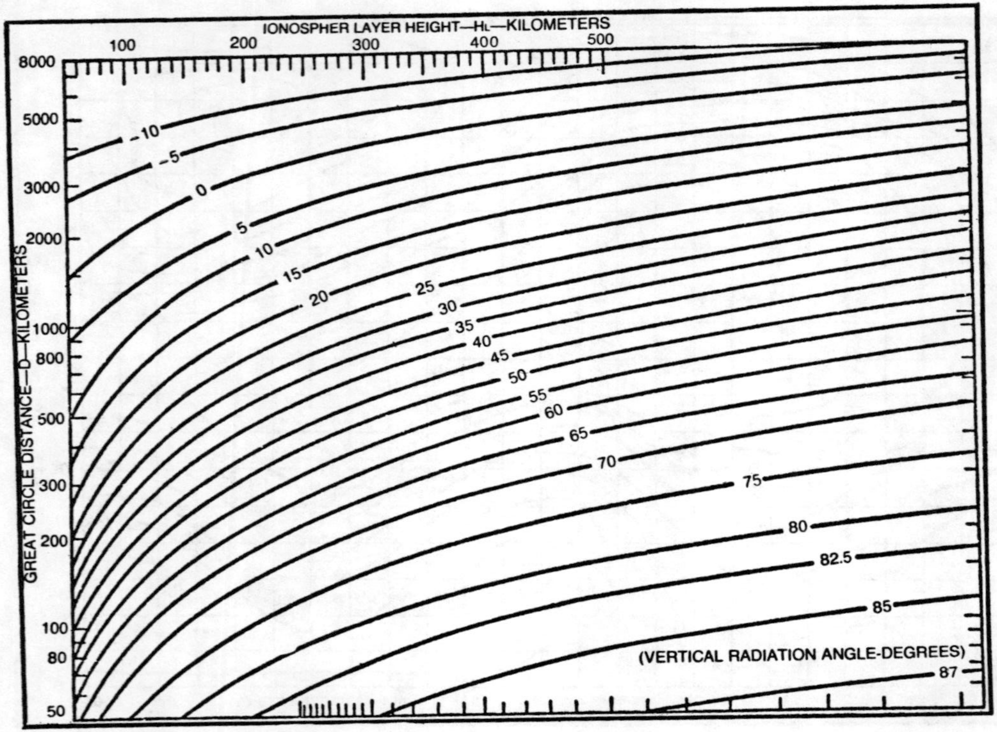

Fig. 3-13. Radiation angle (Δ) as a function of great circle distance and ionospheric layer height.

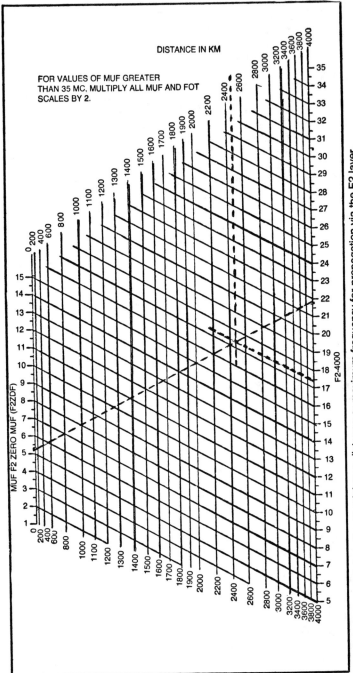

Fig. 3-14. Nomograph to estimate minimum distance or maximum frequency for propagation via the F2 layer.

scale. Last, parallel to the previous two lines, draw a straight line from their intersection to the F2 (4000) MUF scale. Here, we read the path F2 MUF of 17.3 MHz. (It is best, as before, to use a thin transparent plastic overlay and a grease pencil.)

16. Determine the path F2 FOT. FOT means optimum working frequency, from the international French phrase "Frequence Optimale de Travail." This number is found by multiplying the MUFs of column 12 by 0.85, for each two-hour time slot. Record the results in column 13.

Recall at this point that the F2 MUF is the frequency supported between two HF terminals by the ionosphere's F2 layer for 50% (median) of the time of any given month. Now, since the F2 layer is not the most stable and "normally" in a state of ionized turmoil, it follows that the frequencies the layer would reflect would also vary. If the frequency is sufficiently high, it might penetrate the "porous" F2 layer and pass out into space to be lost. To prevent this, empirical considerations and calculations have shown that multiplying the MUF F2 by 0.85 and operating at this lower frequency aids considerably in allowing for the F2-layer instability and increases the chances of ionospheric reflection (propagation) of the wave. The 0.85 factor is an arithmetical convenience (actually this may vary anywhere between 0.6 and 0.92). The F2 FOT frequency increases the chances of F2 layer support from 50% to 90% of the days of the month. It is important to know that this 90% does not mean 90% circuit reliability, except in the case where one might be interested in the percentage of time alone that the ionosphere will support the carrier. This will become clear as we proceed step-by-step through the procedure.

17. Determine the E-layer MUF. To do this, simply use the nomogram (Fig. 3-15), entitled *Nomogram For Converting The MUF (2000) E To MUFs At Other Distances Including F1 Region Effects At Distances Greater Than 2000 KM.* Enter the nomogram with path distance (recall that in the E-layer case we needed two hops so that our path distance for this consideration is 1250 kilometers per hop) and the MUF (2000) E values from worksheet column 7. The results are read from middle line of the graph, MUF MHZ. Log the results in worksheet column 14. The example shown in Fig. 3-15 is that of an E-layer path of 500 kilometers and an MUF (2000) E of 20.0 MHz. The E-layer path MUF in MHz is 8.5. Recall that the shorter the path, the higher our vertical wave angle, and this higher angle, in turn, requires a lower frequency to prevent layer penetration and consequent loss of wave energy.

This nomogram automatically takes into account some effects of the F1 layer. The F layer, during local daytime, separates into two distinct layers, the F2 above and the F1 below. At night, when the sun is not acting upon the ionosphere, the two F layers coalesce into one. It has become customary in HF-propagation work to refer to the "active" portion of the total F-layer as F2. It is interesting to know that since the E layer is quite stable, its MUF and FOT values may be taken to be identical. Paradoxically, therefore. the E MUF is equal to the E FOT.

18. Determine the path LUHF (sometimes also called LUF). This is the lowest usable high frequency, meaning the lowest usable frequency in the high frequency band, or simply, lowest usable frequency. To ac-

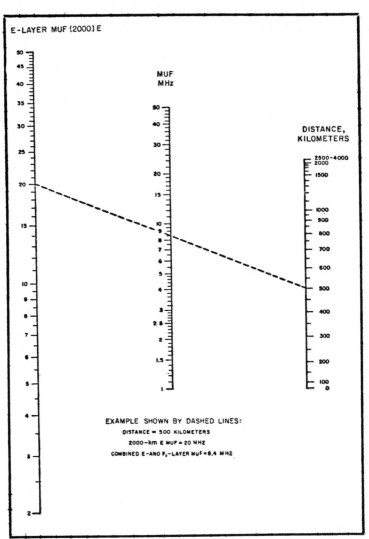

Fig. 3-15. Nomograph for converting the MUF(2000)e to MUFs at other distances including F1-region effects at distances greater than 2000 km.

complish this, there are two simple methods depending upon the path length. These methods are as follows:

METHOD 1 (Path length less than 2,000.0 kilometers great circle distance). Multiply the path E MUF/FOT values of worksheet column 14 by 0.75 and log these values in the corresponding time slots in column 15.

METHOD 2 (Path is between 2,000.0 and 4,000.0 kilometers great circle distance). Multiply the MUF (2000) E values of column 7 by 0.75 and log these values in the corresponding worksheet time slots in column 15.

As our path length in this example is 2,500.0 kilometers, we employ METHOD 2.

Since there is an MUF and an FOT for an HF-propagation path, there is also an LUHF. This sets the lowest frequency limits, below which one should not operate, just as the MUF/FOT sets the higher frequency limits throughout the day for a given month of calculation. The LUHF (or LUF) sets minimum frequencies which will not be absorbed excessively in the ionospheric D-layer, the lower the operating frequency, the higher the absorption. While this LUHF calculation method is not quite as good as the manual method described in NBS Circular 462, which takes into account such values as receiver location noise levels, it is infinitely easier to compute, while rendering quite useful results (akin to that of the Australian method called ALF—Absorption Limiting Frequency). The NBS Circular 462 method of manually calculating the LUHF is formidable, and has all but been abandoned.

Another way to calculate LUHF is to make a series of path calculations using lower and lower frequencies until a frequency is found giving unacceptable S/N (signal-to-noise) ratios for a given path and type of service. This type of manual LUHF calculation, even for educational purposes, is much too time consuming to be practical. Considering then, that one normally operates as near as possible to the FOT, the methods of LUHF calculation described in METHODS 1 and 2, are quite useful, serving as a "flag." Working below the LUHF exacts quite a penalty in terms of loss (absorption), of in the order of 10.0 dB of transmitted power per 1.5 MHz "LUHF trespass." When one is already using a transmitter power of some 100.0 kilowatts (as do, for example, Voice Of America, Radio Free Europe, British Broadcasting Corporation, etc.) this might mean going to one-million watts, a tremendous expense indeed!

19. Calculate the "EFFECTIVE FREQUENCY MHZ," for both the E- and F2-ionospheric layers. The effective frequency is defined as operating frequency plus gyro-frequency. To compute the effective frequency for the F2-ionospheric layer, add worksheet column 13 "PATH F2 FOT (MHz)" values to our gyro frequency (1.51 MHz) logged at the top of the worksheet. Record the effective frequency results for each corresponding 2-hour slot in column 16. To compute the effective frequency for the E-ionospheric layer, add worksheet column 14 "PATH E MUF (FOT)" values to the gyro frequency. Record the results for each corresponding two hours in column 17 "EFFECTIVE FREQ. (MHZ) E." (If your available operating frequencies are legally, financially, or otherwise limited, or different from the calculated FOT, use your actual frequencies in the computation of effective frequency and not the calculated FOTs which reflect the "optimum" frequencies.)

20. We are now ready to compute the ionospheric absorption loss. To do this, use the "ellipse-shaped" nomogram entitled *Ionospheric Absorption Nomogram* (see Fig. 3-16). Simply enter this nomogram with the Ionospheric Absorption Index (I), which we previously recorded in worksheet column 10, and the effective F2 and E frequencies of columns 16 and 17. The example in the nomogram is self explanatory. First, use the Ionospheric Absorption Index (I), column 10, with column 16 (EFFECTIVE FREQ. (MHz) F2 and the antenna radiation angle for the F2 layer,

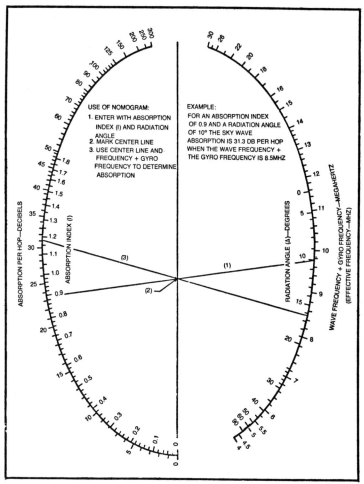

Fig. 3-16. Ionospheric absorption nomograph.

column 11, recording the results in worksheet column 18, "IONOSPHERIC ABSORPTION FOR F2 PROPAGATION (dB)." The, use the Ionospheric Absorption Index (I) of worksheet column 10 with the "EFFECTIVE FREQUENCY (MHZ) E" of column 17 and radiation angle for the E layer indicated at the top of the worksheet. Record these results in column 19, "IONOSPHERIC ABSORPTION FOR E PROPAGATION (dB)."

It is important at this point to recall that in the case of the E-layer our circuit necessitated two hops. Since the ionospheric-absorption losses we have just derived are for one hop, we must double the value because the ray experienced two ionospheric reflections (up and down excursions through the absorptive D layer). The F2-layer values of ionospheric absorption remain the same since there was only one F2-layer hop.

The mechanism of the D-layer absorption loss is interesting. The D-layer, as are the other ionospheric layers, is ionized by solar radiation into positive ions and electrons. The electrons are excited into "oscillatory modes" by incident electromagnetic energy. In the oscillation process, the electrons are afforded greater opportunity for recombination with the free, positive, relatively stationary, ions since the D-layer is lower and therefore denser. Now, since the oscillatory electron motions received their energy to oscillate from the incident signal wave, and since this energy is lost in heat during the recombination process, there is a net subtraction (loss) of energy from the wave.

21. Next, in the case of multiple-hop (more than one) ionospheric propagation, we determine the earth surface reflection losses. To accomplish this procedural step, we enter the appropriate graph, *"Land Reflection Loss"* (Fig. 3-17A) or *"Sea Water Reflection Loss"* (Fig. 3-17B), depending upon whether the earth reflection point is on land or sea with our operating frequency and reflection angle (antenna-take-off angles the earth-reflection angle). Observe that for F2-layer propagation, there were no earth reflections. Consequently, in worksheet column 20, we enter 0.0 for all time slots.

It is possible that a multiple-hop propagation mode for the F2 layer might show less attenuation than single hop. This is due to the steeper radiation angles with the wave spending less time and distance in the absorptive D-layer. (In this case, there would be one or more ground reflections.) In computer programs, all (or most) such modes are examined. While in the manual method, this would be out of the question due to problem complexity and excessive time consumption. Therefore, in the manual method, we consider only the simplest modes.

Continuing with our example, we have one earth reflection in our two-hop E-propagation mode. Since we previously noted at the top of our worksheet that our earth reflection point is land, we'll use Fig. 3-17A and enter it with the E-layer radiation angle and the path E-MUF (FOT) from column 14. Read the decibels per ground reflection and log the results in column 21. As shown, in this particular case, the losses are practically constant at 2.0 to 2.5 dB. In the case of multiple earth (or terrestrial) reflections, the dB losses of each such reflection would be summed and these summed results logged in the appropriate column.

22. Determine the transmission loss due to ray-path distance. Enter Fig. 3-18 (*Nomogram of Transmission Loss Due to Ray-Path Distance*) with the radiation angles previously logged in column 11 for the F2-layer propagation, the antenna radiation angle for E-layer propagation as logged at the worksheet top (6°) the great circle distance (2500 kilometers) and path FOT (for F2 propagation use column 13 or for the E-layer column 14). As before, if known, substitute actual frequencies for the computed FOTs. Log the results for the F2-layer propagation mode in column 22 (RAY PATH DISTANCE LOSS F2-dB) and the results for the E-layer ionospheric propagation mode in column 23 (RAY PATH DISTANCE LOSS E-dB). (Ray-path distance is the same as the so-called "Free-Space-Loss" or FSL of other types of radio-propagation calculations). The example given in Fig. 3-18 is self-explanatory. The inclusion of radiation angle in the nomographic solution of Fig. 3-18 "Free-Space-Loss" encountered by

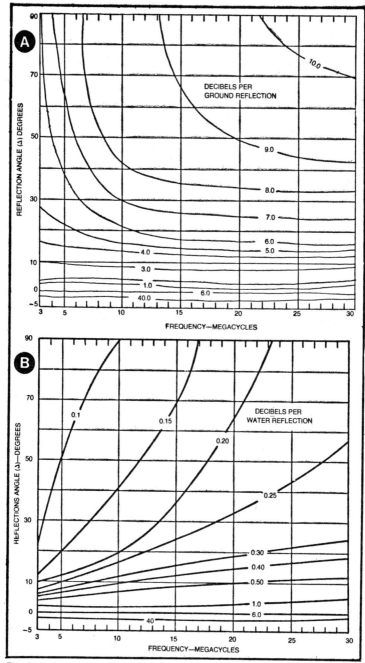

Fig. 3-17. Land reflection loss (A) Sea water reflection loss (B).

the actual earth-ionosphere-earth ray trajectory, as opposed to the great circle path distance. Please keep in mind that this loss is that between isotropic antennas.

23. Next, we determine what in HF work is called, "QUASI MINIMUM SYSTEM LOSS". It is not as formidable as it may sound, being simply the lowest hourly median system loss that can normally be expected at a given hour for any day within the month of calculation (in our case June). It is simply calculated by adding Ionospheric-Absorption Loss, Terrestrial-Reflection Loss, and Ray-Path Distance Loss. For the case of F2-layer propagation, simply add the dB values of worksheet columns 18, 20, and 22 and record these results in column 24, "QUASI MINIMUM SYSTEM LOSS F2 (dB)." Similarly, for the case of E-layer propagation, add the dB values of worksheet columns 19, 21, and 23 and record the results in column 25, "QUASI MINIMUM SYSTEM LOSS E (dB)." This quantity "QUASI MINIMUM SYSTEM LOSS" may be considered, conceptually, as a sort of basic propagational loss.

24. We now adjust the loss values of columns 24 and 25 to "MONTHLY MEDIAN OF HOURLY MEDIAN" values. This id done by entering Fig. 3-19 (*Typical Probability Distribution of Hourly Median Sky-Wave System Loss*) at the 50.0% (median) abscissa value and the reading the correction value from the ordinate via the appropriate curve. In our case, temperate circuits are used since our path control point (path midpoint, M) is located in the temperate zone. Log this at the worksheet top under "SYSTEM LOSS ABOVE QUASI-MINIMUM-dB," as shown. Apply this correction to the values in worksheet column 24 and log the results in column 26. Likewise, apply this same corrections value to the column 25 values and log the results, as shown, in cloumn 27 (SYSTEM LOSS MONHLY MEDIAN OF HOURLY MEDIANS E-dB).

25. Log transmitting and receiving antenna gains in dBi at the top of the worksheet. Correct the column 26 and 27 values by the combined gains of the transmitting and receiving antennas (in this case 20.0 dB) and log the results, respectively, in worksheet columns 28 and 29. These columns, 28 and 29, represent the total system loss from transmitter antenna output to receiving antenna input.

It is considered good HF engineering practice to use directive gain antennas, interference abatement. Antenna "gain" at the transmitting end(s) tends to concentrate the transmitter power toward its intended destination, instead of "spewing it all over the countryside" to cause waste of power and interference. In the receiving case, antenna directionality tends toward reduction of receiver interference and noise.

26. Assume a tentative transmitter output (one kilowatt is a convenient quantity) for the sake of signal-strength calculation. (Later on this power may be modified up or down as required.) Log this assumed transmitter power output at the top of the worksheet as shown. In our example, it is 1.0 kilowatt or + 30.0 dBw.

27. From the transmitter power output (+ 30.0 dBw), subtract column 28 system loss values for the F2 layer and log the results, in dBw, in column 30. Similarly, from the transmitter power output (+ 30.0 dBw) subtract the column 29 system loss values for the E layer and log the results, in dBw, in column 31. Observe that columns 30 and 31 refer to

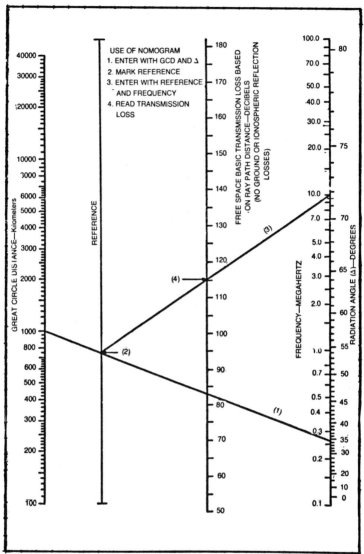

Fig. 3-18. Nomograph of transmission loss due to ray-path distance.

terminals A or B since both terminals have transmitters and receivers, this being a two-way circuit. These columns 30 and 31 values are our tentative receiver input signals, for F2 and E-layer propagation respectively. While the received signals in both directions will not be exactly reciprocal in strength, the additional computational complications are not warranted.

28. Next, we determine the noise levels, at terminals A and B, against which the signals must compete. At HF (3.0 to 30.0 MHz) the noise is considered external to the receiver and comes in three forms which must

be determined, logged, and then summed. These three forms of noise are as follows:

Atmospheric Noise—This noise type is generated by a great number of overlapping lighting discharges, mainly around the areas of the equator. This noise, which varies with the season and time, may be propagated by the ionosphere, just as radio waves of like frequencies, to various parts of the globe.

Cosmic Noise—This comes from the galaxy and outer space. It is also sometimes called galactic noise. Galactic (or cosmic) noise has the characteristics of thermal noise (see Appendix 14 entitled, *"White Noise"*). It is generally of low value in the HF band, in comparison with the levels of the other two noise components, since the ionosphere tends to shield the earth from this noise at a given frequency and angle of incidence.

Man-Made Noise—This type of noise arises from a tremendous number of sources, some of which may be power lines, motors, electrical appliances, or auto ignition, etc. Generally, as shown by Akima in U.S. Dept of Commerce Technical Report IER-34-ITSA-34, the level of man-made noise decreases with frequency at a rate of 28.0 dB per decade of frequency.

Returning to our procedure, to accomplish this step of determining the total noise, we consult Appendix 3 and 4 (CCIR Noise Maps and Standard Man-Made Noise and Galactic Noise in a 1.0 Hz Bandwidth), and proceed as follows:

a. Determine the atmospheric noise at terminal A (Denver, Colorado) and terminal B (Philadelphia, Pennsylvania). Recall that this noise is given in terms of dB above kT_ob. In Appendix 3, turn to the summer (June in Northern Hemisphere) and the appropriate local time blocks covering the terminal A local standard times. In this case, Fig. A3-17A of the CCIR Noise Maps is our beginning point. Using our terminal A geographic coordinates, enter the chart and read the 1.0-MHz noise (atmospheric noise for summer at local standard time block 1600 to 2000 hours) as 78.0 dB above kT_ob. Next, enter graph A3-17B at 1.0 MHz and proceed upward until intersection with the 78.0-dB curve (interpolate when necessary). Then, proceed along this 78.0-dB curve to the frequency logged in column 13 for terminal A's local standard time of 1700 hours. Read 43.0 dB above kT_ob on the F_{am} (dB above kT_ob) scale. Proceed to extrapolated σ F_{am} (standard deviation of values of F_{am}) curve in Fig. A3-18C and enter with path F2 FOT MHz. Read 7.2 dB. Combining the 43.0 and 7.2 dB values renders 50.2 dB above kT_ob for our atmospheric noise component of the total noise. Recalling that $kT_ob = -204.0$ dBw, our atmospheric noise value becomes $50.2 - 204.0 = -153.8$ dBw. Log this value in the proper time slot (in this case 1700 hours standard local) in worksheet column 32. Again please recall our earlier admonition that if you do not actually use the FOT frequencies, then make the noise computations herein at the frequencies you actually employ for your HF traffic. Continue this procedure for the rest of the local standard times in the 24 hour daily period, being sure to use the correct season and local standard time block values of noise maps, and log results in column 32.

b. Repeat the entire above (a) procedure for worksheet column 14 (PATH E MUF-FOT-MHz) frequencies and log results corresponding to these frequencies and local standard times of column 2, in column 33 entitled, "TERMINAL A ATMOSPHERIC NOISE (dBw) FOR PATH E MUF/FOT MHZ FREQUENCY".

c. Read the value of galactic noise in the B graph, as represented by the dashed line, for the appropriate PATH F2 FOTs MHZ of column 13 (or actual operating frequencies). Since this galactic noise is also in relation to $kT_o b$, subtract 204.0 dB as above. Record results in column 34 (TERMINAL A GALACTIC NOISE-dBw-FOR PATH F2 FOT-MHZ-FREQ.), for the appropriate local standard times and frequencies.

d. Repeat the above galactic noise procedure for the PATH E MUF (FOT) MHZ FREQUENCIES of column 14, or actual frequencies used, and log results in column 35.

e. From Appendix 4 (Standard Man-Made Noise and Galactic Noise in 1.0-MHz bandwidth), determine the standard median values of man-made noise at terminal A for the appropriate frequencies. Appendix 4 is also in terms of 1.0 Hz bandwidth. (Note—in this example, we have selected

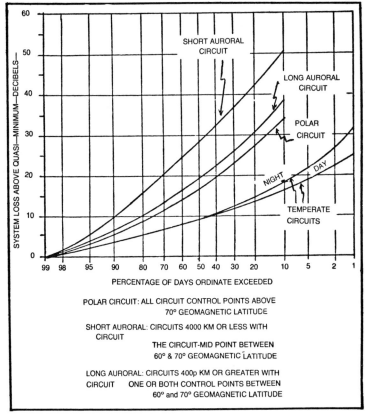

Fig. 3-19. Typical probability distribution of hourly median sky-wave system loss.

arbitrary man-made noise values as representing the results of actual in situ noise measurements. If meaningful measured values are not available, however, use Appendix No. 4 values). Log results directly in dBw in worksheet column 36 (TERMINAL A MAN MADE NOISE dBw FOR PATH F2 FOT MHZ FREQUENCY), or for actual frequencies used.

f. Repeat step e for PATH E MUF-FOT MHZ FREQUENCY of column 14 (or actual frequencies used). Log results in column 37 (TERMINAL A MAN MADE NOISE-dBw-FOR PATH E MUF/FOT MHZ FREQUENCY). (Note—In this step we have also logged in column 37 arbitrarily man-made values).

g. Sum up the respective noise types into a final noise value, against which the signal will have to compete, at terminal A receiving site by employing Appendix 5 (dB POWER ADDITION AND SUBTRACTION CURVES) which is self explanatory. (Note that when any dB values to be added have a difference of the order of 10.0 dB or more, the result will be close to the larger value. Therefore, by inspection it is at times possible to make mental calculations to determine such answers with sufficient accuracy.) Add columns 32, 34, and 36 by means of Appendix 5 and log results in column 38.

h. Perform the above step for columns 33, 35, and 37 and record results in columns 39. (Note—please keep in mind that in all cases where the calculated FOTs are not actually used in traffic, base the calculations on the actual frequencies.) In this example we have used the calculated FOTs.

i. Repeat the above entire procedure (a through h) for terminal B (recall that this is a two-way circuit for traffic handling and therefore the noise values as they affect reception should be determined for both receiving terminals, A and B). Use columns number 40, 41, 42, 43, 44, 45, 46, and 47 as appropriately labelled in the worksheet.

29a. Calculate the tentative S/N (signal-to-noise) ratio at terminal A for the F2 propagation mode. We do this by algebraically subtracting, consecutively, the values of column 38 (TERMINAL A TOTAL NOISE-dBw-FOR PATH F2 FOT-MHZ-FREQ.) from the column 30 values (TENTATIVE RECEIVER INPUT SIGNAL-F2-dBw-TERMINAL A OR B). Log your results, correspondingly, in worksheet column 48 (TENTATIVE TERMINAL A S/N RATIO-dB-FOR F2 PROPAGATION (MEDIAN-50%)).

29b. Repeat the above procedure for worksheet columns 31, 39, and 49.

29c. Repeat the above procedure for worksheet columns 30, 46, and 50.

29d. Repeat the above procedure for worksheet columns 31, 47, and 51. Note—The results of this step are in terms of total receiver signal power to noise in a 1.0-Hz bandwidth ratio. These results may be used directly as S/N ratio requirements for various modes (voice, TTY, Facsimile, etc.), later given in these same terms. However, if you should desire, for any reason (e.g. to use with S/N requirements given in terms of S/N where N (noise) is for the entire receiver bandwidth occupied by the signal), simply subtract the value, 10 Log$_{10}$ BW$_{Hz}$ or, in other words, ten times the logarithm of the occupied bandwidth in Hz. For example, the S/N ratio is 50.0 dB with the noise in a 1.0 Hz bandwidth. The occupied (by the signal) bandwidth is 1.0 KHz (1,000.0 Hz). Simply take 10 Log₁₀ 1,000.0 = 30.0 dB and subtract this value from 50.0 dB to obtain 20.0 dB. Thus, the S/N ratio with the noise considered for a bandwidth of 1.0 KHz is 20.0 dB. The noise-bandwidth may be considered to be analogous to a sliding door, the more it is opened, the wider the aperture for accepting noise.

30. Next, if we employ diversity reception, we must add its S/N advantage to the values in worksheet columns 48, 49, 50, and 51. (Please see graph of Appendix 12. Assuming dual diversity and maximal ratio combining, we note that the S/N improvement is 3.0 dB. Log this value at the top of the worksheet and add it to columns 48, 49, 50, and 51. Log results, respectively, in columns 52, 53, 54, and 55.

31. At this point, we decide our required S/N ratio for our given type of service. This information, for the various types of HF service is given in

Type Of Radio Service	Grade Of Service			
	Operator To Operator		Good Commercial Quality	
	No Diversity	Dual Diversity	No Diversity	Dual Diversity
6A3 (Double sideband—AM)	51.0	48.0	75.0	70.0
3A3 (Single sideband—AM) (3A3a—reduced carrier)	49.0	46.0	73.0	68.0
3A3 (Single sideband—AM) (3A3j—suppressed carrier)	48.0	45.0	72.0	67.0
6A3 (Independent sideband—AM) (6A3b (Two voice channels)	50.0	47.0	74.0	69.0
9A3 (Independent sideband—AM) (9A3b (Three voice channels)	50.0	47.0	74.0	69.0
12A3 (Independent sideband—AM) (12A3b (Four voice channels)	51.0	48.0	75.0	70.0
FACSIMILE (SSB FM subcarrier) (2800 Hz Bandpass Filter)	50.0	47.0	60.0	55.00
Manual C. W. (Morse)	36.0	33.0	45.0	40.0

(Note(1) 5 unit code. no error control schemes. Note (2) power divided equally among channels.	Character Error Rate (Note 1)					
	10^{-2}		10^{-3}		10^{-4}	
	No Div.	Dual Div.	No Div.	Dual Div.	No Div.	Dual Div.
1.1F1 (FSK) (60 WPM) 1500 Hz. Filter (START-STOP)	56.0	51.0	63.0	59.0	69.0	64.0
1.1F1 (FSK) (60 WPM) 150p Hz. Filter (SYNCHRONOUS)	51.0	47.0	59.0	55.0	66.0	61.0
3A7J, suppressed carrier. 16 TTY subchannels-each 42.5 Hz dev. FSK, 110 Hz filter, 100 WPM 5-unit START-STOP (SSB)	62.0	59.0	69.0	65.0	75.0	70.0
Same as above but SYNCHRONOUS	57.0	53.0	65.0	61.0	72.0	67.0
6A9b ISB 1-voice chan. & 16-teletype subchans., each 42.5 Hz dev, FSK, 110 Hz filter. START-STOP. 100 WPM.	65.0	60.0	72.0	66.0	78.0	71.0
Same as above but SYNCHRONOUS	60.0	56.0	68.0	64.0	75.0	70.0
12A9b, 2 voice chan, ISB, 32 TTY subchan, 42.5 Hz Dev, FSK, 11p Hz Filter. 10p WPM. Strt/STOP	66.0	62.0	73.0	68.0	79.0	73.0
Same as above but SYNCHRONOUS	61.0	57.0	69.0	65.0	76.0	71.0

Fig. 3-20. Required S/N ratio occupied bandwidth relative to noise in a 1-Hz bandwidth (dB).

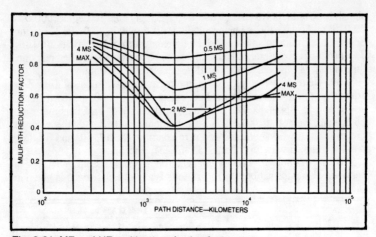

Fig. 3-21. MF and HF multipath reduction factors.

Fig. 3-20. The Fig. 3-20 table is self explanatory. Let us assume that in our example we are using a facsimile SSB FM sub-carrier in an occupied bandwidth of 2,800.0 Hz and that we desire good commercial quality grade of service. We see from the table (Fig. 3-20) that for our case we require a 55.0 dB S/N (noise in a 110 Hz bandwidth). Log the type of service and required S/N ratio at the top of the worksheet. (Note—Regarding Fig. 3-20, there are general limits to the rates at which intelligence can be transmitted by HF sky-wave mode of propagation since these signals may arrive at the receiver by several different paths resulting in what is termed "multipath." This multipath, as pointed out by Akima of the U.S. Dept. Of Commerce in publication IER-34-ITSA-34, coupled with irregular ionosphere motions, distorts the signal in both amplitude, phase, and time of arrival, all of which conspire to limit the rate-of-information. Longer signal elements such as RTTY (radio-teletype), with element pulse lengths of 22.0 to 32.0 milliseconds, would suffer less by an ionosphere component delay than would shorter pulses. Time delay difference between the first and last arriving signal components, due to multipath, depends upon path length and ratio of operating frequency to the MUF. A good way to reduce multipath problems is to operate as near to the MUF as possible; this might necessitate many frequency changes during a 24 hour operating period. This tack tends to reduce the number of ray paths available for multipath distortion. Fig. 3-21 shows expected time delay differences as related to HF path length (great circle distance) and MRF (multipath reduction factor), the MRF being the ratio between the operating frequency and the MUF of the appropriate E or F layers. For example, MUF = 10.0 MHz and actual frequency in use = 7.0 MHz, path length = 5,000.0 kilometers. MRF = 0.7. Enter Fig. 3-21 with these numbers (5,000.0 kilometers and 0.7). This renders a maximal delay of 1.0 millisecond. As can be appreciated, this might not harm a 22.0 millisecond RTTY pulse, but might totally "obliterate" a series of 1.0 millisecond pulses).

It is important to be cognizant that the required S/N ratios above are median values (50.0%); that is, with these S/N ratios the circuit may be expected to render acceptable quality and results around half the number of days of the month of computation, or stated another way, the probability of satisfactory performance on any given day will be 50.0% (provided, of course, that ionospheric support on that day obtains). The probability of attaining this given S/N ratio at a given hour is defined as circuit reliability—in this case, 50.0%.

32. We now determine our required S/N ratios for the E and F2 layer propagation for our required percent reliability. Let's assume that in our case we desire a reliability of 99.0% (Note—It is very important to realize that when we say 99.0% reliability, this must of necessity be 99.0% of 90.0% (or 89.1%) since, as you will recall, the ionosphere for our FOT calculations supports propagation for only 90.0% of the days of the month.) We select the proper chart of Figure 3-22 (from charts 3-22A through

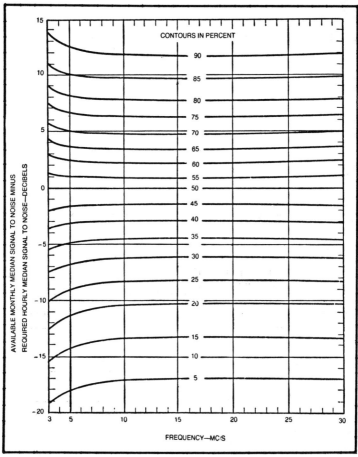

Fig. 3-22A. Chart to estimate reliability of polar circuits (all circuit control points above 70 degrees geomagnetic latitude).

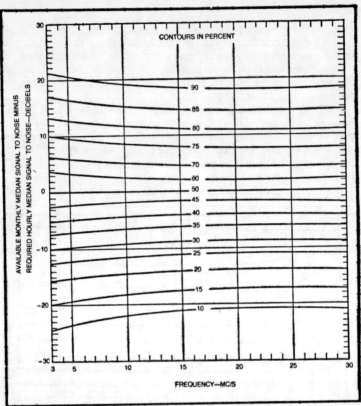

Fig. 3-22B. Chart to estimate reliability of short auroral circuits (circuits 4000 km or less with circuit mid point between 60 and 70 degrees geomagnetic latitude).

3-22E). Since our ionospheric control point (path midpoint M) is at a geomagnetic (dipole) latitude of 50°N, we select the two charts of Fig. 3-22C and 3-22D. We enter these respective graphs according to the condition of daytime or nighttime at the path midpoint, M, with guidance from worksheet column 4 giving path midpoint local standard time. For our example, we arbitrarily assumed path midpoint, M, to be in local standard daytime from its local standard time 0600 hours to 1800 hours and nighttime from 1800 hours local standard time to 0600 hours local standard time. For these graphs, we also use operating frequencies (calculated FOTs or other frequencies actually used). Read decibel values via the 99.0% contour (in this case) and add these values, correspondingly, to our 55.0 dB S/N ratio (50.0% median requirement). The results will be in terms of S/N required for 99.0% *service reliability* per frequency and corresponding time. Log results in appropriate columns 56 or 57 labelled respectively, "S/N RATIO REQUIRED FOR 99% RELIABILITY FOR F2 PROPAGATION" and "S/N (dB) RATIO REQUIRED FOR 99% RELIABILITY FOR E PROPAGATION". Please keep in mind that these S/N (dB) ratios are in terms of signal power in its occupied bandwidth to noise in a 1.0-Hz bandwidth.

33. We now check our actual calculated (available) S/N ratio for deficits for 99.0% service reliability for the F2 and E ionospheric layer propagation for both terminals A and B. Simply perform the following:

a. To check for terminal A S/N ratio deficit for F2 propagation for 99.0% reliability, take column 56 values and subtract from them the column 52 values. Log results at corresponding time slots in column 58.

b. To check for A S/N ratio deficit for E propagation for 99.0% reliability, take the column 57 values and subtract from them the corres-

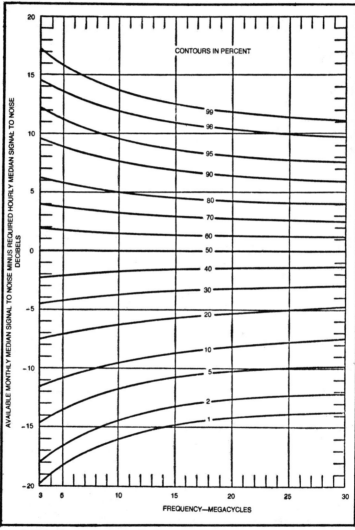

Fig. 3-22C. Chart to estimate daytime reliability of sky-wave circuits below 60 degrees geomagnetic latitude.

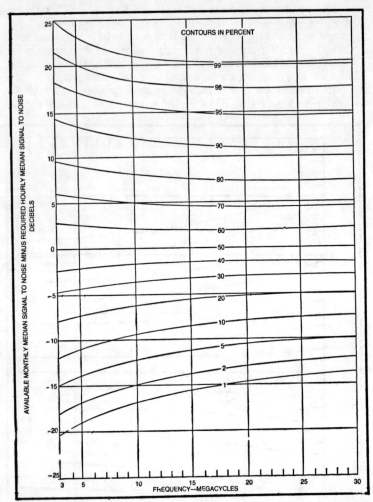

Fig. 3-22D. Chart to estimate nighttime reliability of sky-wave circuits below 60 degrees geomagnetic latitude.

ponding values from column 53. Log results in appropriate time slots in column 59.

c. To check for terminal B S/N ratio deficit for F2 propagation for 99.0% reliability, subtract from the column 56 values those corresponding values of column 54. Log results in appropriate time slots in column 60.

d. To check for terminal B S/N ratio deficit for E propagation for 99.0% reliability, subtract from the column 57 values those corresponding values of column 55. Log results in appropriate time slots in column 61.

34. Inspection of the logged results in columns 58, 59, 60, and 61 will readily show that the E ionospheric layer propagation mode, in this case, exacts a far greater system gain augmentation (antenna gain and/or trans-

mitter power) than does that of the F2 layer. For example, column 59 shows that for the 00 UT time slot the signal-to-noise ratio deficit is 36.7 dB at the receivers at terminal A, while column 61 shows for the same time slot a deficit of 36.7 dB also at the terminal B receivers. If these deficits were made up, respectively, at transmitter terminals B and A, we would need inordinate transmitters powers. To make up these gains by antennas would also be very costly. Each HF calculation, of course, is unique. The value of the traffic must be weighed against the benefits of alternate transmission modes as well as economic factors.

No solution given herein could possibly serve as a universal panacea. Let us, however, for the sake of our example, say that we decide to abandon the E-layer for the F layer, having considered all our important parameters. This means that we would use (or try to approach) the appropriate frequencies throughout the operating day, for a given month of calculation (June herein) shown in worksheet column 13, and antennas providing the wave angles of column 11. In addition, the F2 frequencies are further above the LUHF for a greater period of time than the E frequencies,

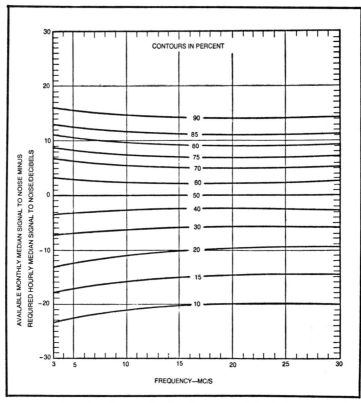

Fig. 3-22E. Chart to estimate reliability of long auroral circuits (circuits greater than 4000 km with one or both control points between 60 and 70 degrees geomagnetic latitude).

thus providing a larger "pad" against ionoshperic D-layer absorption. Please see Fig. 3-23; it is not unusual in HF work to plot the worksheet column 12, 13, 14, and 15 values of MUF, FOT, and LUHF as shown. Additionally, one might plot his actually used frequencies instead of the calculated ones, if this should be necessary, to ascertain their graphical position with respect to the calculated curves. That is to be sure they fall between the FOT and LUHF limits, or if they do not so do, to be prepared for other remedial measures or accept reliabilities less than those calculated.

In our example, herein, we have decided upon a compromise, recouping 5.0-dB gain at each circuit terminal. This seems an acceptable solution (maybe we cinched a good buy on antennas), plus the fact that antennas afford passive gain, requiring no AC power. At any rate, this tack should help in abating interference effects. So 5.0 dB more antenna gain at both transmitters and receivers amounts to 10.0 dB less system loss. Thus, we see that we could "wipe out" the 99.0% reliability deficits of some of the values in worksheet columns 58 and 60. For all values of deficits beyond 10.0 dB, which are not compensatable by this antenna tack, perhaps your traffic loads might allow less reliability at the corresponding "bad time slots". One could concentrate his more important (priority) traffic in the time slots corresponding to the higher percentage reliabilities. In our example, with the added 10.0 dB gain, we might recompute, with the aid of Fig. 3-22C and 3-22D, new percentage reliabilities. See columns 62, 63, 64, and 65. A negative S/N ratio deficit value, of course, is a S/N ratio "excess," while the positive deficit values are still deficits.

In closing this short path procedure (you can now exhale), we might state that as you gain experience and knowledge, you might elect to modify the worksheet forms herein to suit your own needs and fancies. Good luck and wear your "Baptism-of-HF fire button" proudly. Now that we have completed the short path procedure and mastered it, let's go on to the long path computation.

HF Long Path (Greater Than 4,000.0 Kilometers) Procedure

This long-path manual procedure has many points of similarity with that of the short path one and it is presumed that the reader has familiarized himself with the latter so that herein we might obviate unnecessary repetition. So let's get right into the procedure.

1. Enter the names and/or geographic coordinates of the HF circuit terminals at the top of the long HF path worksheet (please see Fig. 3-24). In our long path example, the HF end terminals are A = Washington, D.C. and B = London.

2. Place your transparency over World Map of Fig. 3-2.

3. On the transparency above, trace the world equator line and the 0° and 180° meridians and label them. Mark and label also the A and B end terminals (please see Fig. 3-25).

4. Transfer your transparency to the *"Great Circle Chart Centered on Equator for Modified Cylindrical Projection World Map"* of Fig.3-4. Align the equators, and maintaining them in alignment, slide the transparency horizontally until the terminals A and B lie on the same great circle (solid line) or both terminals lie the same proportionate distance between two solid

line great circles. Trace this great circle path onto the transparency. Estimate the great circle path distance, A to B, using the distances between dot-dashed lines as 1,000 kilometers and those between dot-dashed and dotted lines as 500 kilometers. Enter this great circle path distance value at the top of the worksheet (please see Fig. 3-24) as shown. Mark the path midpoint great circle distance, M. On the transparency, mark the two ionospheric control points A_1 and B_1 along the great circle path, respectively 2,000 kilometers each from terminals A and B. See Fig.

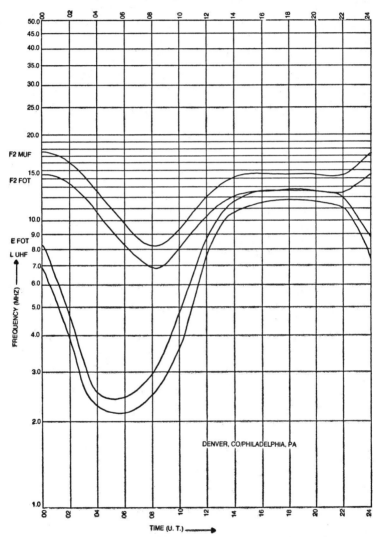

Fig. 3-23. Plot of the F2 MUF, F2 FOT, E FOR, AND LUHF for the Denver to Philadelphia path.

WORKSHEET FOR H.F. IONOSPHERIC PROPAGATION PATH CALCULATION (LONG PATH—OVER 4,000 KILOMETERS)

TERMINAL "A": WASHINGTON, D.C. — 38.9° N — 77° W
TERMINAL "B": LONDON — 51.5° N, 0.0° W
CONTROL POINT "A" GEOGRAPHICAL COORDINATES: 48° N, 51° W
CONTROL POINT "B" GEOGRAPHICAL COORDINATES: 51° N, 29° W
PATH MIDPOINT "M" GEOGRAPHICAL COORDINATES: 50° N–40° W
PATH DISTANCE KILOMETERS: 5900
MONTH AND YEAR: NOVEMBER, 1981
RSSN (R_{12}): PREDICTED 110.7
PATH MIDPOINT ("M") GYRO FREQUENCY (MHz): 1.35
NUMBER OF F2 HOPS: 2
NUMBER OF EARTH REFLECTIONS: 1
CONTROL POINT A: GEOMAGNETIC LATITUDE 58°N
CONTROL POINT B: GEOMAGNETIC LATITUDE 59°N
PATH MIDPOINT M GEOMAGNETIC LATITUDE 60°N

(1) UT, GMT, OR ZULU TIME	(2) TERMINAL "A" LOCAL STANDARD TIME	(3) TERMINAL "B" LOCAL STANDARD TIME	(4) CONTROL POINT "A" IONOSPHERIC LOCAL STANDARD TIME	(5) CONTROL POINT "B" IONOSPHERIC LOCAL STANDARD TIME	(6) PATH MIDPOINT "M" LOCAL STANDARD TIME	(7) CONTROL POINT "A" MUF (4000) F2 MHz	(8) CONTROL POINT "B" MUF (4000) F2 MHz	(9) CONTROL POINT "A" MUF (2000) E MHz	(10) CONTROL POINT "B" MUF (2000) E MHz	(11) TERMINAL "A" FOT MHz F2	(12) TERMINAL "B" FOT MHz F2	(13) PATH F2 FOT MHz	(14) COLUMN #9 VALUE × 0.75	(15) COLUMN #10 VALUE × 0.75	(16) PATH LUHF MHz	(17) SUN'S ZENITH ANGLE CONTROL POINT "A" (DEGREES)	(18) SUN'S ZENITH ANGLE CONTROL POINT "B" (DEGREES)	(19) SUN'S ZENITH ANGLE PATH MIDPOINT "M"	(20) IONOSPHERIC ABSORPTION INDEX (i) CONTROL POINT "A"	(21) IONOSPHERIC ABSORPTION INDEX (i) CONTROL POINT "B"	(22) IONOSPHERIC ABSORPTION INDEX (i) PATH MIDPOINT "M"	(23) AVERAGE IONOSPHERIC ABSORPTION INDEX (i)	(24) EFFECTIVE FREQUENCY (MHz)	(25) PER HOP IONOSPHERIC ABSORPTION (dB)
00	19	00	21	22	21	18	14	4	4	15.3	11.9	11.9	3.0	3.0	3.0	\\\	\\\	\\\	0.0	0.0	0.0	0.0	13.3	0.0
02	21	02	23	24	23	17	15	4	4	14.5	13.6	13.6	3.0	3.0	3.0	\\\	\\\	\\\	0.0	0.0	0.0	0.0	15	0.0
04	23	04	01	02	01	17	15.5	6	3.3	14.5	13.2	13.2	2.7	2.5	2.6	\\\	\\\	\\\	0.0	0.0	0.0	0.0	14.6	0.0
06	01	06	03	04	03	15	13	3.6	3.5	12.8	9.8	9.8	2.7	2.6	2.6	\\\	100°	\\\	0.0	0.02	0.0	0.007	11.6	0.0
08	03	08	05	06	05	14	17	8.5	7.5	11.9	14.5	11.9	6.4	5.6	5.6	94°	85°	90°	0.1	0.24	0.16	0.17	13.3	2.0
10	05	10	07	08	07	20	29	13	12	17.0	24.7	17.0	9.8	9.0	9.0	80°	74°	76°	0.33	0.45	0.55	0.4	18.4	1.5
12	07	12	09	10	09	34	39	16	14.5	29.0	33.2	29.0	12.0	10.9	10.9	67°	70°	69°	0.59	0.53	0.56	0.56	30.4	1.0
14	09	14	11	12	11	42	42	15.5	15	35.7	35.7	35.7	11.6	11.25	11.6	67°	70°	72°	0.58	0.39	0.47	0.42	37.1	1.0
16	11	16	13	14	13	42	39.5	13	12	35.7	33.6	33.6	9.8	9.0	9.8	93°	77°	85°	0.1	0.39	0.25	0.48	35	1.5
18	13	18	15	16	15	39	33	9	9.5	33.2	28.0	28.0	6.8	6.8	6.8	\\\	92°	101°	0.0	0.12	0.0	0.26	29.4	1.0
20	15	20	17	18	17	33	23	5	5	28.0	19.6	19.6	3.8	4.1	3.8	\\\	\\\	\\\	0.0	0.0	0.0	0.04	15.9	0.5
22	17	22	19	20	19	25	17	4	4	21.3	14.5	14.5	3.0	3.0	3.0	\\\	\\\	\\\	0.0	0.0	0.0	0.0	15.9	0.0
24	19	24	21	22	21	18	14	4	4	15.3	11.9	11.9	3.0	3.0	3.0	\\\	\\\	\\\	0.0	0.0	0.0	0.0	13.3	0.0

TRANSMITTER ANTENNA GAIN (dBi) __10.0__
RECEIVER ANTENNA GAIN (dBi) __10.0__
NATURE OF EARTH REFLECTION POINT __SEA WATER__
TRANSMITTER TENTATIVE POWER __1.0 KILOWATT (+30,000 dBW)__
REQUIRED S/N (dB) RATIO (MEDIAN, 50 %) __48.0 dB__
REQUIRED RELIABILITY (%) __99.3 %__

(26)	(27)	(28)	(29)	(30)	(31)	(32)	(33)	(34)	(35)	(36)	(37)	(38)	(39)	(40)	(41)	(42)	(43)	(44)	(45)	(46)	(47)	(48)	(49)	(50)
TOTAL IONOSPHERIC ABSORPTION (dB)	EARTH REFLECTION LOSS (dB)	RAY PATH DISTANCE (dB)	QUASI-MINIMUM SYSTEM LOSS (dB)	SYSTEM LOSS, MONTHLY MEDIAN OF HOURLY MEDIANS (dB)	COMBINED TRANSMITTER AND RECVR ANT. GAINS (dB) CORRECTED FOR COLUMN # 30	TENTATIVE RECEIVER INPUT SIGNAL (dBW) TERMINAL "A" OR "B"	ATMOSPHERIC NOISE TERMINAL "A" (dBW)	GALACTIC NOISE TERMINAL "A" (dBW)	MAN-MADE NOISE TERMINAL "A" (dBW)	TOTAL NOISE TERMINAL "A" (dBW)	ATMOSPHERIC NOISE TERMINAL "B" (dBW)	GALACTIC NOISE TERMINAL "B" (dBW)	TOTAL NOISE TERMINAL "B" (dBW)	TOTAL NOISE TERMINAL "B" (dBW)	TENTATIVE TERMINAL "A" S/N RATIO (dB) (MEDIAN, 50 %)	TENTATIVE TERMINAL "A" S/N RATIO (dB) (MEDIAN, 50%)	COLUMN # 41 S/N (dB) VALUES CORRECTED FOR DIVERSITY	COLUMN #42 S/N (dB) VALUES CORRECTED FOR DIVERSITY	MARGIN AVAILABLE TERMINAL "A" (dB)	MARGIN AVAILABLE TERMINAL "B" (dB)	MARGIN REQUIRED TERMINAL "A" (dB)	MARGIN REQUIRED TERMINAL "B" (dB)	MARGIN EXCESS TERMINAL "A" (dB)	MARGIN EXCESS TERMINAL "B" (dB)
0.0	6.0	129	135	144	124	−94	−162.7	−176.0	−168.0	−161.3	−173.8	−176.0	−168.0	−166.3	67.3	72.3	70.3	75.3	+22.3	+27.3	+21.0	+21.0	+1.3	+6.3
0.0	6.0	130	136	145	125	−95	−164.4	−177.0	−169.0	−164.1	−180.3	−177.0	−169.0	−168.1	69.1	73.1	72.1	76.1	+24.1	+28.1	+21.0	+21.0	+3.1	+7.1
0.0	6.0	130	136	145	125	−95	−165.5	−177.0	−169.0	−163.6	−180.4	−177.0	−169.0	−168.1	68.6	73.1	71.6	76.1	+23.6	+28.1	+21.0	+21.0	+2.6	+7.1
0.0	6.0	127	133	142	122	−92	−158.0	−175.0	−165.0	−157.0	−167.5	−175.0	−165.0	−163.2	65.0	71.2	68.0	74.2	+20.0	+26.2	+21.0	+21.0	−2.6	+4.2
0.0	6.0	130	136	145	125	−95	−158.9	−179.0	−171.0	−158.7	−174.1	−179.0	−171.0	−165.7	63.7	70.7	66.7	74.1	+20.0	+26.1	+22.0	+22.0	−2.0	+4.1
4.0	6.0	132	142	151	131	−101	−178.3	−179.0	−171.0	−168.7	−174.1	−179.0	−171.0	−168.7	67.7	67.7	70.7	70.7	+26.2	+26.1	+21.0	+21.0	+5.2	+5.2
3.0	6.0	137	146	156	135	−105	−206.5	−186.0	−181.0	−177.3	−197.0	−186.0	−181.0	−176.4	72.3	72.3	75.3	75.3	+27.3	+26.4	+11.0	+11.0	+10.7	+9.7
2.0	6.0	137	147	156	135	−105	−215.0	−187.0	−180.0	−179.2	−216.4	−187.0	−181.0	−180.2	74.2	74.2	77.2	77.2	+29.2	+29.2	+11.0	+11.0	+16.3	+15.4
3.0	6.0	139	147	158	136	−106	−207.0	−186.0	−181.0	−180.2	−197.0	−187.0	−181.0	−180.2	73.2	74.2	76.2	77.2	+28.2	+29.2	+10.5	+10.5	+16.3	+18.7
2.0	6.0	137	145	154	134	−104	−192.0	−186.0	−177.0	−176.8	−196.3	−186.0	−177.0	−176.4	73.4	72.4	76.4	75.4	+27.3	+27.4	+11.0	+10.6	+17.6	+17.6
1.0	6.0	138	141	150	130	−100	−171.7	−182.0	−173.0	−166.2	−184.8	−182.0	−173.0	−172.2	66.2	72.2	69.2	75.2	+21.9	+27.2	+11.0	+11.0	+16.3	+16.3
1.0	6.0	131	137	146	126	−96	−164.1	−179.0	−170.0	−162.9	−173.3	−179.0	−170.0	−167.8	66.9	71.8	69.9	74.8	+21.9	+26.8	+12.0	+12.0	+9.2	+15.8
0.0	6.0	129	135	144	124	−94	−162.7	−176.0	−168.0	−161.3	−173.8	−176.0	−168.0	−166.3	67.3	72.3	70.3	75.3	+22.3	+27.3	+20.5	+20.5	+1.4	+5.8
0.0	6.0	129	135	144	124	−94	−162.7	−176.0	−168.0	−161.3	−173.8	−176.0	−168.0	−166.3	67.3	72.3	70.3	75.3	+22.3	+27.3	+21.0	+21.0	+1.3	+6.3

Fig. 3-24. Worksheet for the HF ionospheric propagation path calculations (long path).

3-25. Estimate the number of F2 hops for this path, recalling that for the long path manual calculation method, this layer is most important. The number of F2 hops may be estimated by dividing the total path great circle distance by 4,000 kilometers (the nominal maximum F2 layer hop distance). Then take the next higher integer. In this case, this would be 2. Record this information at the worksheet top as shown in Fig. 3-24. Record at the top of the worksheet the number of earth reflections which is the number of hops minus one. In our case, this would be $(2 - 1 = 1)$, or one earth reflection.

5. Transfer the transparency again to the world map of Fig. 3-2 and determine the nature of the earth reflection point. This is either water or land. (Note—It can be seen by inspection, in this case, that our earth reflection point is at midpoint position, M). Record this at the worksheet top as shown (Fig. 3-25). Also record at the top of the worksheet, the geographical coordinates or control point A, control point B, and path midpoint M.

6. Enter month, year, and RSSN (Running smoothed sunspot number) for this calculation at worksheet top. For our example, we arbitrarily choose November 1981. Consulting Fig. 3-5 for this date yields a predicted value of smoothed sunspot number 110.7 with 90.0% confidence limits of 45. Herein, for our example, we shall use the middle value of 110.7. (Note—As in our short-path calculation, the long-path claculations must be made for each month of a sunspot cycle).

7. Transfer the transparency to the World Map of Geomagnetic Latitudes (Fig. 3-7). With the equators and 0° and 180° meridians aligned, read the geomagnetic latitudes of control point A, A_1, control point B, B_1, and path midpoint M. Record these values at top of the long-path worksheet.

8. Opposite the UT times in column 1 of the worksheet (Fig. 3-24), write in the corresponding local standard times of terminals A and B, ionospheric control points A_1 and B_1, and also those of path midpoint M. This may be easily determined by Appendix 16. Record these local times, correspondingly, in worksheet columns 2,3,4,5, and 6.

9. Determine the following for ionospheric control point A_1 and B_1.
  a. Monthly median MUF (4000)F2 MHz.
  b. Monthly median MUF(2000)E MHz.

These values are for an ideal 0° take-off and landing wave angle. Actually, in many cases, the earth dissipates radiated energy below about 3°. However, the 0° value appears to fit empirical HF propagation results. Since our predicted RSSN (Running smoothed sunspot number) is 110.7, we may use VOL. 3 of our OT/TRER-13 set, which is for an RSSN of 110.0. Remember that if our sunspot number, for which we are calculating, should be a value between those which are covered by OT/TRER-13, our MUFs must be worked out from the two "bracketing" volumes and then linearly interpolated. For our present problem, we use OT/TRER-13 VOL. 3, pages 363, 365, 367, 369, 371, 373, 375, 377, 379, 381, 383, and 385 for monthly median MUF(4000)F2 MHz. Similarly for monthly median MUF(2000)E MHz, use OT/TRER-13 VOL. 3, pages 386 through 397. Simply overlay the transparency (Fig. 3-25) consecutively on each of the above OT/TRER-13 plots, being certain to align the equators and meri-

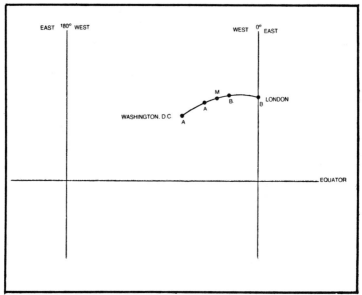

Fig. 3-25. Overlay for the Washington to London path.

dians at each one, and read appropriate values of frequency in MHz for ionospheric control points A_1 and B_1 and record these values in appropriate worksheet columns 7, 8, 9, and 10. See Fig. 3-8B and 3-8C for samples of these MUF value isoplethic presentations.

10. Compute the terminal A F2 FOT values by multiplying the column 7 values by 0.85. Record the results in column 11. Similarly compute terminal B F2 FOT values by multiplying the column 8 values by 0.85. Record results in column number 12.

11. Compare worksheet columns 11 and 12. For each time slot, select the lowest F2 FOT MHz value and record in corresponding time slots in column 13 (PATH F2 FOT MHz).

12. Compute the path LUHF values. This is done by multiplying the column 9 values by 0.75 and then the column 10 values by 0.75 and then recording the results respectively in columns 14 and 15. Next select the higher values of columns 14 and 15 for each time slot and record correspondingly in column 16 which is the LUHF for the path in MHz.

13. Determine the sun's zenith angle at ionosphere control points A_1 and B_1 and also at the path midpoint, M. Use Fig. 3-10K for our calculation month of November. Align the transparency and Fig. 3-10K equators and slide the transparency horizontally so that the 0° meridian of the transparency covers, in turn, all the local standard times in the usual two-hour increments of worksheet column 1. Thus, the solar angles will be a function of UT. Record results in appropriate columns 17, 18, and 19.

14. With the worksheet values of columns 17, 18, and 19 and the RSSN of 110.7, which we already recorded at the top of our worksheet, enter nomogram entitled, *"Nomogram of ionospheric absorption index (I)"* of Fig. 3-11. Log results correspondingly in columns 20, 21, and 22.

15. Take the average of the ionospheric absorption index values of columns 20, 21, and 22. Log corresponding average values in worksheet column 23.

16. Determine the path midpoint, M, gyro-frequency (Mhz) by overlaying your transparency (Fig. 3-25) over Fig. 3-12 (World Map Of E-Region Gyro-Frequency) with equators and 0° and 180° meridians aligned as usual. Record this value at top of worksheet as shown.

17. Calculate the effective frequency by adding the gyro-frequency to the operating frequency. Herein we use as our operating frequencies, those logged in column 13 (Path F2 FOT MHz). If you must employ other frequencies for one reason or another (e.g. regulatory), then use those frequencies actually employed for this calculation. Record results in column 24.

18. Determine the ionospheric absorption per hop by entering *"Ionospheric Absorption Nomogram"* of Fig.3-16 with the effective frequencies of column 24, the radiation (wave) angle (0°) for our F2 propagation for paths in excess of 4,000 kilometers and the average absorption indices logged in column 23. Record results in appropriate time slots in column 25.

19. Determine the total ionospheric absorption loss (dB) by multiplying the per-hop value of step 18 above (column 25) by the number of hops (2 hops in this case). Record the results in worksheet column 26.

20. Determine the earth reflection loss. In our present example, we have only one earth reflection, this reflection occurring on the Atlantic Ocean. We use Fig. 3-17B (Sea Water Reflection Loss). Entering this graph with the reflection angle (Δ) degrees (0° in this example) and the path FOT (or your actual frequencies of operation) of column 13, read the corresponding losses and record results in column 27.

21. Determine the path distance loss due to ray path distance. Enter *"Nomogram of Transmission Loss Due to Ray-Path Distance"* (Fig.3-18) with our radiation angle (0°), path great circle distance (in our case 5,900 kilometers), and operating frequency (in our case, we use the column 13 path F2 FOT MHz). Record results in column 28 (Ray path Distance Loss, dB).

22. Determine the "Quasi-Minimum System Loss" for the HF circuit. This is accomplished by simply summing the decibel loss values of "Total Ionospheric Absorption(dB)" of column 26, "Earth Reflection Loss(dB)" of column 27, and "Ray Path Distance Loss(dB)" of column 28 and entering the results in appropriate time slots, as usual, in column 29 ("Quasi-Minimum System Loss, dB).

23. We next adjust the "Quasi-Minimum System Loss" of column 29 to "Monthly Median of Hourly Medains" value. Enter Fig. 3-19 (*"Typical Probability Distribution of Hourly Median Sky Wave System Loss"*) at the 50% (median) abscissa value and read correction value at the ordinate. Apply this correction (higher loss) to the column 29 values and log results in column 30 ("System Loss, Monthly Median Of Hourly Medians, dB").

24. Log transmitter and receiver site antenna gains at top of worksheet. As in our short-path problem, we employ 10.0-dBi antenna gains for each. Next, correct the column 30 values by the combined gains of the transmitting receiving antennas (i.e. subtract 20.0 dB from the column 30 values and log results in column 31).

25. Assume a tentative transmitter power (one kilowatt is usually a convenient number). Log this assumed transmitter power value at the top of the worksheet.

26. From + 30.0 dBw (our transmitter power of 1000.0 watts), algebraically subtract the system losses of column 31. Log results in appropriate time slots of column 32. This, then, represents the signal inputs in dBw at the receivers at Washington, D.C. and London.

27. We now determine the total noise values at the receiver inputs at both Washington, D.C. and London. Use the basic methods outlined in step 28 for our SHORT PATH problem, along with Appendices 3, 4, and 5. For the man-made noise component of the total noise, let's assume that both of the receiving sites are in a rural area. Log results in appropriate columns 33, 34, 35, 36, 37, 38, 39, and 40.

28. Determine the tentative (i.e. with 1.0 kilowatt transmitter power) S/N ratio at both HF terminals, A and B. For terminal A receiver site, take the difference between worksheet columns 32 and 36. Log results in worksheet column 41. Similarly for terminal B receiver site, take the difference between 32 and 40 and log the results in worksheet column 42.

29. Consult Appendix 12. Accordingly correct worksheet column 41 and log the results in column 43. Similarly correct column 42 and log the results in column 44. Assume dual diversity reception using maximal ratio type combining. In our example, the correction is 3.0 dB, so our S/N ratio gains this amount.

30. From Fig. 3-20, determine the required median (50%) S/N ratio (dB) in occupied bandwidth relative to noise in a 1.0-Hz bandwidth. Recall that our noise was computed in a 1.0-Hz bandwidth. Assuming operator-to-operator grade of service for a four-channel independent sideband signal, we find our S/N ratio requirement to be, in this case, 48.0 dB. Log results at top of worksheet as shown.

31. Log required reliability (99%) at the worksheet top as shown. Recall that this is NOT, in reality, 99% circuit reliability even though it may be so called by convention, but 0.99 × 0.90 or 89.1% since our FOT will afford ionospheric propagation support for only a nominal 90% of the days of the month.

32. We next determine our margin. This is simply the available monthly median S/N ratio (dB) minus the required hourly median S/N ratio (dB) and is found as follows:

a. For terminal A, column 43 minus 48.0 dB. Log results in column 45.

b. For terminal B, column 44 minus 48.0 dB. Log results in column 46.

33. Determine the required margins (dB) for terminals A and B and log correspondingly in worksheet columns 47 and 48. Since our ionospheric control points are located at 60° or less in geomagnetic latitude, we use Fig. 3-22C and 3-22D for the points A_1 and B_1 depending upon whether they are in local standard daytime or nighttime. Recall that these figures, 3-22C and 3-22D are entered with our path FOT from column 13 (unless other frequencies are actually used). Arbitrarily we herein designate 0600 to 1800 hours local as daytime and 1800 to 0600 hours local as nighttime at the ionospheric control points. Actually sunrise and sunset at the ionospheric control points might be better criteria.

34. Determine our terminal A and B S/N ratio margin excess (or deficit). Subtract margin required values from the margin available values

as follows: For terminal A, subtract column 47 values from those of column 45 and log results in column 49. Similarly for terminal B, subtract column 48 values from those of column 46 and log results in column 50. Thus, we can see that there are excesses of S/N ratio based upon our 1.0 kilowatt of transmitter power and 10.0 dB antenna gains. For energy conservation and concomitant cost savings, along with abatement of HF spectrum pollution, we are obliged to reduce our e.i.r.p. (effective isotropic radiated power). This is basically a cost effectivity item. However, decreasing antenna directivity should not generally be resorted to in the interests of spectrum pollution. Instead, other factors being equal, we could reduce our transmitter power by the decibel amounts of the margin excesses.

There are many possible variations of the worksheet formats used herein and these may be tailored by the user to fit his particular computational convenience. The worksheet samples shown herein have worked well for the writer, but as the reader becomes more familiar with the "caprices" of HF propagation, he can design his own forms by modifying those herein. He might, for example, log his actual frequencies used, actual transmitter powers used, and so on to compare actual results with predicted values and thus contribute to the propagational science.

Now, as our parting shot, we plot, in a manner not unlike that of the short path example, our path F2 FOT MHz of column 13 and path LUHF MHz of column 16 (see Fig. 3-26).

By way of acknowledgements, we wish to extend our gratitude to the following at the National Telecommunications and Information Administration (NTIA), Institute for Telecommunication Science (ITS), Boulder, Colorado, for their concomitantly listed works, which were drawn upon for our herein presented short and long HF path calculation procedures.

a. NBS Report 7249, "Technical Considerations In The Selection Of Optimum Frequencies For High Frequency Sky-wave Communication Services" by Mr. George Haydon, Mr. Donald L Lucas, and Mr. Rodney A. Hanson.

b. OT/TRER-13, Telecommunications and Engineering Report-13. "Ionospheric Predictions" by Mr. William Roberts and Mr. Rayner Rosich.

c. "Monograph 80", U. S. Department Of Commerce, by Mr. Kenneth Davies.

HF PROPAGATION PREDICTION BY COMPUTER

This information regarding computerization of HF propagation is excerpted from the latest available information on the subject, contained in work "OT REPORT 76-102" entitled, "The Performance Of High Frequency Sky-Wave Telecommunications Systems (The Use Of The HFMUFES-4 Program) by Mr. George W. Haydon, Ms. Margo Leftin, and Mr. Rayner Rosich of National Telecommunication and Information Administration (N.T.I.A.) Institute For Telecommunication Sciences (I.T.S.) Boulder, Colorado, USA. The author is indebted to all the above, as well as to Mr. Vaughn Agy of the department, but especially to Dr. Rosich whose knowledge and patience during the wiriting of this HF chapter were tried to the limit.

It is probable that the reader will not be doing manual HF path computations as a regular activity, except under unusual circumstances. He may perform several examples in order to better understand the problem and/or work out an occasional required HF path calculation, but for any greater volume of commercial work, he will doubtless rely upon modern computer techniques. The U.S.A. Department Of Commerce, N.T.I.A./ I.T.S. (National Telecommunication and Information Administration/ Institute for Telecommunication Science) at Boulder, Colorado has a variety of latest computer programs (HFMUFES-4) on hand. Costs generally vary around the values shown as follows: Optimum Frequencies only, approximately $2.00 per circuit-month. If circuit reliability and/or LUHF and/or S/N ratios, etc., are requested, costs increase up to about $25.00 per circuit-month depending upon antenna complexity, number of hours per

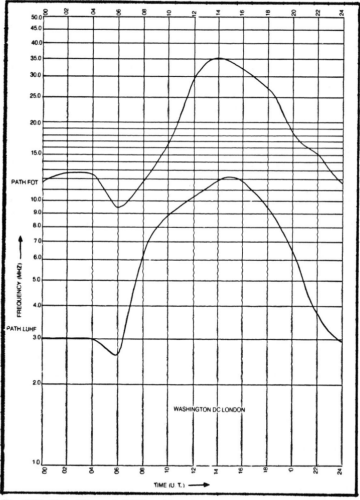

Fig. 3-26. Plot of the path FOT and LUHF.

day, frequency complements, etc. In all cases, there exists a minimum charge of $100.00. If you have a large scale computer, NTIA/ITS is able to provide the program and necessary additional material on a computer tape for a fee of $600.00. These prices were effective as of May 24, 1978. Write to: NTIA/ITS
    Ionospheric Transmission Technology Group
    Applied Electromagnetic Science Division
    U.S.A. Department Of Commerce
    Boulder, Colorado 80303
    U.S.A.

To obtain computerized HF propagation predictions service, simply complete the forms of Fig. 3-27A and 3-27B. Let us quickly run through an example service request form:

General Information:

(Box-1) The name, address, etc. are self explanatory.

Item A: Give the year of the required prediction and also the known or predicted RASSN (Running Average Sun Spot Number) and period of the solar cycle (high, low, or medium). RASSN low is typically 10.0 while RASSN 110.0 to 130.0 is considered typically high. If the RASSN is not given, NTIA/ITS can choose a number for you from the other data. Please be certain never to use daily or monthly solar activity indicies as only the 12-month expected running averages (RASSN) are acceptable for meaningful computer results.

Item B: Circle the appropriate month(s) for which your HF propagation predictions are required (June and December are typical seasonal extremes for long range planning while in other cases specific months may be given as desired). Recall that the RASSN is predicted and/or calculated on a monthly basis. List the day of the month if it is desired that your computerized predictions be centered around a day other than the 15th of the month.

Item C: In this item list your transmitter antenna site radio-geological data, ground conductivity (δ) in mohos/meter (or otherwise known as Siemens/meter), and the dielectric constant (ϵ). (Please see Appendix 1). As a second best alternative, you may list the earth characteristics in grosser terms such as excellent ground (sea water), good ground, fair ground, or poor ground as shown. Typical "standard" ground classifications in this context are given in the following table.

Ground Classification	Typical of Classification
Good Ground	Low hills with unforested rich soil, or, flat, wet, coastal region
Fair Ground	Medium hills or forested heavy soil.
Poor Ground	Rock steep hills, sandy dry coastal regions and city industrial areas.

Item D: Repeat above for this item, but for receiving antenna site.

1. GENERAL INFORMATION

NAME:________; COMPANY:________; ADDRESS:________;
CITY:________; STATE:________; ZIP CODE:________; AND TELEPHONE NUMBER:________.
(A) YEAR ________ OR PERIOD OF SOLAR CYCLE________; SUNSPOT NUMBER________.
(B) MONTHS (Circle) J F M A M J J A S O N D; DAY OF MONTH________.

			CHECK ONE*				
				SEA WATER	GOOD GROUND	FAIR GROUND	POOR GROUND
(C) TRANSMITTER SITE CHARACTERISTICS:	σ________MHOS/m,	ϵ________	OR	________	________	________	________.
(D) RECEIVER SITE CHARACTERISTICS:	σ________MHOS/m,	ϵ________	OR	________	________	________	________.

* Each antenna requires specification of its local ground electrical characteristics per numbers 1(C) and 1(D).

2. PROGRAM CONTROL INFORMATION

(A) Type of tabulation (method) desired 1 2 3 4 5 6 7 8 9 10.
(B) Type of graph display desired 0 1 2 3 4 5 6 .
(C) Beginning hour of calculation (UT)________(do not use zero UT).
(D) Ending hour of calculation (UT)________.
(E) Every________hours
(F) Circut Reliability________ OR Service Probability________.
(G) Antenna Pattern: None______ Transmitter______ Receiver______ Both______.
(H) Great Circle Path: Short______ OR Long______ .

3. FREQUENCY COMPLEMENT

________ ________ ________ ________ ________ ________ ________ ________ ________ ________ ________ MHz

A typical complement covering the frequency range is 2, 3, 5, 7, 9, 12, 15, 22, 26, 30 MHz.

4. CIRCUIT AND SYSTEM PERFORMANCE

(A) Transmitter Location Name________, Latitude________, Longitude________. (C) Minimum Takeoff Angle________°.
(B) Receiver Location Name________, Latitude________, Longitude________. (D) Transmitter Power________kw.
(E) Required S/N Ratio________dB OR Required Service________.
(F) Manmade Noise at Receiving Site ______DBW/MHZ at 3 MHz OR Urban _____, Suburban_____, Rural_____.
(G) Circuit Reliability Required for LUF or Service Probability Computation ______%. (H) Is Multipath Computation Required: Yes____ No____.
(I) Minimum Tolerable Power Ratio Between Multiple Modes ________dB. (J) Maximum Tolerable Time Delay Between Multiple Modes ________ms.

6. TRANSMITTING ANTENNAS			7. RECEIVING ANTENNAS		
TYPE	FREQUENCY - MHz		TYPE	FREQUENCY - MHz	
	Begin	End		Begin	End

Fig. 3-27A. Input data required for HF system performance predictions.

5. ANTENNA PARAMETERS (TRANSMITTING OR RECEIVING) - All dimensions must be in meters, wavelengths, or degrees unless otherwise specified.

Type 1 - rhombic	Type 3 - horizontal 1/2λ dipole	Type 5 - vertical dipole	Type 7 - terminated sloping vee	Type 8 - inverted "L"
feed height_____ leg length_____ tilt angle_____ true bearing_____	feed height_____ additional gain_____ dB true bearing_____	feed height_____ length_____ additional gain_____ dB true bearing_____	feed height_____ leg length_____ semi-apex angle_____ true bearing_____ termination height_____	vertical height_____ horizontal length_____ true bearing_____

Type 9 - terminated sloping rhombic	Type 11 - sloping wire	Type 2 - vertical monopole	Type 14 - arbitrary tilted dipole	Type 15 - half rhombic
feed height_____ leg length_____ semi-apex angle_____ true bearing_____ termination height_____	feed height_____ length_____ slope_____ true bearing_____	length_____ additional gain_____ dB Type 12 - constant gain gain_____ dB	feed height_____ length_____ angle of element from horizontal_____ true bearing_____	leg length_____ angle between leg and ground_____ true bearing_____

TYPE 4 - horizontal Yagi	Type 6 - curtain	Type 13 - horizontal LPA	Type 16 - dual rhomboid
feed height_____ driven element length_____ reflector element length_____ director element length_____ true bearing_____ number of elements_____ director spacing_____ reflector spacing_____	lowest element height_____ length of element_____ number of bays_____ true bearing_____ no. of elements in each bay_____ bay separation_____ vertical spacing of elements_____ reflector spacing_____	feed height_____ rear element length_____ array slope_____ true bearing_____ unloaded transmission line impedance_____ angle between array axis and element tips_____ geometric ratio of element length_____ number of elements_____	feed height_____ short leg length_____ angle between principle antenna axis and: one rhombic major axis_____ short leg_____ long leg_____ true bearing_____ long leg length_____ termination height_____

Type 17 - vertical with ground system
length_____
length of radials_____
radius of radials_____ m
number of radials_____

REMARKS

Fig. 3-27B. Antenna data required for HF system performance predictions.

Program Control Information:

(Box-2).

Item A: Type of tabulation (method) desired. Select the output(s) required and then choose the simplest tabulation providing it (them), the simplest one usually being the one with the fewest outputs. Computer output tabulation samples are as follows:

Tabulation-1: (Please see Fig. 3-28). Upper useful freuquency limit tabulation (MUF and FOT). You will recall that the MUF is an estimate of a frequency supported by the ionosphere for 50.0% or more of the days of the month, while FOT is the freuqency supported for 90.0% or more of the days of the month. These data are plotted hourly (while in our manual method we plotted every two hours) in UT (or GMT). The first line at the top of the page lists the month, day, and the year and also the solar activity level as indicated by both methods, i.e. 12-month running average Zurich sunspot number and the 12-month average solar 10.0 centimeter flux radiation density in units of 10^{-22} watts per square meter per 1.0-Hz bandwidth. The second and third lines contain the names and coordinates (geographical) of the transmitting and receiving sites, the azimuthal bearings (in degrees East of true North) of the receiving site from the trasmitting site, and vice versa, and also the great circle path distance in statute miles and kilometers. The fourth line indicates the minimum vertical (wave) angle used in the propagation mode selected (e.g. one-hop F2, two-hop E, etc.). The table body itself, is the predicted MUF and FOT in MHz for each hour of universal time (UT or GMT). Should the MUF involve the sporadic-E (E_s) propagation mechanism, the MUF value is followed by an asterisk.

Tabulation-2: Please see Fig. 3-29. As the FOT (Optimum Traffic Frequency) is an estimate of the highest frequency supported by the ionosphere for 90.0% (or better) of the days of the month, the LUF (or

1					(HF MUFE S4 75/10/31)	
	JANUARY 15, 1970	10 CM FLUX 145		(SSN 100)		
BOULDER, COLO.	TO ST. LOUIS, MO.		AXIMUTHS		MILES	KM.
40.03N – 105.27W	38.67N – 90.25W		91.86 281.42		807.1	1298.8
	MINIMUM ANGLE	2.0 DEGREES				
UT	MUF	FOT	UT	MUF	FOT	
01	14.0	11.1	07	8.5*	6.2	
02	11.5	9.1	08	8.4*	6.8	
03	9.5	7.5	09	8.3	6.7	
04	8.0*	6.2	10	7.9*	6.1	
05	8.2*	5.7	11	7.8*	5.2	
06	8.4*	5.6	12	7.9*	4.9	
UT	MUF	FOT	UT	MUF	FOT	
13	8.6*	6.7	19	21.8	18.6	
14	11.7	10.0	20	21.7	18.6	
15	16.1	13.7	21	21.4	18.4	
16	19.0	16.1	22	20.7	17.8	
17	20.3	17.5	23	19.3	16.6	
18	21.1	18.2	24	16.8	14.5	

Fig. 3-28. Sample upper useful frequency limit tabulation (MUF-FOT).

```
                    4         (HF MUFE S4 75/10/31)

        JANUARY 15, 1970    10 CM FLUX 145    (SSN 100)
  BOULDER, COLO.      TO ST. LOUIS, MO.         AXIMUTHS     MILES      KM.
  40.03N – 105.27W    38.67N – 90.25W        91.86   281.42   807.1   1298.8
                    MINIMUM ANGLE  0.0 DEGREES
    XMTR 2.0 TO 30.0    VERTICAL    H  –0.00 L   35.00 A  –0.0 OFF AZ   0.0
    RCVR 2.0 TO 30.0    VERTICAL    H  –0.00 L   –.50 A   –0.0 OFF AZ   0.0
  POWER= 30.00KW    3 MHZ NOISE=–148.6DBW    REQ. REL.=.90    REQ. S/N=55.0DB
```

UT	FOT	LUF	UT	FOT	LUF
01	11.1		07	6.2	
02	9.1	–2.0	08	6.8	2.9
03	7.5		09	6.7	
04	6.2	–2.0	10	6.1	2.9
05	5.7		11	5.2	
06	5.6	2.5	12	4.9	–2.0

UT	FOT	LUF	UT	FOT	LUF
13	6.7		19	18.6	
14	10.0	4.6	20	18.6	8.5
15	13.7		21	18.4	
16	16.1	7.5	22	17.8	6.6
17	17.5		23	16.6	
18	18.2	8.8	24	14.5	3.8

Fig. 3-29. Sample useful frequency range tabulation (FOT-LUF).

LUHF) is an estimate of the lowest usable frequency rendering an acceptable S/N ratio for the required time period. The frequency range between the FOT and LUF is the useful range. The first line in the heading of the computer printout (see Fig. 3-29) gives the month, day, and year and the solar activity level in 10.0 centimeter flux density in units of 10^{-22} watts per square meter per 1.0-Hz bandwidth, as well as in terms of the 12-month running average Zurich sunspot number. The second and third lines give the names and geographical coordinates of the transmitter and receiver site locations, the azimuthal bearings in degrees East of true North, from the transmitter to the receiver sites, and vice versa, as well as the path great circle distance in statute miles and kilometers. The fourth line indicates the lowest vertical (wave) angle considered in the mode selection process. The fifth and sixth lines describe the physical parameters of the transmitting and receiving antennas and the orientation of the antennas main beams relative to the great circle path. It might be mentioned here that in all methods involving antennas up to three transmitting and/or receiving antennas may be considered. It is necessary, in this case, to specify the frequency range for each antenna; however, these frequency ranges may not overlap. The seventh line contains the transmitter power output, the man-made noise at 3.0 MHz in a bandwidth of 1.0 Hz at the receiving location, the required reliability (which is needed only in an LUHF (LUF), or service probability computation), and the hourly median S/N ratio required to provide desired grade of service. Note—The concept of the term "Service Probability" may be appreciated by: the actual results of any calculated path will vary from path to path, even if all the parameters entered into calculation were identical. This is a result of the program's or formula's inability to account, in the present state of the art, for the indefinite (and their mixes) of the various propagational parameters and effects entering into the real-life propagational problem. Stated in other words, real life propagation just isn't as "neat" as the propagational formula nor the computer program. For example, say an HF ionospheric (or sky-wave) path, or any propagational path for that matter, is calculated to provide a given grade of service for a given percentage of time and the system designed is installed on one hundred different paths having (or appearing to have) identical parameters, as far as we are able to assess them. These one hundred paths, over the long term, would reveal that only a certain number of them meet the design critieria. This number, divided by the total number of paths (one hundred) is the service probability. Statistically the actual results obtained on the 100 paths describe a normal distribution specified by a mean and a standard deviation. The mean is the calculated or predicted value and the standard

```
                                       6            (HF MUFES4 75/10/31)
             JANUARY 15, 1970       10 CM FLUX 145    (SSN 100)
BOULDER, COLO.       TO ST. LOUIS, MO.          AZIMUTHS     MILES    KM.
40.03N - 105.27W     38.67N -  90.25W         91.86  281.42   807.1  1298.8
                     MINIMUM ANGLE  0.0 DEGREES
XMTR  2.0 TO 30.0         RHOMBIC    H  20.00 L 114.00 A  70.0 OFF AZ  0.0
RCVR  2.0 TO 30.0         RHOMBIC    H  23.00 L 120.00 A  68.0 OFF AZ  0.0
POWER= 30.00KW   3 MHZ NOISE=-148.6DBW   REQ. REL.=.90   REQ.S/N=55.0CB
MULTIPATH POWER TOLERANCE=10.0 DB      MULTIPATH DELAY TOLERANCE= .85 MS.
                          FREQUENCIES IN MHZ
UT   MUF    2.0   3.0   5.0   7.5  10.0  12.5  15.0  17.5  20.0  25.0  30.0

01   14.0  K(0)MUF=14.0     K(5)MUF=14.0      ESMUF=  8.1   ESHPF=16.9
      1F    3F    2F    2F    1F    1F    1F    1F    1F    -     -     -   MODE
     23.9  46.6  34.6  35.0  17.3  18.1  19.8  23.9  23.9   -     -     -   ANGLE
      5.0   6.5   5.4   5.5   4.7   4.7   4.8   5.0   5.0   -     -     -   DELAY
      337   232   234   243   242   254   277   337   337   -     -     -   VIRT HT
      .50   .99   .39   .99   .99   .97   .73   .36   .12   -     -     -   F. DAYS
      129   122   112   105    94    89    96   138   150   -     -     -   LOSS CB
       53    34    44    50    60    65    64    48    37   -     -     -   DBU
      -84   -78   -67   -60   -49   -45   -51   -93  -105   -     -     -   SIG.DBW
     -164  -144  -149  -154  -159  -162  -163  -162  -162   -     -     -   NOI.DBW
       80    66    81    94   110   117   113    69    57   -     -     -   S/N DB
      .98   .86   .99   .99   .99   .99   .99   .90   .59   -     -     -   F. S/N
      .49   .85   .98   .99   .99   .96   .73   .33   .07   -     -     -   REL.
       -    .99   .99    -     -     -     -     -     -    -     -     -   MF PROB

02   11.5  K(0)MUF=11.5     K(5)MUF=11.5      ESMUF=  7.8   ESHPF=18.1
      1F    3F    2F    1F    1F    1F    1F    1F    -     -     -     -   MODE
     24.2  46.6  34.9  17.3  19.1  19.8  24.2  24.2   -     -     -     -   ANGLE
      5.0   6.5   5.5   4.7   4.7   4.8   5.0   5.0   -     -     -     -   DELAY
      342   238   241   242   253   277   342   342   -     -     -     -   VIRT HT
      .50   .99   .99   .99   .99   .78   .34   .09   -     -     -     -   F. DAYS
      105   123   112   105    93    91   113   140   -     -     -     -   LOSS DB
       59    34    43    52    60    64    57    46   -     -     -     -   DBU
      -61   -78   -67   -60   -49   -46   -69   -95   -     -     -     -   SIG.DBW
     -163  -143  -148  -153  -158  -162  -164  -162   -     -     -     -   NOI.CBW
      102    65    80    93   109   116    95    67   -     -     -     -   S/N DB
      .99   .84   .99   .99   .99   .99   .99   .89   -     -     -     -   F. S/N
      .56   .83   .38   .99   .99   .77   .33   .08   -     -     -     -   REL.
       -    .99   .99   .99    -     -     -     -    -     -     -     -   MF PROB

03    9.5  K(0)MUF= 9.5     K(5)MUF= 9.5      ESMUF=  7.8   ESHPF=19.4
      1F    3F    2F    1F    1F    1F    1F    -     -     -     -     -   MODE
     24.7  47.4  35.7  18.0  19.4  24.7  24.7   -     -     -     -     -   ANGLE
      5.0   6.6   5.5   4.7   4.8   5.0   5.0   -     -     -     -     -   DELAY
      350   247   250   252   273   350   350   -     -     -     -     -   VIRT HT
      .50   .39   .99   .99   .90   .40   .08   -     -     -     -     -   F. DAYS
       97   122   112   104    93    98   117   -     -     -     -     -   LOSS DB
       61    34    44    53    61    61    56   -     -     -     -     -   DBU
      -52   -78   -67   -60   -48   -54   -72   -     -     -     -     -   SIG.DBW
     -161  -142  -147  -153  -158  -162  -164   -     -     -     -     -   NOI.DBW
      109    65    80    93   110   109    92   -     -     -     -     -   S/N DB
      .99   .83   .98   .93   .99   .99   .99   -     -     -     -     -   F. S/N
      .50   .82   .38   .99   .90   .39   .08   -     -     -     -     -   REL.
       -    .99   .39   .91    -     -     -    -     -     -     -     -   MF PROB
```

Fig. 3-30. Sample system performance predictions.

deviation is a measure of the prediction uncertainty. The body of the table provides the FOT (optimum traffic frequency—90.0%) at each hour of universal time (UT) and the LUF (or LUHF)—lowest usable high frequency) determined for any interval of universal time required.

Tabulation-3: (see Fig. 3-30). The first line of this tabulation, at the top of the computer read-out sheet, gives the month, day, and the year and the solar activity level in terms of the 12-month moving average Zurich sunspot number and 10.0 cm. solar flux. The second and third lines give the names and geographical coordinates of the transmitter and receiver locations, the aximuth bearings (in degrees East of true North) of the receiving site from the transmitting site, and vice versa, as well as the circuit path great circle distance in statute miles and kilometers. The fourth line gives the lowest vertical angle (minimum angle) considered in the mode selection process, while the fifth line and sixth line describe the physical parameters of each terminal's antenna as well as the antenna great circle path orientation. The seventh line indicates transmitter power output, the man-made noise at 3.0 MHz in a 1.0-Hz bandwidth at the receiving site location, the required circuit reliability (e.g. 0.9 = 90.0%) and the required S/N ratio. The eighth line indicates tolerances used in multipath determination while the ninth and tenth lines list the frequencies considered in the table body. The first line in the body of the table gives the universal time (UT or GMT—or Zulu) and the MUF (maximum usable frequency-50.0%). This is followed by an estimate of predicted MUF for a local magnetic K-Index of 0.0 and 5.0 (the expected MUF for other values of K-Index may be determined by linear interpolation or extrapolation) and the inclusion of sporadic-E (E_s) propagation modes in the MUF calculation. (Note—The K-Index is a measure of variation or disturbance in the earth's magnetic field. The K figures range from 0.0 (very quiet) to 9.0 (extremely disturbed). For each frequency in line 10, the body of the tabulation contains MODE (the mode having the greatest reliability, e.g. 2E means two-hop E-Layer reflected), ANGLE (the vertical take-off and arrival angles in degrees associated with the corresponding mode), DELAY (time delay in milliseconds), VIRT. HT.(the virtual height of the reflecting ionosphere in kilometers), F-DAYS fraction of days that any sky-wave mode exists, transmission loss in dB, dBu (median field strength at receiving location in decibels relative to 1.0 microvolt per meter, SIG, dBw (median signal power available at the receiver antenna terminals in decibels relative to 1.0 watt), NOI. dBw (median value of total noise at the receiving site location in terms of decibels relative to 1.0 watt), S/N dB—(median signal-to-noise ratio, i.e. signal in occupied bandwidth and noise in a bandwidth of 1.0 Hz) for days during which the ionosphere supports propagation, F. S/N (the number of days of a given month, expressed as a fraction, in which the required S/N ratio is predicted to attain, REL—the number of days of a given month, expressed as a fraction, that ionosphere support will be present and the required S/N ratio will be attained, and MP PROB—probability of multiple propagation paths within the power and delay tolerance indicated in the computer print-out heading. Note—Symbols denote propagation modes as follows: E (E-ionospheric layer, F (F2 ionospheric layer, E_s (sporadic E propagation modes). The time blocks are repeated for each hour interval of universal time (UT).

```
                JANUARY 15, 1970    10 CM FLUX 145       (SSN 100)      (HFMUFES4 75/10/31)
BOULDER, COLO.     TO ST. LOUIS, MO.              AZIMUTHS        MILES        KM.
40.03N - 105.27W    38.67N - 90.25W         91.86    281.42      807.1      1298.8
                                MINIMUM ANGLE  0.0 DEGREES
XMTR  2.0 TO 30.0 ARBITARY DIPOLE H  -.25  L  -.50  A  45.0 OFF AZ   0.0
RCVR  2.0 TO 30.0  CONSTANT GAIN   H   5.00 L  -0.0  A  -0.0 OFF AZ   0.0
POWER= 30.00KW   3 MHZ NOISE=-148.6DBW    REQ. REL.=.70   REQ.S/N=70.0DB
                         FREQUENCIES IN MHZ
UT    MUF   2.0    3.0    5.0    7.5   10.0   12.5   15.0   17.5   20.0   25.0   30.0

01    14.0  K(0)MUF=14.0        K(5)MUF=14.0       ESMUF=  8.1     ESHPF=16.9
       1F    2F     1F     1F    1F     1F     1F    1F     1F     -      -     -   MODE
      23.9  35.4   17.9   17.1  17.3   18.1   19.8  23.9   23.9    -      -     -   ANGLE
       .50   .99    .99    .99   .99    .97    .73   .36    .12    -      -     -   F. DAYS
        48    44     48     48    48     48     48    48     48    -      -     -   DBU
        85    78     83     84    85     86     86    85     84    -      -     -   S/N DB
       .48   .64    .91    .94   .95    .95    .71   .35    .11    -      -     -   S. PROB

02    11.5  K(0)MUF=11.5        K(5)MUF=11.5       ESMUF=  7.8     ESHPF=18.1
       1F    1F     1F     1F    1F     1F     1F    1F     -      -      -     -   MODE
      24.2  18.2   17.3   17.3  18.1   19.8   24.2  24.2    -      -      -     -   ANGLE
       .50   .99    .99    .99   .99    .78    .34   .09    -      -      -     -   F. DAYS
        48    47     47     47    48     48     48    48    -      -      -     -   DBU
        86    80     81     83    84     86     86    85    -      -      -     -   S/N DB
       .50   .88    .93    .96   .98    .77    .33   .08    -      -      -     -   S. PROB

03     9.5  K(0)MUF= 9.5        K(5)MUF= 9.5       ESMUF=  7.8     ESHPF=19.4
       1F    1F     1F     1F    1F     1F     1F    -      -      -      -     -   MODE
      24.9  17.9   17.6   19.0  19.4   24.7   24.7   -      -      -      -     -   ANGLE
       .50   .99    .99    .99   .90    .40    .08   -      -      -      -     -   F. DAYS
        48    47     47     48    48     48     48   -      -      -      -     -   DBU
        86    80     81     82    84     87     86   -      -      -      -     -   S/N DB
       .50   .84    .90    .95   .89    .39    .09   -      -      -      -     -   S. PROB

04     7.9  K(0)MUF= 7.9        K(5)MUF= 7.9       ESMUF=  8.0     ESHPF=20.4
       1F    1F     1F     1F    1F     1F     -     -      -      -      -     -   MODE
      25.3  18.3   18.2   19.0  22.3   25.3    -     -      -      -      -     -   ANGLE
       .50   .99    .99    .99   .60    .11    -     -      -      -      -     -   F. DAYS
        48    47     47     48    48     48    -     -      -      -      -     -   DBU
        85    79     80     82    84     87    -     -      -      -      -     -   S/N DB
       .49   .82    .88    .94   .60    .11    -     -      -      -      -     -   S. PROB

05     6.9  K(0)MUF= 6.9        K(5)MUF= 6.9       ESMUF=  8.2     ESHPF=20.8
       1F    1F     1F     1F    1F     -      -     -      -      -      -     -   MODE
      25.8  19.2   19.1   20.2  25.8    -      -     -      -      -      -     -   ANGLE
       .50   .99    .99    .99   .31    -      -     -      -      -      -     -   F. DAYS
        48    48     48     48    48    -      -     -      -      -      -     -   DBU
        84    80     81     82    85    -      -     -      -      -      -     -   S/N DB
       .49   .81    .88    .94   .31    -      -     -      -      -      -     -   S. PROB

06     6.8  K(0)MUF= 6.8        K(5)MUF= 6.8       ESMUF=  9.4     ESHPF=20.5
       1F    1F     1F     1F    1F     -      -     -      -      -      -     -   MODE
      25.9  20.2   19.7   20.7  25.9    -      -     -      -      -      -     -   ANGLE
       .50   .99    .99    .99   .29    -      -     -      -      -      -     -   F. DAYS
        48    48     48     48    48    -      -     -      -      -      -     -   DBU
        85    80     81     82    85    -      -     -      -      -      -     -   S/N DB
       .49   .81    .87    .94   .29    -      -     -      -      -      -     -   S. PROB
```

Fig. 3-31. Sample condensed system performance predictions showing service probability.

Tabulation-4: (See Fig. 3-31). The first line gives the date (month, day, and year) and the solar activity level in terms of both 12-month moving average 10-centimeter flux density (in units of 10^{-22} watts per square meter per Hz) and 12-month running average Zurich sunspot number. Second and third lines are names and geographic coordinates of transmitter and receiver site locations, the azimuth bearings (in degrees East of true North) of the receiver location from the transmitter location, and vice versa, plus the length of the HF ionospheric circuit path (great circle length) in statute miles and kilometers. The minimum angle on the fourth line indicates the lowest vertical angle considered in the mode selection process. The fifth and sixth lines describe the physical parameters of the antenna system for each terminal and the orientation of the antenna main lobe relative to the great circle path in all computer methods involving antennas. Up to three transmitting and/or receiving antennas may be considered, but the frequency ranges of these antennas may not overlap. The seventh line contains transmitter power output, the 3.0 MHz. man-made noise in a 1.0-Hz bandwidth assumed for the receiving site location, the required circuit reliability needed only in LUHF and service probability computations and the hourly median S/N ratio required to provide a given type of service. The eighth and ninth lines list the frequencies considered in the table body. The first line in each UT (GMT or Zulu) time block lists the MUF, estimate of expected MUF for local magnetic K-Index of 0.0 and 5.0, and MUF-E_s when sporadic E is included in the MUF calculations (expected MUF for other values of K-Index may be found by linear interpolation or extrapolation). For each frequency listed, the tabulation body contains MODE (the mode having the greatest reliability), ANGLE (vertical angle in degrees associated with this mode), F-DAYS (fraction of days during which a regular sky-wave mode is expected to exist, dBu (median incident field strength for the above mode, available at the receiving site location in terms of decibels relative to 1.0 microvolt per meter), S/N dB (median S/N ratio—signal in occupied bandwidth to noise in a 1.0-Hz bandwidth at the receiver input, in decibels, for the days during which sky-wave (ionospheric support) exists), and S. PROB (liklihood that the circuit reliability requirement will be met). Symbols denoting propagation modes are as follows: E-layer (E), F2-layer (F), E-F2-layers (X)—e.g. 3X = one-hop E and two-hops F2. Sporadic E (E_s) modes not included.

Tabulation-5: (see Fig. 3-21). First line at top of page (below the page number) gives month, day, and year with corresponding solar activity level in terms of 12-month moving average 10 centimeter flux density in units of 10^{-22} watts per square meter per Hz and also 12-month running average Zurich sunspot number. The second and third lines give the names and geographical coordinates of the transmitter and receiver site locations, azimuth bearings (in degrees East of true North) of receiver from transmitter, and vice versa, as well as the circuit great circle distance in statute miles and kilometers. The minimum angle on heading line 4 indicates the lowest vertical angle considered in the mode selection process. The fifth and sixth lines contain the list of frequencies considered in the table body. The first line in each UT time block in the table body gives the MUF, estimated MUF for local magnatic k-index of 0.0 and 5.0 and last MUF when sporadic-E (E_s) are included in the computerization (expected MUF for other values of K-Index may be determined by linear interpolation or extrapolation). For each listed frequency the tabulation body contains MODE (the mode having the greatest reliability—number of hops and layer), ANGLE (vertical angle in degrees associated with this mode), DELAY (time delay in milliseconds), VIRT. HT. (virtual height of ionospheric reflection in kilometers), and F. DAYS (fraction of days that any sky-wave mode is expected to exist). Symbols used to denote propagation modes are as follows: E-layer (E), F2-layer (F), E-F2(X)—e.g. 3X describes one E-Layer hop and two hops by the F2-layer. Sporadic-E (E_s) modes not included.

```
                                      21            (HFMUFES4 75/10/31)
             JANUARY 15, 1976     10 CM FLUX 145   (SSN 100)
BOULDER, COLO.    TO ST. LOUIS, MO.          AZIMUTHS     MILES      KM.
40.03N - 105.27W    38.67N - 90.25W       91.86  281.42   807.1    1298.8
                         MINIMUM ANGLE  0.0 DEGREES
                              FREQUENCIES IN MHZ
UT    MUF   2.0   3.0   5.0   7.5  10.0  12.5  15.0  17.5  20.0  25.0  30.0

01   14.0   K(0)MUF=14.0         K(5)MUF=14.0    ESMUF= 8.1    ESHPF=16.9
      1F    1E    1E    1F    1F    1F    1F    1F    1F    1F    -    -    MODE
     23.9   5.2   5.6  17.1  17.3  18.1  19.6  23.9  23.9  23.9   -    -    ANGLE
      5.0   4.4   4.7   4.7   4.7   4.7   4.8   5.0   5.0   5.0   -    -    DELAY
      337   93    98   235   242   254   277   337   337   337    -    -    VIRT HT
      .50   .99   .99   .99   .99   .97   .73   .36   .12   .03   -    -    F. DAYS

02   11.5   K(0)MUF=11.5         K(5)MUF=11.5    ESMUF= 7.8    ESHPF=18.1
      1F    1E    1F    1F    1F    1F    1F    1F    1F    -     -    -    MODE
     24.2   5.7  17.3  17.3  18.1  19.8  24.2  24.2  24.2   -     -    -    ANGLE
      5.0   4.4   4.7   4.7   4.7   4.8   5.0   5.0   5.0   -     -    -    DELAY
      342   99   237   242   253   277   342   342   342    -     -    -    VIRT HT
      .50   .99   .99   .99   .99   .78   .34   .09   .01   -     -    -    F. DAYS

03    9.5   K(0)MUF= 9.5         K(5)MUF= 9.5    ESMUF= 7.8    ESHPF=19.4
      1F    1F    1F    1F    1F    1F    1F    -     -     -     -    -    MODE
     24.7  17.8  17.6  19.0  19.4  24.7  24.7   -     -     -     -    -    ANGLE
      5.0   4.7   4.7   4.7   4.8   5.0   5.0   -     -     -     -    -    DELAY
      350  242   244   252   273   350   350    -     -     -     -    -    VIRT HT
      .50   .99   .99   .99   .90   .40   .08   -     -     -     -    -    F. DAYS

04    7.9   K(0)MUF= 7.9         K(5)MUF= 7.9    ESMUF= 8.0    ESHPF=20.4
      1F    1F    1F    1F    1F    1F    -     -     -     -     -    -    MODE
     25.3  18.3  19.2  19.0  22.3  25.3   -     -     -     -     -    -    ANGLE
      5.1   4.7   4.7   4.8   4.9   5.1   -     -     -     -     -    -    DELAY
      359  250   254   266   314   359    -     -     -     -     -    -    VIRT HT
      .50   .99   .99   .99   .60   .11   -     -     -     -     -    -    F. DAYS

05    6.9   K(0)MUF= 6.9         K(5)MUF= 6.9    ESMUF= 8.2    ESHPF=20.8
      1F    1E    1F    1F    1F    1F    -     -     -     -     -    -    MODE
     25.8   6.5  19.1  20.2  25.8  25.8   -     -     -     -     -    -    ANGLE
      5.1   4.4   4.8   4.8   5.1   5.1   -     -     -     -     -    -    DELAY
      367  108   264   282   367   367    -     -     -     -     -    -    VIRT HT
      .50   .99   .99   .99   .31   .02   -     -     -     -     -    -    F. DAYS

06    6.8   K(0)MUF= 6.8         K(5)MUF= 6.8    ESMUF= 8.4    ESHPF=20.5
      1F    1E    1F    1F    1F    1F    -     -     -     -     -    -    MODE
     25.9   5.8  19.7  20.7  25.9  25.9   -     -     -     -     -    -    ANGLE
      5.1   4.4   4.8   5.1   5.1   5.1   -     -     -     -     -    -    DELAY
      368  100   272   289   368   368    -     -     -     -     -    -    VIRT HT
      .50   .99   .99   .99   .29   .02   -     -     -     -     -    -    F. DAYS
```

Fig. 3-32. Sample tabulation of propagation path geometry.

Tabulation-6: (see Fig. 3-33). The first line in the heading (below the page number), shows the date (month, day, and year) and the solar activity (12-month moving average 10.0 centimeter flux density in units of 10^{-22} watts per square meter per Hz bandwidth) and also 12-month running average Zurich sun spot number (ZSSN). The second and third lines give the names and geographic coordinates of transmitter and receiver site locations, azimuth bearing (in degrees East of true North) of the receiver site from the transmitter site, and vice versa, and the great circle HF path circuit distance in both statute miles and kilometers. The fourth line indicates the minimum vertical antenna angle (firing angle) considered in the mode-selection process. The fifth and sixth lines in the heading of the computerized readout describe the physical parameters of the antenna systems for each of the terminals (transmitting and receiving) and orientation of the antenna main lobe relative to the HF circuit great circle path (in all methods involving antennas, up to three transmitting and/or receiving antennas may be considered; the frequency range for each such antenna must be specified and no frequency range may not overlap). The seventh

```
                                    26
                                                    (HF MUFES4 75/10/31)

              JANUARY 15, 1970    10 CM FLUX 145    (SSN 100)
       BOULDER, COLO.    TO ST. LOUIS, MO.      AZIMUTHS     MILES     KM.
       40.03N - 105.27W    38.67N - 90.25W      91.86      281.42    807.1   1298.8
                         MINIMUM ANGLE  0.0 DEGREES
       XMTR 2.0 TO 30.0   INVERTED L     H  10.00 L 21.30 A  -0.0 OFF AZ    0.0
       RCVR 2.0 TO 30.0 TERM. SLOPING V H  15.20 L 121.90 A  22.5 OFF AZ    1.4
    POWER= 30.00KW   3 MHZ NOISE=-148.6DBW    REQ. REL.=.90    REQ. S/N=55.0DB
                                  RELIABILITIES
```

UT	MUF	2.0	3.0	5.0	7.5	10.0	12.5	15.0	17.5	20.0	25.0	30.0	MUF
01	14.0	.54	.83	.96	.99	.96	.72	.35	.12	-	-	-	.49
02	11.5	.53	.80	.96	.99	.77	.33	.08	-	-	-	-	.49
03	9.5	.50	.78	.96	.90	.39	.08	-	-	-	-	-	.50
04	7.9	.48	.78	.96	.60	.11	-	-	-	-	-	-	.50
05	6.9	.49	.79	.96	.31	-	-	-	-	-	-	-	.49
06	6.8	.49	.79	.96	.29	-	-	-	-	-	-	-	.49
07	7.5	.48	.78	.96	.49	.06	-	-	-	-	-	-	.50
08	8.2	.47	.78	.96	.70	.14	-	-	-	-	-	-	.50
09	8.3	.46	.77	.96	.74	.05	-	-	-	-	-	-	.50
10	7.5	.48	.78	.95	.51	-	-	-	-	-	-	-	.50
11	6.4	.53	.81	.90	.10	-	-	-	-	-	-	-	.49
12	6.1	.62	.85	.87	.04	-	-	-	-	-	-	-	.50
13	7.9	.42	.78	.97	.66	-	-	-	-	-	-	-	.50
14	11.7	-	.36	.95	.98	.89	.25	-	-	-	-	-	.48
15	16.1	-	-	.76	.98	.98	.94	.71	.17	-	-	-	.47
16	19.0	-	-	.61	.98	.98	.96	.94	.73	.28	-	-	.48
17	20.3	-	-	-	.88	.99	.97	.98	.88	.55	-	-	.49
18	21.1	-	-	-	.86	.99	.97	.98	.92	.68	.04	-	.49
19	21.6	-	-	-	.86	.99	.97	.98	.94	.74	.06	-	.48
20	21.7	-	-	-	.90	.99	.98	.99	.95	.75	.07	-	.49
21	21.4	-	-	.66	.93	.99	.98	.99	.94	.72	.05	-	.49
22	20.7	-	-	.82	.99	.99	.98	.98	.91	.62	-	-	.49
23	19.3	-	-	.89	.99	.99	.98	.96	.78	.36	-	-	.49
24	16.8	.30	.75	.97	.99	.99	.97	.84	.35	.04	-	-	.48

Fig. 3-33. Sample tabulation of maximum usable frequency and circuit reliability.

line shows the transmitter power output, the 3.0 MHz man-made noise in a 1.0-Hz bandwidth at the receiving site location, the circuit reliability required (required circuit reliability needed only in the LUHF (LUF) and service probability computations), and the hourly median S/N ratio required to provide the type of service requested. The eighth line gives the subject heading while the ninth line lists the frequencies (MHz) considered. The body of the table reflects the hour in UT, the MUF in MHz, and the reliabilities for the frequencies in the frequency complement, including the circuit reliability at the MUF A dash (—) indicates that the reliability is less than 0.01 or 1.0%.

Tabulation-7: (see Fig. 3-34A and 3-34B). This computerized output shows in detail the various parameters involved in the system performance predictions by tabulating the ray path geometries used, the ionospheric characteristics involved, and the expected performance of each ionospheric

```
                                       27                   (HFMUFES4 75/10/31)
               JANUARY 15, 1970      10 CM FLUX 145    (SSN 100)
BOULDER, COLO.      TO ST. LOUIS, MO.           AZIMUTHS        MILES      KM.
40.03N - 105.27W       38.67N -  98.25W       91.86   281.42    807.1    1298.8
                             MINIMUM ANGLE   0.0 DEGREES
XMTR  2.0 TO 30.0           RHOMBIC      H  20.00 L 114.00 A  70.0 OFF AZ   0.0
RCVR  2.0 TO 30.0           RHOMBIC      H  23.00 L 120.00 A  68.0 OFF AZ   0.0
POWER= 30.00KW     3 MHZ NOISE=-148.60BW    REQ. REL.=.90   REQ.S/N=55.0LB

                              UT = 12

                        REFLECTION AREA DATA
                            1         2         3         4         5
DISTANCE FROM TX        649.397   324.699   974.096   649.397   649.397
LATITUDE                 39.592    39.873    39.190    39.592    39.592
LONGITUDE                97.686   101.466    93.944    97.686    97.686
GEOMAGNETIC LAT          49.403    49.228    49.404    49.403    49.403
TIME IN HOURS             5.488     0.000     0.000     0.000     0.000
ABSORP. FACTOR             .010     0.000     0.000     0.000     0.000
E-LAYER CRITICAL           .805     0.000     0.000     0.000     0.000
F-LAYER BOTTOM          266.082     0.000     0.000     0.000     0.000
HEIGHT OF FMAX          337.978     0.000     0.000     0.000     0.000
GYRO-FREQUENCY            1.352     0.000     0.000     0.000     0.000
F-LAYER CRITICAL          3.417     0.000     0.000     0.000     0.000
ES MEDIAN                 1.737     0.000     0.000     0.000     0.000

TIME AT RECEIVER    5.98   ABSORP. FACTOR      .01   E-LAYER CRITICAL     .81
GYRO-FREQUENCY      1.52   HEIGHT OF FMAX   336.86   SEMI-THICKNESS     71.90
F-LAYER CRITICAL    3.42   MUF              6.10    FOT    4.94   HPF    7.14
EXCESS SYS LOSS     9.30   ESM              1.71    ESL    1.23   ESU    3.48
   K=0    K=1    K=2    K=3    K=4    K=5    K=6    K=7    K=8    K=9
  6.10   6.10   6.10   6.10   6.10   6.10   6.10   6.10   6.10   6.10

                              FREQUENCY =   6.101 MHZ
NOISE-ATMOSPHERIC   GALACTIC   MAN-MADE   ADJUSTED   EFFICIENCY   CONTROLLING
     -158.80        -171.78    -157.66      0.00       -1.70        -156.79
                    E-MODES               F-MODES                  EF-MODES
                        1        2        1        2        3        5        6
NUMBER OF HOPS          1        2        1        2        3        -        -
TAKE-OFF ANGLE        .00      .00    26.86    49.12    62.11        -        -
VIRTUAL HEIGHT        .00      .00   381.90   400.76   440.18        -        -
TIME DELAY IN MS        -        -     5.14     7.04        -        -        -
SKY WAVE LOSS           -        -   111.31   114.64        -        -        -
ABSORPTION LOSS         -        -      .12      .15        -        -        -
GROUND REF. LOSS        -        -      .00     7.52        -        -        -
XMTR ANT. GAIN          -        -    12.40   -10.00        -        -        -
RCVR ANT. GAIN          -        -    13.81   -10.00        -        -        -
TRANSMISSION LOS        -        -    95.12   151.01        -        -        -
FIELD STRENGTH          -        -    58.75    26.07        -        -        -
SIGNAL POWER            -        -   -50.35  -100.84        -        -        -
S/N IN DB               -        -   106.44    43.25        -        -        -
FRACTION OF DAYS        -        -      .50      .00        -        -        -
FRACTION OF S/N         -        -      .99      .18        -        -        -
RELIABILITY             -        -      .50      .00        -        -        -
SERVICE PROBABLE        -        -      .50      .00        -        -        -
```

Fig. 3-34. Sample tabulation of system performance of individual modes.

```
                      FREQUENCY =   2.000 MHZ
NOISE-ATMOSPHERIC  GALACTIC   MAN-MADE   ADJUSTED   EFFICIENCY   CONTROLLING
    -150.65        -204.00    -143.42      0.00       -1.70        -144.27
                   E-MODES              F-MODES                  EF-MODES
NUMBER OF HOPS        1          2         1          2        3     0       0
TAKE-OFF ANGLE      5.33       17.22     21.80      39.09    51.34   -       -
VIRTUAL HEIGHT     94.93      110.78    270.91     276.58   282.10   -       -
TIME DELAY IN MS    4.41       4.61      4.86       5.82     7.24    -       -
SKY WAVE LOSS     100.89      101.29    101.73     103.31   105.20   -       -
ABSORPTION LOSS     .97        1.01       .42        .53      .65    -       -
GROUND REF. LOSS    .06        3.94       .00       4.55     9.18    -       -
XMTR ANT. GAIN    -10.00      -10.00    -10.00     -2.92      .26    -       -
RCVR ANT. GAIN    -10.00      -10.00     -8.87      -.38     2.34    -       -
TRANSMISSION LOS  131.06      135.54    130.33     120.99   121.73   -       -
FIELD STRENGTH     38.93       32.45     36.54      37.39    33.92   -       -
SIGNAL POWER      -86.29      -90.77    -85.56     -76.21   -76.96   -       -
S/N IN DB          56.28       51.83     58.71      68.05    67.31   -       -
FRACTION OF DAYS    .99         .99       .99        .99      .99    -       -
FRACTION OF S/N     .56         .35       .65        .39      .38    -       -
RELIABILITY         .55         .35       .65        .38      .87    -       -
SERVICE PROBABLE    .00         .00       .01        .42      .36    -       -


                      FREQUENCY =   3.000 MHZ
NOISE-ATMOSPHERIC  GALACTIC   MAN-MADE   ADJUSTED   EFFICIENCY   CONTROLLING
    -152.67        -204.00    -148.60      0.00       -1.70        -148.86
                   E-MODES              F-MODES                  EF-MODES
NUMBER OF HOPS        1          2         1          2        3     0       0
TAKE-OFF ANGLE      6.15        .00     20.44      40.38    53.00    -       -
VIRTUAL HEIGHT    104.66        .00    276.26     294.90   312.41    -       -
TIME DELAY IN MS   -.02         -        4.82       5.95     7.69    -       -
SKY WAVE LOSS     104.44         -     105.18     107.02   109.25    -       -
ABSORPTION LOSS     .62         -        .33        .38      .47     -       -
GROUND REF. LOSS    .00         -        .00       5.54    11.34     -       -
XMTR ANT. GAIN    -10.00        -      -3.76       4.79     4.42     -       -
RCVR ANT. GAIN    -10.00        -       -.54       6.80     5.32     -       -
TRANSMISSION LOS  134.36        -     119.11     110.65   120.62     -       -
FIELD STRENGTH     37.16        -       42.95     44.07    39.58     -       -
SIGNAL POWER      -89.59        -      -74.34    -65.38   -75.84     -       -
S/N IN DB          57.58        -       74.53     82.98    73.02     -       -
FRACTION OF DAYS    .99         -        .99        .97      .91     -       -
FRACTION OF S/N     .61         -        .95        .39      .94     -       -
RELIABILITY         .60         -        .95        .96      .86     -       -
SERVICE PROBABLE    .00         -        .90        .97      .76     -       -
```

Fig. 3-34. (Continued). Sample Tabulation of System Performance of Individual Modes (Tabulation Number 7)

mode for selected frequencies and times. Beginning at the top of the printout sheet, (below the page number) the first line, as usual, shows month, day, year and the solar activity level as indicated by the 12-month moving average 10-centimeter flux in units of 10^{-22} watts per square meter per one Hz of bandwidth and the 12-month running average Zurich sun spot number (ZSSN). The second and third lines contain the name and geographical coordinates (in degrees East of true North) of the receiver site location from that of the transmitter site location, and vice versa, and the HF circuit great circle distance in both statute miles and kilometers. The fourth heading-line indicates the minimum vertical antenna radiation angle considered in the mode selection process. The fifth and sixth lines in the heading describe the physical parameters of the antenna system for each terminal receiver and transmitter) and the orientation of the antenna main beam relative to the great circle path (in methods requiring antennas, up to three receiving and/or transmitting antennas may be considered, but the frequency ranges of these antennas may not overlap and must be specified). The seventh line gives the transmitter power, the 3.0 MHz man-made noise in a 1.0 -Hz bandwidth at the receiving site location, the required circuit reliability (needed only in a LUHF (or LUF) or service probability computation) and the hourly median S/N ratio required to provide the requested service.

The table body proper of the computer readout sheet is divided into time blocks, one block for each hour of UT as desired (the UT = 12 time block is shown in our example). These time blocks are further divided into two types of tabulation concerning ionospheric reflection data followed by a tabulation for selected frequencies which always include the MUF (maximum usable frequency for 50.0% of the time or days of the month). Each major time block is headed by the UT (GMT or Zulu) time.

Following this major heading, the first tabulation shows the ionospheric characteristics for each sample area used to describe the characteristics of the sky-wave propagation path (REFLECTION AREA DATA). The sample areas, identified by numbers from 1 to 5, appear as table captions. For each sample area, the following are tabulated: DISTANCE FROM TRANSMITTER (distance in kilometers along the great circle path from transmitter toward the receiver to each ionospheric reflection (actually refraction) are where the ionosphere is sampled), LATITUDE (the latitude in degrees of sample location, South latitudes being identified by a preceding minus (−) sign), LONGITUDE (longitude of sample location, East logitude being identified by a preceding minus (−) sign), GEOMAGNETIC LATITUDE (the latitude related to the geomagnetic instead of the geographic equator), TIME IN HOURS (local standard time at the sample location), ABSORPTION FACTOR (index of ionospheric absorption depending primarily upon the solar zenith angle), F-LAYER BOTTOM (height of lower part of the ionospheric E-region in kilometers), HEIGHT OF F-MAX (height of F-region in kilometers, GYRO FREQUENCY (gyro-frequency in MHz at a 100.0 kilometer height) F-LAYER CRITICAL (critical frequency of the F-region in MHz), E_s MEDIAN (monthly median frequency supported by sporadic-E at vertical incidence. Note—The F-heights have not been reduced by E-layer retardation.

Following the "Reflection Area Data," there is additional computerized tabulated data. These include TIME AT RECEIVER (local standard time at the receiver site location), ABSORB FACTOR (absorption for the total path derived from the absorption factor at sample areas) F-LAYER CRITICAL (path E-region critical frequency used in calculating mixed modes), GYRO-FREQUENCY (path gyro-frequency for calculation of ionospheric absorption), HEIGHT OF F-MAX (the path F-region height at maximum ionization), SEMI-THICKNESS (path semi-thickness of the F-region, F-LAYER CRITICAL (path F-Layer critical used to calculate active ionospheric transmission modes), MUF (path monthly median of the maximum frequency propagated via the ionosphere), FOT (path lower percentile of this maximum frequency), HPF (path upper percentile of this maximum frequency), EXCESS SYS LOSS (estimate of propagation losses in dB which are not specifically included in the computations), ESM (monthly median sporadic-E in MHz for the path), ESL (lower decile of sporadic-E) ESU (upper decile of sporadic-E). (Note—These heights *have* been reduced by the E-layer retardation factor.)

The balance of this second part of the computerized tabulation shows the expected variation of the MUF as a function of the daily magnetic K-Index. The next portion of the tabulation considers the expected performance of the various propagation modes for a specific frequency, e.g. the MUF. The frequency is shown as a caption for this part of the table. Immediately following the frequency caption, the noise level at the receiving station is shown: NOISE ATOMSPHERIC (expected monthly median atmospheric noise level for the frequency being considered, in dB referenced to one watt of power, for a bandwidth of 1.0 Hz), GALACTIC (expected galactic noise as above), MAN MADE (expected man-made noise as above), ADJUSTED (no longer used), EFFICIENCY (efficiency of the receiving antenna system), and CONTROLLING (a summation of the external noise adjusted for receiving antenna efficiency). The controlling noise is expressed in dBw in a 1.0-Hz bandwidth. The balance of this tabulation shows parameters associated with the expected modes of propagation (E-MODES, F-MODES, and E-F MODES). For each mode, the tabulation includes NUMBER OF HOPS (the number of ionospheric reflections considered for each mode), TAKE-OFF ANGLE (the vertical take-off angle or angle of departure and arrival corresponding to the number of hops, VIRTUAL HEIGHT (apparent height of ionospheric reflection for the mode), TIME DELAY IN MS (propagation time for the particular mode in milliseconds) SKY-WAVE LOSS (loss in dB due to free-space spreading of the transmitted energy as it travels via the sky-wave), ABSORPTION LOSS (loss in dB due to absorption in the lower ionosphere), GROUND REF LOSS (loss in dB due to ground reflection(s) (TRANSMITTER ANTENNA GAIN (gain at the vertical angle of the transmitting antenna in dBi), RECEIVER ANTENNA GAIN (corresponding receiver antenna gain in dBi), TRANSMISSION LOSS (summation of the above losses including the excess system loss and antenna gain), SIGNAL POWER (estimate of receiver signal power in dBw at the receiver input), S/N n dB (total signal power relative to the noise in a bandwidth of 1.0 Hz at the receiver input), FRACTION OF DAYS (percentage of days within the month during which this mode of propagation is expected), FRACTION OF S/N (fraction of days that the required S/N ratio is expected to be equalled or exceeded), RELIABILITY (percentage of days within the month that the required S/N ratio is expected to be equalled or exceeded and the mode is expected to be supported by the ionosphere, i.e. the product of the FRACTION OF DAYS and FRACTION OF S/N), SERVICE PROBABILITY (liklihood that the H.F. circuit reliability will be met). The portion of the table below the frequency caption is repeated for additional frequencies as required.

Tabulation-8: (see Fig. 3-35). This tabulation shows the expected MUF change as a function of Magnetic K-Index (K_p) from 0.0 to 9.0. A description of the HF circuit parameters used in the calculations of the system performance predictions is shown in the heading of each tabulation. Beginning with the first heading line below the page number is the date (month, day, year) and the solar activity phase (12-month moving average 10.0 centimeter flux density in units of 10^{-22} watts per square meter per 1.0-Hz bandwidth and the 12-month running average Zurich sun spot number). The second and third lines in the heading contain the names and the geographical coordinates of the transmitting site and the receiving site locations, azimuth bearings (in degrees East of true North) of the receiver site location from the transmitter site location, and vice versa, and the great circle path distance of the HF circuit in statute miles and also in kilometers.

```
                                    30                    (HFMUFES4 75/10/31)
            JANUARY 15, 1976    10 CM FLUX 145      (SSN 100)
BOULDER, COLO.    TO ST. LOUIS, MO.           AZIMUTHS      MILES      KM.
40.03N - 105.27W    38.67N - 90.25W     91.86   281.42     807.1    1298.8
                            MINIMUM ANGLE  0.0 DEGREES
              PREDICTED MUF MODIFIED FOR LOCAL MAGNETIC K-INDEX
UT    MUF     K=0   K=1   K=2   K=3   K=4   K=5   K=6   K=7   K=8   K=9

01    14.0   14.0  14.0  14.0  14.0  14.0  14.0  14.0  14.0  14.0  14.0
02    11.5   11.5  11.5  11.5  11.5  11.5  11.5  11.5  11.5  11.5  11.5
03     9.5    9.5   9.5   9.5   9.5   9.5   9.5   9.5   9.5   9.5   9.5
04     7.9    7.9   7.9   7.9   7.9   7.9   7.9   7.9   7.9   7.9   7.9
05     6.9    6.9   6.9   6.9   6.9   6.9   6.9   6.9   6.9   6.9   6.9
06     6.8    6.8   6.8   6.8   6.8   6.8   6.8   6.8   6.8   6.8   6.8
07     7.5    7.5   7.5   7.5   7.5   7.5   7.5   7.5   7.5   7.5   7.5
08     8.2    8.2   8.2   8.2   8.2   8.2   8.2   8.2   8.2   8.2   8.2
09     8.3    8.3   8.3   8.3   8.3   8.3   8.3   8.3   8.3   8.3   8.3
10     7.5    7.5   7.5   7.5   7.5   7.5   7.5   7.5   7.5   7.5   7.5
11     6.4    6.4   6.4   6.4   6.4   6.4   6.4   6.4   6.4   6.4   6.4
12     6.1    6.1   6.1   6.1   6.1   6.1   6.1   6.1   6.1   6.1   6.1
13     7.9    8.6   8.3   8.1   7.8   7.6   7.3   7.1   6.8   6.6   6.3
14    11.7   12.8  12.4  12.0  11.7  11.3  10.9  10.6  10.2   9.8   9.4
15    16.1   17.5  17.0  16.5  16.0  15.4  14.9  14.4  13.9  13.4  12.9
16    19.0   20.0  19.5  19.0  18.5  18.0  17.5  17.0  16.5  16.0  15.4
17    20.3   21.5  20.9  20.4  19.8  19.3  18.8  18.2  17.7  17.1  16.6
18    21.1   22.3  21.7  21.2  20.6  20.0  19.5  18.9  18.3  17.8  17.2
19    21.6   21.6  21.6  21.6  21.6  21.6  21.6  21.6  21.6  21.6  21.6
20    21.7   21.7  21.7  21.7  21.7  21.7  21.7  21.7  21.7  21.7  21.7
21    21.4   21.4  21.4  21.4  21.4  21.4  21.4  21.4  21.4  21.4  21.4
22    20.7   20.7  20.7  20.7  20.7  20.7  20.7  20.7  20.7  20.7  20.7
23    19.3   19.3  19.3  19.3  19.3  19.3  19.3  19.3  19.3  19.3  19.3
24    16.8   16.9  16.8  16.8  16.8  16.8  16.8  16.8  16.8  16.8  16.8
```

Fig. 3-35. Sample tabulation of maximum usable frequency showing expected MUF change with local magnetic K index.

The minimum angle on the fourth line indicates the lowest vertical angle (antenna firing angle) considered in the mode selection process. The body of the table is the expected MUF and its variation as a function local magnetic K-Index for each hour of UT (universal time).

Tabulation-9: This tabulation displays the vertical HF antenna pattern in terms of dBi. The first line in the heading identifies the tabulation, i.e. ANTENNA PATTERN. The second and third lines describe FREQUENCY RANGE (range of frequencies considered), ANTENNA TYPE (physical type of antenna), HEIGHT (antenna height in meters or when prefixed by a minus sign antenna height in wavelengths), LENGTH (length of the antenna in meters or when prefixed by a minus sign the antenna

```
                                                    ANTENNA PATTERN
 FREQUENCY RANGE   ANTENNA TYPE      HEIGHT    LENGTH       ANGLE
  2.0 TO   30.0      VERTICAL        0.000     10.673       0.000
              2     3     4     5     6     7     8     9    10    11
         90 -10.0 -10.0 -10.0 -10.0 -10.0 -10.0 -10.0 -10.0 -10.0 -10.0
         88 -10.0 -10.0 -10.0 -10.0 -10.0 -10.0 -10.0 -10.0 -10.0 -10.0
         86 -10.0 -10.0 -10.0 -10.0 -10.0 -10.0 -10.0 -10.0 -10.0 -10.0
         84 -10.0 -10.0 -10.0 -10.0 -10.0 -10.0 -10.0 -10.0 -10.0 -10.0
         82 -10.0 -10.0 -10.0 -10.0 -10.0 -10.0 -10.0 -10.0 -10.0 -10.0
         80 -10.0 -10.0 -10.0 -10.0 -10.0 -10.0 -10.0 -10.0 -10.0 -10.0
         78  -9.0  -9.2  -9.4  -9.7 -10.0 -10.0 -10.0 -10.0 -10.0 -10.0
         76  -7.7  -7.8  -8.1  -8.4  -8.7  -9.2  -9.8 -10.0 -10.0 -10.0
         74  -6.6  -6.7  -6.9  -7.2  -7.6  -8.0  -8.7  -9.5 -10.0 -10.0
         72  -5.6  -5.7  -5.9  -6.2  -6.5  -7.0  -7.6  -8.4  -9.4 -10.0
      E  70  -4.7  -4.8  -5.0  -5.3  -5.6  -6.1  -6.7  -7.4  -8.4  -9.8
      L  68  -3.9  -4.0  -4.2  -4.5  -4.8  -5.2  -5.8  -6.5  -7.5  -8.9
      E  66  -3.2  -3.3  -3.5  -3.7  -4.1  -4.5  -5.0  -5.7  -6.5  -7.9
      V  64  -2.5  -2.6  -2.8  -3.1  -3.4  -3.8  -4.3  -5.0  -5.8  -7.1
      A  62  -1.9  -2.0  -2.2  -2.4  -2.7  -3.1  -3.6  -4.2  -5.1  -6.3
      T  60  -1.4  -1.5  -1.6  -1.9  -2.1  -2.5  -3.0  -3.6  -4.4  -5.5
      I  58   -.9  -1.0  -1.1  -1.3  -1.6  -1.9  -2.4  -2.9  -3.7  -4.7
      O  56   -.4   -.5   -.6   -.8  -1.1  -1.4  -1.8  -2.4  -3.1  -4.0
      N  54    .0   -.0   -.2   -.4   -.6   -.9  -1.3  -1.8  -2.4  -3.3
         52    .5    .4    .2    .1   -.2   -.4   -.8  -1.3  -1.9  -2.7
      A  50    .8    .7    .6    .5    .3    .0   -.3   -.7  -1.3  -2.1
      N  48   1.2   1.1   1.0    .9    .7    .4    .1   -.3   -.6  -1.5
      G  46   1.5   1.4   1.3   1.2   1.0    .8    .6    .2   -.3   -.9
      L  44   1.8   1.8   1.7   1.5   1.4   1.2   1.0    .6    .2   -.3
      E  42   2.1   2.0   2.0   1.9   1.7   1.6   1.3   1.0    .7    .2
         40   2.4   2.3   2.2   2.2   2.0   1.9   1.7   1.4   1.1    .7
      I  38   2.6   2.6   2.5   2.4   2.3   2.2   2.0   1.8   1.5   1.2
      N  36   2.9   2.8   2.8   2.7   2.6   2.5   2.3   2.2   1.9   1.6
         34   3.1   3.0   3.0   2.9   2.9   2.8   2.6   2.5   2.3   2.0
      D  32   3.3   3.2   3.2   3.2   3.1   3.0   2.9   2.8   2.7   2.4
      E  30   3.5   3.4   3.4   3.4   3.3   3.3   3.2   3.1   3.0   2.8
      G  28   3.6   3.6   3.6   3.6   3.5   3.5   3.4   3.4   3.3   3.2
      R  26   3.8   3.8   3.7   3.7   3.7   3.7   3.7   3.6   3.6   3.5
      E  24   3.9   3.9   3.9   3.9   3.9   3.9   3.9   3.9   3.8   3.9
      E  22   4.0   4.0   4.0   4.0   4.0   4.0   4.1   4.1   4.1   4.1
      S  20   4.2   4.1   4.1   4.2   4.2   4.2   4.2   4.3   4.3   4.3
         18   4.3   4.2   4.3   4.3   4.3   4.3   4.4   4.4   4.5   4.5
         16   4.3   4.3   4.3   4.4   4.4   4.4   4.5   4.6   4.6   4.7
         14   4.4   4.4   4.4   4.4   4.5   4.5   4.6   4.7   4.8   4.9
         12   4.5   4.5   4.5   4.5   4.5   4.6   4.7   4.8   4.9   5.0
         10   4.5   4.5   4.5   4.5   4.6   4.6   4.7   4.8   4.9   5.1
          8   4.5   4.5   4.5   4.5   4.6   4.6   4.7   4.8   4.9   5.1
          6   4.5   4.4   4.4   4.4   4.5   4.5   4.6   4.7   4.9   5.0
          4   4.4   4.3   4.3   4.3   4.3   4.4   4.4   4.5   4.6   4.8
          2   4.0   3.8   3.7   3.6   3.6   3.6   3.6   3.7   3.8   3.9
          0 -10.0 -10.0 -10.0 -10.0 -10.0 -10.0 -10.0 -10.0 -10.0 -10.0
              2     3     4     5     6     7     8     9    10    11
                                           FREQUENCIES IN MEGAHERTZ
```

Fig. 3-36. Sample antenna pattern tabulation.

length is in terms of wavelengths), ANGLE (antenna orientation relative to the earth's tangential plane in degrees), AZIMUTH (the azimuth upon which the pattern was calculated relative to the antenna in degrees—the extra descriptors EX(1), EX(2), EX(3) and EX(4) are used to describe extra (additional) antenna parameters if required.). The last two entries on heading lines 2 and 3 describe the conductivity (CONDUCT) in Mhos/Meter (also known as Siemens/Meter) and the dielectric constant

(HFMUFES4 75/10/31)

```
AZIMUTH    EX(1)     EX(2)     EX(3)     EX(4)   CONDUCT.   DIELECT.
 0.000     0.000     0.000     0.000     0.000     5.000     80.000
   12       13        14       16        18        20       22       24        26        28       30
 -10.0    -10.0    -10.0    -10.0    -10.0    -10.0    -10.0    -10.0    -10.0    -10.0    -10.0  90
 -10.0    -10.0    -10.0    -10.0    -10.0    -10.0    -10.0    -10.0    -10.0    -10.0    -10.0  88
 -10.0    -10.0    -10.0    -10.0    -10.0    -10.0    -10.0    -10.0    -10.0    -10.0    -10.0  86
 -10.0    -10.0    -10.0    -10.0    -10.0     -8.6     -9.5    -10.0    -10.0    -10.0    -10.0  84
 -10.0    -10.0    -10.0    -10.0     -9.5     -6.1     -7.0    -10.0    -10.0    -10.0    -10.0  82
 -10.0    -10.0    -10.0    -10.0     -7.6     -4.2     -5.0     -7.9    -10.0    -10.0    -10.0  80   E
 -10.0    -10.0    -10.0    -10.0     -6.2     -2.6     -3.4     -6.2    -10.0    -10.0    -10.0  78   L
 -10.0    -10.0    -10.0    -10.0     -4.9     -1.3     -2.0     -4.8     -9.2    -10.0    -10.0  76   E
 -10.0    -10.0    -10.0    -10.0     -3.9      -.2      -.8     -3.4     -7.7    -10.0    -10.0  74   V
 -10.0    -10.0    -10.0    -10.0     -3.0       .8       .2     -2.3     -6.3    -10.0    -10.0  72   A
 -10.0    -10.0    -10.0    -10.0     -2.3      1.6      1.1     -1.2     -5.0    -10.0    -10.0  70   T
 -10.0    -10.0    -10.0    -10.0     -1.7      2.3      2.0      -.2     -3.7    -10.0    -10.0  68   I
  -9.8    -10.0    -10.0    -10.0     -1.2      3.0      2.8       .7     -2.6     -9.3    -10.0  66   O
  -8.9    -10.0    -10.0    -10.0      -.8      3.6      3.4      1.5     -1.5     -7.4    -10.0  64   N
  -7.9    -10.0    -10.0    -10.0      -.5      4.0      4.1      2.3      -.4     -5.6    -10.0  62
  -7.1     -9.6    -10.0    -10.0      -.3      4.4      4.6      3.0       .5     -3.9    -10.0  60
  -6.2     -8.5    -10.0    -10.0      -.2      4.8      5.1      3.7      1.4     -2.4    -10.0  58   A
  -5.4     -7.5    -10.0    -10.0      -.2      5.0      5.5      4.3      2.3     -1.0    -10.0  56   N
  -4.6     -6.5     -9.9    -10.0      -.3      5.2      5.9      4.8      3.1       .2     -7.2  54   G
  -3.8     -5.6     -8.6    -10.0      -.6      5.3      6.2      5.3      3.8      1.4     -4.3  52   L
  -3.1     -4.7     -7.3    -10.0      -.9      5.4      6.4      5.7      4.4      2.4     -2.0  50   E
  -2.4     -3.8     -6.1    -10.0     -1.5      5.3      6.6      6.0      5.0      3.3      -.1  48
  -1.7     -3.0     -5.0    -10.0     -2.2      5.2      6.7      6.3      5.4      4.2      1.5  46   I
  -1.1     -2.2     -4.0    -10.0     -3.3      4.9      6.7      6.5      5.8      4.9      2.9  44   N
   -.5     -1.5     -3.0    -10.0     -4.7      4.5      6.8      6.6      6.2      5.5      4.1  42
    .1      -.8     -2.0     -8.7    -6.6      4.0      6.4      6.6      6.4      6.0      5.0  40   D
    .7      -.1     -1.2     -6.5    -9.4      3.3      6.1      6.6      6.5      6.3      5.8  38   E
   1.2       .6      -.4     -4.0   -10.0      2.4      5.7      6.4      6.6      6.6      6.4  36   G
   1.7      1.2       .4     -3.0   -10.0      1.2      5.2      6.2      6.5      6.8      6.8  34   R
   2.2      1.7      1.1     -1.6   -10.0      -.3      4.5      5.3      6.4      6.6      7.1  32   E
   2.6      2.3      1.8      -.4     -9.8     -2.3      3.7      5.3      6.1      6.7      7.2  30   E
   3.0      2.8      2.4       .7     -5.8     -5.2      2.6      4.7      5.6      6.4      7.2  28   S
   3.4      3.2      3.0      1.7     -3.0     -9.9      1.2      3.8      5.1      6.0      7.0  26
   3.8      3.7      3.5      2.6      -.9   -10.0      -.5      2.3      4.3      5.5      6.5  24
   4.1      4.0      4.0      3.4       .8   -10.0     -2.8      1.6      3.4      4.7      5.9  22
   4.4      4.4      4.4      4.1      2.2     -7.7     -6.0       .0      2.3      3.8      5.0  20
   4.6      4.7      4.8      4.7      3.3     -3.8   -10.0     -1.9      1.0      2.6      3.9  18
   4.8      5.0      5.1      5.2      4.3     -1.2   -10.0     -4.3      -.7      1.2      2.4  16
   5.0      5.2      5.4      5.6      5.1       .7   -10.0     -7.6     -2.7      -.6       .6  14
   5.2      5.4      5.6      6.0      5.7      2.1     -9.2   -10.0     -5.2     -2.7     -1.6  12
   5.3      5.5      5.7      6.2      6.2      3.1     -5.9   -10.0     -8.3     -5.3     -4.5  10
   5.3      5.5      5.8      6.4      6.5      3.8     -3.9   -10.0   -10.0     -8.4     -8.1   8
   5.2      5.5      5.8      6.4      6.7      4.2     -2.7   -10.0   -10.0   -10.0   -10.0    6
   5.0      5.2      5.5      6.2      6.5      4.2     -2.2   -10.0   -10.0   -10.0   -10.0    4
   4.0      4.3      4.5      5.2      5.5      3.2     -3.0   -10.0   -10.0   -10.0   -10.0    2
 -10.0    -10.0    -10.0    -10.0   -10.0   -10.0    -10.0    -10.0   -10.0   -10.0   -10.0    0
   12       13       14       16       18       20       22       24       26       28       30
```

(DIELECT) of the earth area over whch the antenna is erected. The body of the table is antenna gain in dBi as a function of elevation angle and operating frequency

Tabulation-10: (see Fig. 3-37). This computer output is designed to provide detailed information concerning the frequency dependence of system performance by considering each frequency in the 2.0 to 30.0 MHz HF range. A description of the circuit parameters is shown in the heading. Starting at the top of the page, below the page number, the heading is as follows: The first line gives the date (month, day, year) and the solar

```
                      JANUARY 15, 1970     10 CM³³FLUX 143
             BOULDER, COLO.    TO ST. LOUIS, MO.              AZIMUTHS
             40.03N - 105.27W   38.67N - 90.25W          91.86  281.42
                                      MINIMUM ANGLE  0.0 DEGREES
             XMTR  2.0 TO 10.0        RHOMBIC    H  20.00 L 114.00
             XMTR 10.0 TO 20.0     INVERTED L    H  10.00 L  21.30
             XMTR 20.0 TO 30.0 ARBITARY DIPOLE   H   -.25 L   -.50
             RCVR  2.0 TO 10.0        RHOMBIC    H  23.00 L 120.00
             RCVR 10.0 TO 20.0 TERM. SLOPING V   H  15.20 L 121.90
             RCVR 20.0 TO 30.0  CONSTANT GAIN    H   5.00 L   -0.00
             POWER=  30.00KW    3 MHZ NOISE=-148.60BM   REQ. REL.=.90
             MULTIPATH POWER TOLERANCE=10.0 DB        MULTIPATH DELAY
                                                     FREQUENCIES IN MHZ
  MUF  FOT    2    3    4    5    6    7    8    9   10   11   12   13   14   15
 11.5  9.1                                                UT 02
   1F   1F   3F   2F   2F   1F   1F   1F   1F   1F   1F   1F   1F   1F   1F   1F
   24   19   47   35   35   17   18   18   18   19   20   21   24   24   24   24
  .50  .90  .99  .99  .99  .99  .99  .99  .98  .91  .78  .00  .42  .27  .16  .09
   41   63   34   43   48   52   56   59   62   63   64   62   39   41   44   44
   79  115   65   88   89   93  101  107  112  115  116   83   77   78   78   77
  .49  .89  .83  .98  .99  .99  .99  .99  .97  .91  .77  .59  .41  .26  .15  .08
    -    -  .99  .99    -  .99    -    -    -    -    -    -    -    -    -    -

  7.9  6.2                                                UT 04
   1F   1F   3F   2F   1F   1F   1F   1F   1F   1F   1F   1F    -    -    -    -
   25   20   49   37   19   19   20   21   25   25   25   25    -    -    -    -
  .50  .90  .99  .99  .99  .99  .99  .93  .74  .47  .24  .11  .04    -    -    -
   61   58   34   44   48   53   57   60   61   61   61   43    -    -    -    -
  109  103   65   88   87   94  102  109  110  110  108   81    -    -    -    -
  .50  .89  .82  .98  .99  .99  .93  .73  .47  .24  .11  .04    -    -    -    -
    -    -  .99  .99  .99  .64    -    -    -    -    -    -    -    -    -    -

  6.8  5.6                                                UT 06
   1F   1F   2F   2F   1F   1F   1F   1F   1F   1F    -    -    -    -    -    -
   26   21   38   39   20   21   22   26   26   26    -    -    -    -    -    -
  .50  .90  .99  .99  .99  .99  .79  .43  .19  .07    -    -    -    -    -    -
   60   57   37   44   49   54   58   60   61   61    -    -    -    -    -    -
  107  101   65   80   86   95  104  109  110  110    -    -    -    -    -    -
  .50  .89  .82  .97  .98  .99  .79  .43  .18  .07    -    -    -    -    -    -
    -    -  .99  .99  .99    -    -    -    -    -    -    -    -    -    -    -

  8.2  6.8                                                UT 08
   1F   1F   2F   2F   2F   1F   1F   1F   1F   1F   1F    -    -    -    -    -
   26   22   39   39   48   21   21   22   24   26   26    -    -    -    -    -
  .50  .90  .99  .99  .99  .99  .93  .85  .55  .29  .14  .06    -    -    -    -
   61   60   38   44   47   54   58   61   62   61   60   43    -    -    -    -
  110  108   65   80   86   95  103  109  112  110  108   82    -    -    -    -
  .50  .89  .81  .97  .99  .99  .93  .84  .55  .29  .14  .06    -    -    -    -
    -    -  .99  .99  .99    -    -    -    -    -    -    -    -    -    -    -
```

Fig. 3-37. Sample tabulation of condensed system performance predictions all integer frequencies 2-30 MHz.

activity level as indicated by the 12-month moving average 10 centimeter flux density in units of 10^{-22} watts per square meter per 1.0 Hz of bandwidth and the 12-month running average ZSSN (Zurich Sun Spot Number). The second and third lines contain the names and geographical coordinates of the transmitter site and the receiver site locations, the azimuthal bearings (in degrees East of true North) of the receiver site location from the transmitting site location, and vice versa, and the great circle path distance of the HF circuit in both statute miles and kilometers. The minimum angle on the fourth line indicates the lowest vertical (firing) angle considered in the mode selection process. The fifth through tenth

```
(SSN 100 (HFMUFES4 73/10/31)
       MILES      KM.
       807.1     1298.8

A  70.0 OFF AZ   0.0
A  -0.0 OFF AZ   0.0
A  45.0 OFF AZ   0.0
A  68.0 OFF AZ   0.0
A  22.5 OFF AZ   1.4
A  -0.0 OFF AZ   0.0
       REQ.S/N=55.0DB
   TOLERANCE= .25 MS.

  16 17  18  19  20  21  22  23  24  25  26  27  28  29  30


  1F  -   -   -   -   -   -   -   -   -   -   -   -   -   -  - MODE
  24  -   -   -   -   -   -   -   -   -   -   -   -   -   -  - ANGLE
 .04  -   -   -   -   -   -   -   -   -   -   -   -   -   -  - F. DAYS
  43  -   -   -   -   -   -   -   -   -   -   -   -   -   -  - DBU
  74  -   -   -   -   -   -   -   -   -   -   -   -   -   -  - S/N DB
 .04  -   -   -   -   -   -   -   -   -   -   -   -   -   -  - REL.
   -  -   -   -   -   -   -   -   -   -   -   -   -   -   -  - MP PROB


   -  -   -   -   -   -   -   -   -   -   -   -   -   -   -  - MODE
   -  -   -   -   -   -   -   -   -   -   -   -   -   -   -  - ANGLE
   -  -   -   -   -   -   -   -   -   -   -   -   -   -   -  - F. DAYS
   -  -   -   -   -   -   -   -   -   -   -   -   -   -   -  - DBU
   -  -   -   -   -   -   -   -   -   -   -   -   -   -   -  - S/N DB
   -  -   -   -   -   -   -   -   -   -   -   -   -   -   -  - REL.
   -  -   -   -   -   -   -   -   -   -   -   -   -   -   -  - MP PROB


   -  -   -   -   -   -   -   -   -   -   -   -   -   -   -  - MODE
   -  -   -   -   -   -   -   -   -   -   -   -   -   -   -  - ANGLE
   -  -   -   -   -   -   -   -   -   -   -   -   -   -   -  - F. DAYS
   -  -   -   -   -   -   -   -   -   -   -   -   -   -   -  - DBU
   -  -   -   -   -   -   -   -   -   -   -   -   -   -   -  - S/N DB
   -  -   -   -   -   -   -   -   -   -   -   -   -   -   -  - REL.
   -  -   -   -   -   -   -   -   -   -   -   -   -   -   -  - MP PROB


   -  -   -   -   -   -   -   -   -   -   -   -   -   -   -  - MODE
   -  -   -   -   -   -   -   -   -   -   -   -   -   -   -  - ANGLE
   -  -   -   -   -   -   -   -   -   -   -   -   -   -   -  - F. DAYS
   -  -   -   -   -   -   -   -   -   -   -   -   -   -   -  - DBU
   -  -   -   -   -   -   -   -   -   -   -   -   -   -   -  - S/N DB
   -  -   -   -   -   -   -   -   -   -   -   -   -   -   -  - REL.
   -  -   -   -   -   -   -   -   -   -   -   -   -   -   -  - MP PROB
```

lines describe the physical parameters of the antenna system for each terminal (transmitting and receiving) and the orientation of the antenna main lobe relative to the great circle path (in computer methods requiring antennas, up to three transmitting and/or receiving antennas may be considered. The frequency range for such antenna must be specified, and such frequency ranges may not overlap). The eleventh line of the heading contains the transmitter power output, the 3.0 MHz man-made noise in a 1.0-Hz bandwidth, at the receiver site location, the required circuit reliability (required circuit reliability needed only in an LUHF (LUF) or Service Probability computation), and the hourly median S/N RATIO required to provide the type of service requested. The twelfth line gives the tolerances used in multipath computation. The thirteenth and the fourteenth lines of the computerized heading give the frequencies considered in the body of the table. The first line of each time block in the body of the table gives the UT (universal time) and the MUF (maximum usable frequency) as well as the fot (frequency which has a 90.0% probability of ionospheric support for HF sky-wave propagation).

For each frequency in the list, the tabulation body contains MODE (the mode having the greatest reliability—the number of hops and the layers), ANGLE (vertical angle in degrees associated with this mode), F-DAYS (fraction of days that any sky-wave mode exists), DBU (median incident field strength for the above mode available at the receiving site location in dBu), S/N DB (median signal-to-noise ratio—signal in occupied bandwidth versus noise in a 1.0-Hz bandwidth, at the input to the receiver, in dB for the days during which the sky-wave exists), REL (fraction of days within the month during which the skywave will be present and the required S/N will be obtained) and MP PROB (probability of multiple propagation paths within the power and delay tolerances shown in the heading). The symbols used to denote propagation modes are as follows, E-Layer(E), F2-Layer(F), E-F2-Layers(X) i.e. 3X denotes one E-Layer hop and two F2-Layer hops. Sporadic-E(e_s) modes not included. The time blocks are repeated for each interval of universal time (UT or GMT) as required.

Item-B: (see Fig. 3-38). Type Of Graph Display Desired.

Graph Display-1: This figure graphically and tabularly displays an estimate of the frequencies having ionospheric support for 50.0% of the time (MUF or maximum usable frequency) and those having ionospheric support for 90.0% of the time (FOT or optimum traffic frequency). These are generally adequate as estimates of upper frequency limits for system planning and are computerized and printed out at selected intervals of frequency and universal time (UT or GMT). A description of the circuit parameters used is shown in the heading, starting at the top of the page, below the page number this heading is as follows. The first line contains month, day, and year and also the solar activity level as indicated by the 12-month moving average 10 centimeter flux density in units of 10^{-22} watts per square meter per 1.0 Hz of bandwidth and also the 12-month running average Zurich sun spot number (ZSSN or RASSN). The second and third lines list the names and geographical coordinates of the transmitter site and the receiver site locations, azimuthal bearing (in degrees East of North) of the receiver site location from the transmitter site location, and vice versa, and the great circle path distance in statute miles and kilometers. The fourth line indicates minimum vertical (firing angle) used in the process of ionospheric mode selection. In the body of the graph the maximum usable frequency is plotted by a locus of asterisks while the FOT (optimum working frequency) is plotted by plus sign. All values equal to or greater than 30.0 MHz are plotted at the top of the graph. All plotted values are tabulated at the right hand side of the graph.

Graph Display-2: (see Fig. 3-39). This is a graphical (and tabular) display of the estimate of the highest frequency having ionospheric support for 90.0% of the days of the month (FOT), usually used as the estimate of the upper frequency limit for system planning. The figure gives an estimate

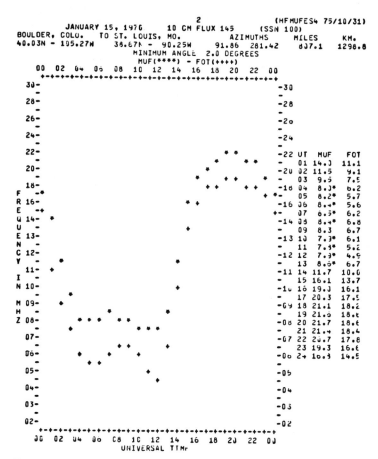

Fig. 3-38. Sample graph of usable frequency limits (MUF-FOT).

of the LUHF or LUF (the lowest frequency rendering an adequate S/N ratio for the required percentage of days. A description of the circuit parameters used is shown in the heading. Starting at the top of the page below the page number, this heading is described as follows: The first line contains month, day, and year and the solar activity level as indicated by 12-month running average 10 centimeter flux density in units of 10^{-22} watts per square meter per a 1.0-Hz bandwidth, as well as the 12-month running average ZSSN (Zurich Sun Spot Number). The second and third lines contain the transmitter site and receiving site locations, the azimuth bearings (in degrees East of true North) of the receiving site location from the transmitting site locations, and vice versa, and the circuit great circle distance in both statute miles and kilometers. The minimum angle on the fourth line indicates the lowest vertical angle (wave angle) considered in the mode selection process. The fifth and sixth lines describe the physical parameters of the antenna systems for each terminal (transmitting and

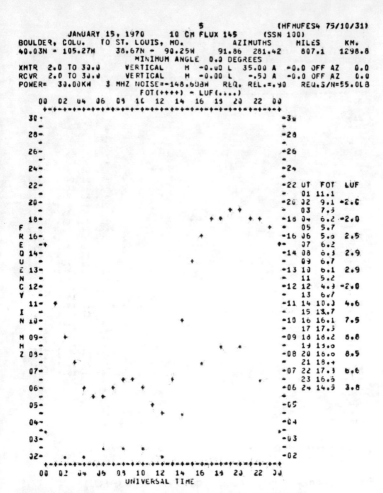

Fig. 3-39. Sample graph of useful frequency limits (FOT-LUF)

receiving) and the orientation of the antennas' main lobes relative to the great circle path (in all methods requiring antennas, up to three transmitting and/or receiving antennas may be considered in this program). The frequency range of each antenna must be specified, but the frequencies may not overlap. The seventh line contains the transmitter power output, the 3.0 MHz man-made noise referred to a 1.0-Hz bandwidth at the receiving site location, required circuit reliability (required circuit reliability needed only for LUHF or service probability computation), and the hourly median S/N ratio required to provide the type of service requested. In the body of the graph, the FOT is plotted as a locus of plus signs while LUF is displayed in dots. When no frequency is expected to render the required reliability, an X is shown at the bottom of the graph. All values equal to or greater than 30.0 MHz are plotted at the top of the graph. All LUF (LUHF) values below

2.0 MHz are plotted at the graph bottom. All values including FOTs above 30.0 MHz are tabulated at the right of the graph. In the tabulation a minus (−) sign or dash corresponds to an X in the graph (e.g. a −2.0 indicates an LUF (LUHF) below 2.0 MHz).

Graph Display-3: (see Fig. 3-40). A description of HF circuit parameters is shown in the heading below the page number as follows: The first line contains the month, day, and year and the solar activity level as indicated by the 12-month moving average 10 centimeter flux density in units of 10^{-22} watts per square meter per 1.0 Hz of bandwidth. The second and third lines contain the names and geographic coordinates of the transmitter site and receiver site locations, the azimuth bearings in degrees East of true North of the receiver location from the transmitter location, and vice

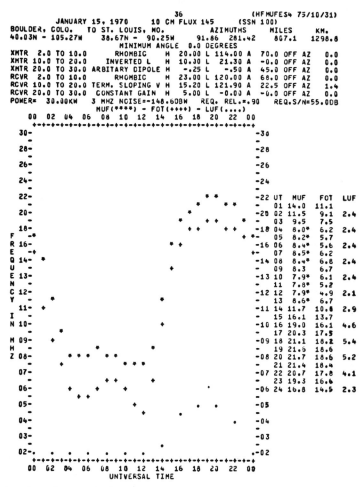

Fig. 3-40. Sample graph of useful frequency limits (MUF-FOT-LUF).

versa, and the circuit great circle distance in statute miles and kilometers. The minimum angle on the fourth line of the heading indicates the lowest vertical (firing or wave) angle considered in the mode selection process. The fifth and sixth lines describe the physical parameters of the antenna system for each terminal (transmitter and receiver) and the orientation of the antenna main lobe relative to the great circle bearing (in all methods requiring antennas, up to three transmitting and/or receiving antennas may be considered). The frequency range of each antenna must be stipulated (but the ranges may not overlap). The seventh line contains the transmitter output power, the man-made noise level at 3.0 MHz in dBw for a bandwidth of 1.0 Hz at the receiving location, the required circuit reliability (required reliability needed only in LUHF and Service Probability computerizations), and the hourly S/N ratio required to provide the type of service requested. In the body of the graph, the maximum usable frequency (MUF) is plotted as a locus of asterisks, the FOT as the locus of plus signs, and the LUF as a locus of dots. When no frequency computes out to obtain a 90.0% reliability, and X is shown at the bottom of the graph, any values equal to or greater than 30.0 MHz are plotted along the top of the graph. An LUF below 2.0 MHz is plotted along the graph bottom. All values, including the possible MUFs above 30.0 MHz, are tabulated at the right of the graph. A minus (−) sign, or dash, in the tabulation corresponds to an X on the graph. For example, a −2.0 indicates an LUHF below 2.0 MHz. The LUHF can be computerized for any interval of universal time (UT) as requred.

Graph Display-4: (see Fig. 3-41). Starting at the top of the page, below the page number, the heading gives information as follows: The first line gives month, day, and year and the solar activity level as indicated by the 12-month moving average 10 centimeter flux density in units of 10^{-22} watts per square meter per 1.0 Hz bandwidth and the 12-month running average ZSSN (Zurich Sun Spot Number). The second and third lines contain the names and geographical coordinates of the transmitting site and the receiving site locations, the azimuthal bearings, in degrees East of true North, of the receiving site location from the transmitter site location, and vice versa, and the circuit great circle distance in statute miles as well as in kilometers. The minimum angle, as the fourth line indicates, is the lowest vertical angle (wave angle) considered in the mode selection process. The fifth and sixth lines describe the physical parameters of the antenna system for each terminal (transmitter and receiver sites) and the orientation of the antennas' main beams relative to the great circle path (in all computerized methods requiring antennas, up to three transmitting and/or receiving antennas may be considered in the programs, but their frequency ranges may not be allowed to overlap). The seventh line contains the transmitter power output, the 3.0 MHz man-made noise in a 1.0-Hz bandwidth at the receiver, the required circuit reliability (required circuit reliability is needed only in an LUF or Service Probability computation), and the hourly median S/N ratio required to provide the type of service requested. The graph proper displays the plotted maximum usable frequency (MUF) as a locus of asterisks and lowest usable high frequency (LUHF) as a series of dots. When no frequency predicts out to render the required reliability, an X is shown at the bottom, of the graph. Any value equal to or greater than 30.0 MHz is plotted on the top line of the graph (30.0 MHz line). An LUHF

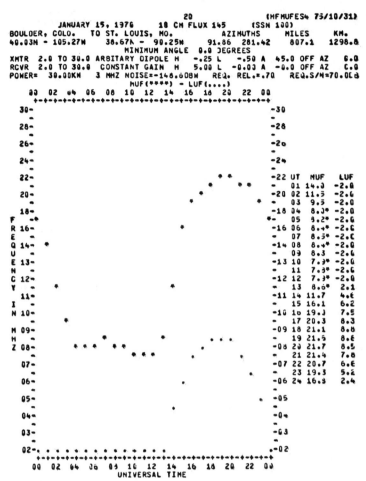

Fig. 3-41. Sample graph of useful frequency limits (MUF-LUF).

(LUF) below 2.0 MHz is plotted at the bottom (2.0 MHz line). All values including the MUF above 30.0 MHz are tabulated at the right hand side of the graph. A dash in the tabulation corresponds to an X on the graph. A −2.0 indicates an LUHF (LUF) below 2.0 MHz. The LUHF can be computed for any interval of UT(GMT) desired.

Tabulation-5: (see Fig. 3-42). In addition to the upper frequency limits (MUF and FOT), it is sometimes useful to know the frequencies above which there is a low probability of ionospheric support. This upper frequency limit is often chosen at a 10.0% probability level and is called the HPF (Highest Probable Frequency). Figure 3-42 shows the HPF in addition to the MUF and the FOT. A description of the circuit parameters is shown in the heading. Starting at the top of the page, below the page number, we have in the first line the month, day, and year and the solar activity level as

109

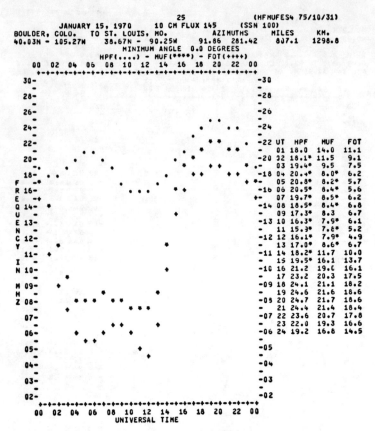

Fig. 3-42. Sample graph of upper frequency limits (HPF-MUF-FOT).

indicated by the 12-month moving average 10 centimeter flux density in units of 10^{-22} watts per square meter per 1.0 Hz of bandwidth and the 12-month running average Zurich sun spot number (ZSSN). The second and third lines give the name and coordinates of the transmittter site and the receiver site locations, the azimuth bearing in degrees East of True North of the receiving site location from that of the receiver site, and vice versa, and the length of the circuit in statute miles and kilometers. The minimum angle on the fourth line indicates the lowest vertical (firing or wave) angle considered in the mode selection process. In the body of the graph the HPF is represented by a locus of dots, the MUF by asterisk, and the FOT by plus sign. All values equal to or greater than 30.0 MHz are plotted at the top of the graph. All values, including those above 30.0 MHz, are tabulated at the right of the graph.

Tabulation-6: (see Fig. 3-43). This is a computerized presentation of MUF, FOT, the most useful mode (MODE) for this MUF/FOT, and the corresponding vertical (wave or firing) angle of radiation or reception. For HF ionospheric circuits operating near the upper useful frequency limit, these vertical angles may be used to assist in the selection or design of appropriate HF antennas. The graph body shows the MUF in asterisks and the FOT in plus signs while the table at the right reflects the universal time (UT), the MUF, the vertical angle for the MODE of propagation, and the MODE proper. A description of the HF circuit parameters is given in the heading. Starting at the top of the page, below the page number, this heading is as follows: The first line gives month, day, and year and the solar activity level as indicated by the 12-month moving average 10 centimeter flux density in units of 10^{-22} watts per square meter per 1.0-Hz bandwidth and the 12-month running average ZSSN. The second and third lines contain the name and geographical coordinates of the transmitter site location and the receiver site location, the azimuth bearings in degrees East of true North of the receiving station from the transmitting station, and vice versa, and the length of the HF circuit in both statute miles and kilometers. The minimum angle on the fourth line indicates the lowest vertical angle considered in selecting the mode (1F, 1E, etc.). In the body of the graph, the diurnal variation of the MUF is plotted as a locus of asterisks while the

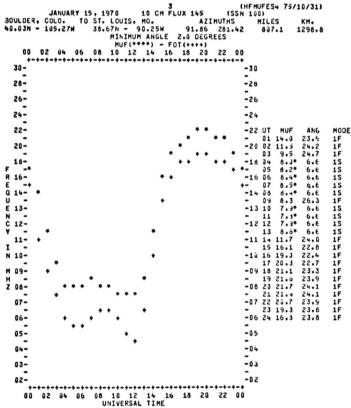

Fig. 3-43. Sample graph of upper usable frequency limits with MODE and VERTICAL ANGLE of the MUF (MUF-FOT. ANG).

Fig. 3-45. Summary of options, tabulations, and graphs associated with the various methods of HFMUFES 4.

An X in the body of the table indicates the input data is required for a given METHOD.

Data #	Description of Input Data	1 MUF-FOT	2 FOT-LUF	3 System Performance	4 Condensed System Performance	5 Most Probable Mode	6 MUF and Circuit Reliability	7 All Modes	8 MUF as a Function of K	9 Antenna Pattern	10 All Frequencies 2-30 MHz
1											
a	Year or Sunspot Number	X	X	X	X	X	X	X	X		X
b	Month or Months	X	X	X	X	X	X	X	X		X
c	Transmitter Site Characteristics		X	X	X		X	X	X	X	X
d	Receiver Site Characteristics		X	X	X		X	X	X	X	X
2											
a	Type of Tabulation	X	X	X	X	X	X	X	X		X
b	Type of Graph or no Graph	X	X	X	X	X	X	X	X		X
c	Beginning Hour	X	X	X	X	X	X	X	X		X
d	Ending Hour	X	X	X	X	X	X	X	X		X
e	Hour Interval	X	X	X	X	X	X	X	X		X
f	Circuit Reliability or Service Probability		X	X	X		X	X			X
g	Antenna Pattern Tabulation?	X	X	X	X	X	X	X	X	X	X
h	Short or Long Path		X	X	X	X	X	X	X		X
3	Frequency Complement	X	X	X	X	X	X	X			X
4											
a	Transmitter Location	X	X	X	X	X	X	X	X	X	X
b	Receiver Location	X	X	X	X	X	X	X	X	X	X
c	Minimum Take-off Angle	X	X	X	X	X	X	X	X	X	X
d	Transmitter Power		X	X	X		X	X			X
e	Required S/N Ratio		X	X	X		X	X			X
f	Man Made Noise		X	X	X		X	X			X
g	Time (circuit reliability) Required		X	X	X		X	X			X
h	Multipath Probability?		X	X	X		X	X			X
i	Maximum Power Difference		X	X	X			X			X
j	Minimum Time Delay		X	X	X			X			X
5	Antenna Parameters		X	X	X		X	X		X	X
6	Transmitter Antenna Type/Types		X	X	X		X	X			X
7	Receiver Antenna Type/Types		X	X	X		X	X			X

Fig. 3-45. continued from page 112

The X in the body of the table shows options available with various methods and the tabulation and/or graphs provided as outputs from these methods. (X) indicates tabulation or graph not both.

		METHOD									
		1 MUF-FOT	2 FOT-LUF	3 System Performance	4 Condensed System Performance	5 Most Probable Mode	6 MUF and Circuit Reliability	7 All Modes	8 MUF as a Function of K	9 Antenna Patterns	10 All Frequencies 2-30 MHz
OPTION:											
Sporadic-E	1		x	x	x		x	x			x
Circuit Reliability	2			x	x		x	x			x
Service Probability	3			x	x		x	x			x
Antenna Pattern	4			x	x		x	x			x
Short/Long Path	5			x	x		x	x			x
Multipath Probability	6			x	x		x	x			x
TABULATIONS:											
MUF-FOT	1	x	x				x	x			
FOT-LUF	2		x				x	x			
System Performance	3			x							
Condensed System Performance	4	(X)	(X)		x						
Path Geometry	5					x					
Circuit Reliability	6						x				
Individual Modes	7							x			
MUF-K Index	8								x		
Antenna Pattern	9									x	
All Frequencies 2-30 MHz	10										x
GRAPHS*:											
MUF-FOT	1			x	x		x	x			x
FOT-LUF	2			x	x		x	x			x
MUF-FOT-LUF	3			x	x		x	x			x
MUF-LUF	4	(X)	(X)	x	x	x	x	x			x
HPF-MUF-FOT	5	(X)	(X)	x	x	x	x	x			x
MUF-FOT-ANG	6	x	x	x	x	x	x	x	x		x

*Only one graph per method per computation.

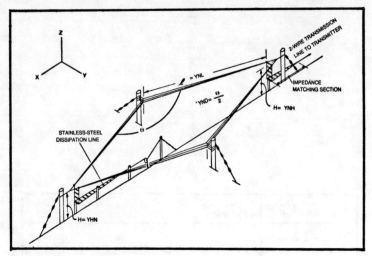

Fig. 3-46A. Geometry of a single three-wire horizontal rhombic antenna.

FOT is plotted as a locus of plus signs. All values equal to or greater than 30.0 MHz are plotted at the top of the graph (on the 30.0-MHz level). The symbols used to denote HF ionospheric propagation modes are: E-Layer(E), F2-Layer(F), E-F2Layers(X) i.e. 3X denotes one hop E and two hop F2, E_s-Layer(S) and E_s-F2-Layers(Y) i.e. 4Y denotes one hop E_s and three hops F2.

Item-C: Beginning hour of computer calculation (UT or GMT). The first hour for which computerized predictions are required expressed in two digits (fractions of hours may not be specified).

Item-D: Ending hour of computer calculation (UT or GMT). The last hour for which computerized predictions are required.

Item-E: Every_____hours. The hourly increment of required computer calculations. For graphs requiring LUHF (Lowest Usable High Frequency), it is recommended that the beginning hour be 02 hours and the ending hour be 24 while the increments be two hours. For all graphs, the MUF, FOT, and HPF are always computerized for every UT hour.

Item-F: Circuit reliability or service probability. In some methods of computer solution of HF propagation problems, a choice must be made between the two. Please see the table of Fig. 3-44 for these computer methods. As already pointed out herein, circuit reliability is an estimate of the percentage of days of a month during which the signal quality will be acceptable. In HF work, this quantity is a product of the fractional percentage of the days of the month (of calculation) during which regular sky-wave is expected (E_s or sporadic E excluded) and the fractional percentage of days during which the S/N ratio is adequate. Service probability, on the other hand, is an estimate of the liklihood that a specified circuit reliability will be met.

Item-G: Antenna Pattern Tabulation. In all computer methods requiring antenna characteristics (or if the method of tabulation-9 is chosen), a tabulation of antenna gains is available.

ANTENNA PATTERN

FREQUENCY RANGE 2.0 TO 30.0 | ANTENNA TYPE RHOMBIC | HEIGHT 16.760 | LENGTH 88.390 | ANGLE 67.500 | AZIMUTH 0.000 | EX(1) 0.300 | EX(2) 0.000 | EX(3) 0.000 | EX(4) 0.000 | CONDUCT. 0.001 | DIELECT. 4.000

ELEVATION ANGLE IN DEGREES	2	3	4	5	6	7	8	9	10	11	12	13	14	16	18	20	22	24	26	28	30	ELEVATION ANGLE IN DEGREES
90	-2.0	-10.0	-10.0	-0.7	-3.3	-10.0	-8.7	-9.1	-10.0	-8.9	-1.3	-10.0	-10.0	-8.5	-10.0	-10.0	-0.8	-10.0	-8.7	-10.0	-10.0	90
88	-1.4	-10.0	-10.0	-0.9	-5.0	-10.0	-10.0	-7.1	-10.0	-10.0	-1.4	-4.4	-10.0	-4.1	-10.0	-6.3	-5.1	-10.0	-6.5	-10.0	-1.4	88
86	-1.0	-10.0	-10.0	-1.3	-3.7	-10.0	-10.0	-6.5	-10.0	-10.0	-3.9	-0.7	-10.0	-4.0	-10.0	-2.2	-10.0	-2.7	-10.0	-9.1	-5.6	86
84	-0.5	-8.4	-10.0	-2.3	-2.0	-10.0	-10.0	-7.1	-6.3	-10.0	-9.8	0.1	-5.0	-8.4	-10.0	-4.7	-10.0	-2.0	-10.0	-4.4	-10.0	84
82	-0.2	-6.1	-10.0	-3.7	-1.0	-10.0	-10.0	-9.2	-4.1	-10.0	-10.0	-1.8	-0.5	-10.0	-7.1	-10.0	-1.7	-10.0	-7.5	-10.0	-4.4	82
80	0.2	-4.2	-10.0	-5.7	-0.4	-7.8	-10.0	-10.0	-3.6	-5.9	-10.0	-7.1	0.6	-10.0	-5.7	-10.0	0.5	-10.0	-4.2	-10.0	-0.8	80
78	0.4	-2.5	-10.0	-8.6	-0.4	-4.7	-10.0	-10.0	-4.6	-2.3	-10.0	-10.0	-1.3	-10.0	-9.4	-8.5	-5.6	-3.9	-10.0	-7.0	-10.0	78
76	0.6	-1.1	-10.0	-10.0	-1.0	-2.6	-10.0	-10.0	-7.4	-0.9	-5.2	-10.0	-7.3	-3.2	-10.0	-2.2	-10.0	1.0	-10.0	-5.1	-10.0	76
74	0.8	0.1	-10.0	-10.0	-2.1	-1.3	-10.0	-10.0	-10.0	-1.4	-0.7	-10.0	-10.0	-0.2	-10.0	-2.7	-10.0	-3.2	-5.7	-10.0	-1.2	74
72	0.9	1.2	-9.9	-10.0	-3.9	-0.6	-6.3	-10.0	-10.0	-3.7	1.1	-3.3	-10.0	-0.9	-10.0	-10.0	0.0	-10.0	-0.5	-10.0	-3.6	72
70	0.9	2.2	-6.3	-10.0	-6.6	-0.7	-3.7	-10.0	-10.0	-8.5	0.8	1.1	-8.4	-6.1	-4.5	-10.0	1.4	-10.0	-5.8	-5.5	-10.0	70
68	0.9	2.9	-3.4	-10.0	-10.0	-1.4	-2.0	-9.2	-10.0	-10.0	-1.7	2.7	-0.5	-10.0	-2.1	-10.0	-4.8	1.7	-10.0	-3.5	-9.8	68
66	0.9	3.6	-1.0	-10.0	-10.0	-2.9	-1.2	-5.6	-10.0	-10.0	-7.0	1.9	3.1	-10.0	-4.2	-5.6	-10.0	2.9	-5.7	-10.0	-2.4	66
64	0.8	4.2	1.0	-10.0	-10.0	-5.3	-1.2	-3.4	-9.2	-10.0	-10.0	-1.4	3.8	-6.3	-10.0	-2.8	-10.0	-4.7	3.1	-10.0	-8.1	64
62	0.6	4.6	2.6	-7.2	-10.0	-9.2	-2.0	-2.1	-5.4	-10.0	-10.0	-8.4	1.9	1.5	-10.0	-4.9	-1.7	-10.0	1.9	-1.1	-10.0	62
60	0.4	4.9	4.0	-3.2	-10.0	-10.0	-3.7	-1.7	-3.3	-6.7	-10.0	-10.0	-3.2	4.7	-10.0	-10.0	0.5	-5.7	-10.0	2.8	-7.7	60
58	0.1	5.2	5.1	-0.1	-10.0	-10.0	-6.7	-2.2	-2.2	-3.3	-8.3	-10.0	-10.0	4.7	-0.5	-10.0	-3.5	2.8	-10.0	-3.8	0.7	58
56	-0.2	5.3	6.1	2.4	-3.2	-10.0	-10.0	-3.7	-2.1	-1.7	-2.9	-10.0	-10.0	1.5	4.3	-10.0	-10.0	3.0	1.9	-10.0	-2.0	56
54	-0.5	5.4	6.8	4.4	-3.2	-10.0	-10.0	-6.5	-2.8	-1.3	-0.4	-2.9	-10.0	-6.6	5.2	-1.0	-10.0	-4.2	5.9	-0.9	-10.0	54
52	-0.9	5.4	7.4	6.0	0.6	-10.0	-10.0	-10.0	-4.6	-2.0	0.4	0.9	-2.9	-10.0	2.5	3.9	-9.4	-10.0	2.2	6.7	-4.2	52
50	-1.4	5.3	7.9	7.3	3.5	-5.3	-10.0	-10.0	-7.8	-3.6	-0.3	2.3	1.9	-10.0	-5.6	4.7	-0.2	-10.0	-10.0	5.4	6.6	50
48	-1.9	5.1	8.2	8.4	5.8	-0.3	-10.0	-10.0	-10.0	-6.3	-2.1	1.9	3.9	-1.8	-10.0	1.1	3.5	-3.7	-10.0	-6.2	6.9	48
46	-2.5	4.8	8.4	9.2	7.7	3.4	-5.3	-10.0	-10.0	-10.0	-5.0	-0.1	3.8	4.2	-10.0	-10.0	2.8	1.0	-3.3	-10.0	-3.2	46
44	-3.1	4.5	8.4	9.9	9.2	6.2	0.2	-10.0	-10.0	-10.0	-9.3	-3.4	2.0	6.8	1.5	-10.0	-3.7	2.1	0.6	-1.6	-10.0	44
42	-3.8	4.0	8.3	10.3	10.3	8.5	4.3	-3.2	-10.0	-10.0	-10.0	-8.2	-1.7	6.9	7.1	-3.3	-10.0	-1.7	0.8	2.9	0.5	42
40	-4.5	3.5	8.2	10.6	11.2	10.2	7.4	2.3	-6.5	-10.0	-10.0	-10.0	-7.2	4.9	9.3	6.1	-9.2	-10.0	-2.0	1.1	5.5	40
38	-5.3	3.0	7.9	10.7	11.8	11.6	9.8	6.3	0.5	-9.3	-10.0	-10.0	-10.0	0.7	8.9	10.2	4.7	-10.0	-10.0	-3.2	2.8	38
36	-6.1	2.3	7.5	10.5	12.2	12.6	11.6	9.3	5.4	-0.9	-10.0	-10.0	-10.0	-6.2	6.1	11.1	10.6	3.7	-10.0	-10.0	-4.5	36
34	-7.0	1.6	7.0	10.4	12.5	13.3	13.0	11.6	9.0	4.8	-1.6	-10.0	-10.0	-10.0	0.7	9.5	12.5	10.9	3.5	-10.0	-10.0	34
32	-8.0	0.8	6.4	10.1	12.5	13.8	14.1	13.4	11.7	8.9	4.7	-1.5	-10.0	-10.0	-8.6	5.1	11.6	13.6	11.4	4.3	-10.0	32
30	-9.0	-0.1	5.7	9.6	12.3	14.0	14.8	14.7	13.8	12.0	9.2	5.1	-0.7	-10.0	-10.0	-2.9	7.9	13.1	14.5	12.4	6.1	30
28	-10.0	-1.1	4.9	9.0	12.0	14.0	15.2	15.6	15.3	14.3	12.6	9.9	6.1	-6.6	-10.0	-10.0	0.6	9.6	14.1	15.4	13.7	28
26	-10.0	-2.1	4.0	8.3	11.5	13.8	15.3	16.2	16.4	16.0	15.0	13.3	10.9	3.0	-10.0	-10.0	-10.0	2.4	10.6	14.8	16.2	26
24	-10.0	-3.3	3.0	7.5	10.9	13.4	15.2	16.4	17.1	17.2	16.8	15.9	14.4	9.3	0.3	-10.0	-10.0	-10.0	3.0	10.8	15.1	24
22	-10.0	-4.5	1.9	6.6	10.1	12.8	14.9	15.4	17.4	17.9	18.0	17.7	16.9	13.8	8.1	-1.4	-10.0	-10.0	-10.0	2.5	10.4	22
20	-10.0	-5.8	0.7	5.5	9.2	12.1	14.4	16.1	17.4	18.3	18.8	18.9	18.6	16.9	13.4	7.7	-1.7	-10.0	-10.0	-10.0	0.6	20
18	-10.0	-7.2	-0.6	4.3	9.1	11.2	13.6	15.6	17.1	18.3	19.1	19.6	19.7	19.1	17.1	13.6	8.0	-0.8	-10.0	-10.0	-10.0	18
16	-10.0	-8.7	-2.1	3.0	6.9	10.1	12.7	14.8	16.6	18.0	19.0	19.8	20.3	20.5	19.6	17.6	14.2	9.1	1.2	-10.0	-10.0	16
14	-10.0	-10.0	-3.6	1.5	5.5	9.8	11.6	13.9	15.7	17.3	18.6	19.6	20.4	21.2	21.1	20.2	18.2	15.1	10.6	3.8	-7.1	14
12	-10.0	-10.0	-5.3	-0.2	4.0	7.4	10.2	12.6	14.6	16.4	17.8	19.0	20.0	21.3	21.9	21.7	20.7	18.9	16.2	12.2	6.3	12
10	-10.0	-10.0	-7.3	-2.0	2.2	5.6	8.6	11.1	13.2	15.1	16.6	18.0	19.1	20.8	21.9	22.2	22.0	21.1	19.5	17.0	13.5	10
8	-10.0	-10.0	-9.5	-4.2	0.0	3.6	6.6	9.1	11.4	13.3	15.0	16.5	17.7	19.8	21.1	21.9	22.2	21.9	21.0	19.5	17.3	8
6	-10.0	-10.0	-10.0	-6.9	-2.6	1.0	4.0	6.7	9.0	11.0	12.8	14.3	15.7	17.9	19.5	20.6	21.2	21.4	21.0	20.1	18.6	6
4	-10.0	-10.0	-10.0	-10.0	-6.2	-2.6	0.5	3.2	5.5	7.6	9.4	11.0	12.5	14.8	16.7	18.0	18.8	19.2	19.2	18.7	17.7	4
2	-10.0	-10.0	-10.0	-10.0	-10.0	-3.6	-5.5	-2.7	-0.4	1.7	3.6	5.2	6.7	9.2	11.1	12.5	13.5	14.0	14.2	13.9	13.2	2
0	-10.0	-10.0	-10.0	-10.0	-10.0	-10.0	-10.0	-10.0	-10.0	-10.0	-10.0	-10.0	-10.0	-10.0	-10.0	-10.0	-10.0	-10.0	-10.0	-10.0	-10.0	0
	2	3	4	5	6	7	8	9	10	11	12	13	14	16	18	20	22	24	26	28	30	

FREQUENCY IN MEGAHERTZ

Fig. 3-46B. Example of horizontal terminated rhombic antenna gains.

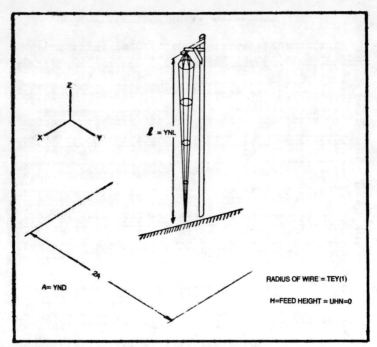

Fig. 3-47A. Geometry of a vertical monopole antenna.

Item-H: Great circle path. Except for very long paths, only the short great circle path is usually considered. If both the long and the short great circle paths are required, separate computer calculations are necessary.

Frequency Compliment (BOX-3):

The frequencies which are available and which should be considered in the computer calculations. A selection of up to eleven frequencies is usual for most formats. Additional frequencies generally may be considered to require additional computer time and concomitant expense.

Circuit and System Parameters (BOX-4):

Item-A: Transmitter site location by name and geographic coordinates (expressed in degrees and decimal parts of degrees)—e.g. 63°30′ = 60.5°.

Item-B: Receiver site location by name and geographic coordinates (expressed in degrees and decimal parts of degrees)—e.g. 20°45′ = 20.75°.

Item-C: Minimum take-off angle. Normally considered as zero (0°) unless the particular antenna performance is expected to be so inferior at low angles that these angles should be eliminated from consideration in the computerized estimation of the upper useful frequency limits, or if the horizon is so obstructed that low take-off (wave or firing) angles are unlikely. Note that this holds reciprocally true for both transmitting and receiving antennas.

(HFMUFES3 74/12/30)
ANTENNA PATTERN

FREQUENCY RANGE 2.0 TO 30.0	ANTENNA TYPE VERTICAL					HEIGHT 0.000	LENGTH -0.250		ANGLE 0.300	AZIMUTH 0.000			EX(1) 0.300	EX(2) 0.000		EX(3) 0.000		EX(4) 0.000	CONDUCT. 0.001		DIELECT. 4.000	
	2	3	4	5	6	7	8	9	10	11	12	13	14	16	18	20	22	24	26	28	30	
90	-10.0	-10.0	-10.0	-10.0	-10.0	-10.0	-10.0	-10.0	-10.0	-10.0	-10.0	-10.0	-10.0	-10.0	-10.0	-10.0	-10.0	-10.0	-10.0	-10.0	-10.0	90
88	-10.0	-10.0	-10.0	-10.0	-10.0	-10.0	-10.0	-10.0	-10.0	-10.0	-10.0	-10.0	-10.0	-10.0	-10.0	-10.0	-10.0	-10.0	-10.0	-10.0	-10.0	88
86	-10.0	-10.0	-10.0	-10.0	-10.0	-10.0	-10.0	-10.0	-10.0	-10.0	-10.0	-10.0	-10.0	-10.0	-10.0	-10.0	-10.0	-10.0	-10.0	-10.0	-10.0	86
84	-10.0	-10.0	-10.0	-10.0	-10.0	-10.0	-10.0	-10.0	-10.0	-10.0	-10.0	-10.0	-10.0	-10.0	-10.0	-10.0	-10.0	-10.0	-10.0	-10.0	-10.0	84
82	-10.0	-10.0	-10.0	-10.0	-10.0	-10.0	-10.0	-10.0	-10.0	-10.0	-10.0	-10.0	-10.0	-10.0	-10.0	-10.0	-10.0	-10.0	-10.0	-10.0	-10.0	82
80	-10.0	-10.0	-10.0	-10.0	-10.0	-10.0	-10.0	-10.0	-10.0	-10.0	-10.0	-10.0	-10.0	-10.0	-10.0	-10.0	-10.0	-10.0	-10.0	-10.0	-10.0	80
78	-10.0	-10.0	-10.0	-10.0	-10.0	-10.0	-10.0	-10.0	-10.0	-10.0	-10.0	-10.0	-10.0	-10.0	-10.0	-10.0	-10.0	-10.0	-10.0	-10.0	-10.0	78
76	-10.0	-10.0	-10.0	-10.0	-10.0	-10.0	-10.0	-10.0	-10.0	-10.0	-10.0	-10.0	-10.0	-10.0	-10.0	-10.0	-10.0	-10.0	-10.0	-10.0	-10.0	76
74	-10.0	-10.0	-10.0	-10.0	-10.0	-10.0	-10.0	-10.0	-10.0	-10.0	-10.0	-10.0	-10.0	-10.0	-10.0	-10.0	-10.0	-10.0	-10.0	-10.0	-10.0	74
72	-10.0	-10.0	-10.0	-10.0	-10.0	-10.0	-10.0	-10.0	-10.0	-10.0	-10.0	-10.0	-10.0	-10.0	-10.0	-10.0	-10.0	-10.0	-10.0	-10.0	-10.0	72
70	-9.1	-9.5	-9.6	-9.7	-9.7	-9.7	-9.7	-9.7	-9.6	-9.6	-9.5	-9.6	-9.6	-9.5	-9.5	-9.5	-9.5	-9.5	-9.5	-9.4	-9.4	70
68	-8.3	-8.6	-8.8	-8.9	-8.9	-8.9	-8.9	-8.9	-8.8	-8.8	-8.8	-8.8	-8.8	-8.7	-8.7	-8.7	-8.7	-8.7	-8.6	-8.6	-8.6	68
66	-7.5	-7.9	-8.1	-8.1	-8.1	-8.1	-8.1	-8.1	-8.1	-8.1	-8.1	-8.0	-8.0	-8.0	-8.0	-8.0	-7.9	-7.9	-7.9	-7.9	-7.9	66
64	-6.9	-7.2	-7.4	-7.5	-7.5	-7.5	-7.5	-7.4	-7.4	-7.4	-7.4	-7.4	-7.4	-7.3	-7.3	-7.3	-7.3	-7.3	-7.2	-7.2	-7.2	64
62	-6.2	-6.6	-6.8	-6.9	-6.9	-6.9	-6.8	-6.8	-6.8	-6.8	-6.8	-6.8	-6.7	-6.7	-6.7	-6.7	-6.7	-6.6	-6.6	-6.6	-6.6	62
60	-5.7	-6.0	-6.2	-6.3	-6.3	-6.3	-6.3	-6.3	-6.3	-6.2	-6.2	-6.2	-6.2	-6.2	-6.1	-6.1	-6.1	-6.1	-6.1	-6.1	-6.1	60
58	-5.1	-5.5	-5.7	-5.9	-5.8	-5.8	-5.8	-5.7	-5.7	-5.7	-5.7	-5.7	-5.7	-5.6	-5.6	-5.6	-5.6	-5.6	-5.6	-5.6	-5.5	58
56	-4.7	-5.0	-5.2	-5.3	-5.3	-5.3	-5.3	-5.3	-5.3	-5.2	-5.2	-5.2	-5.2	-5.2	-5.1	-5.1	-5.1	-5.1	-5.1	-5.1	-5.1	56
54	-4.2	-4.6	-4.8	-4.9	-4.9	-4.9	-4.8	-4.8	-4.8	-4.8	-4.8	-4.8	-4.8	-4.7	-4.7	-4.7	-4.7	-4.7	-4.7	-4.6	-4.6	54
52	-3.8	-4.2	-4.3	-4.4	-4.4	-4.4	-4.4	-4.4	-4.4	-4.4	-4.4	-4.4	-4.4	-4.3	-4.3	-4.3	-4.3	-4.3	-4.3	-4.2	-4.2	52
50	-3.4	-3.8	-4.0	-4.0	-4.1	-4.1	-4.1	-4.1	-4.0	-4.0	-4.0	-4.0	-4.0	-4.0	-3.9	-3.9	-3.9	-3.9	-3.9	-3.9	-3.9	50
48	-3.0	-3.4	-3.6	-3.7	-3.7	-3.7	-3.7	-3.7	-3.7	-3.7	-3.7	-3.7	-3.6	-3.6	-3.6	-3.6	-3.6	-3.6	-3.5	-3.5	-3.5	48
46	-2.7	-3.1	-3.3	-3.4	-3.4	-3.4	-3.4	-3.4	-3.4	-3.4	-3.4	-3.4	-3.3	-3.3	-3.3	-3.3	-3.3	-3.3	-3.2	-3.2	-3.2	46
44	-2.4	-2.8	-3.0	-3.1	-3.1	-3.2	-3.1	-3.1	-3.1	-3.1	-3.1	-3.1	-3.1	-3.0	-3.0	-3.0	-3.0	-3.0	-3.0	-3.0	-3.0	44
42	-2.2	-2.6	-2.8	-2.9	-2.9	-2.9	-2.9	-2.9	-2.9	-2.9	-2.9	-2.8	-2.8	-2.8	-2.8	-2.8	-2.8	-2.7	-2.7	-2.7	-2.7	42
40	-1.9	-2.4	-2.6	-2.7	-2.7	-2.7	-2.7	-2.7	-2.7	-2.7	-2.6	-2.6	-2.6	-2.6	-2.6	-2.6	-2.6	-2.5	-2.5	-2.5	-2.5	40
38	-1.7	-2.2	-2.4	-2.5	-2.5	-2.5	-2.5	-2.5	-2.5	-2.5	-2.5	-2.5	-2.4	-2.4	-2.4	-2.4	-2.4	-2.4	-2.4	-2.3	-2.3	38
36	-1.6	-2.0	-2.2	-2.3	-2.4	-2.4	-2.4	-2.4	-2.3	-2.3	-2.3	-2.3	-2.3	-2.3	-2.3	-2.2	-2.2	-2.2	-2.2	-2.2	-2.2	36
34	-1.4	-1.9	-2.1	-2.2	-2.3	-2.3	-2.3	-2.2	-2.2	-2.2	-2.2	-2.2	-2.2	-2.2	-2.2	-2.1	-2.1	-2.1	-2.1	-2.1	-2.1	34
32	-1.3	-1.8	-2.0	-2.1	-2.2	-2.2	-2.2	-2.2	-2.2	-2.2	-2.1	-2.1	-2.1	-2.1	-2.1	-2.1	-2.1	-2.1	-2.0	-2.0	-2.0	32
30	-1.3	-1.8	-2.0	-2.1	-2.1	-2.2	-2.2	-2.1	-2.1	-2.1	-2.1	-2.1	-2.1	-2.1	-2.1	-2.0	-2.0	-2.0	-2.0	-2.0	-2.0	30
28	-1.2	-1.8	-2.0	-2.1	-2.2	-2.2	-2.2	-2.2	-2.2	-2.1	-2.1	-2.1	-2.1	-2.1	-2.1	-2.1	-2.1	-2.0	-2.0	-2.0	-2.0	28
26	-1.3	-1.8	-2.1	-2.2	-2.2	-2.2	-2.2	-2.2	-2.2	-2.2	-2.2	-2.2	-2.2	-2.2	-2.1	-2.1	-2.1	-2.1	-2.1	-2.1	-2.1	26
24	-1.3	-1.9	-2.2	-2.3	-2.3	-2.3	-2.3	-2.3	-2.3	-2.3	-2.3	-2.3	-2.3	-2.3	-2.3	-2.2	-2.2	-2.2	-2.2	-2.2	-2.2	24
22	-1.5	-2.1	-2.3	-2.3	-2.5	-2.5	-2.5	-2.5	-2.5	-2.5	-2.5	-2.5	-2.5	-2.5	-2.4	-2.4	-2.4	-2.4	-2.4	-2.4	-2.4	22
20	-1.7	-2.3	-2.6	-2.7	-2.8	-2.8	-2.8	-2.8	-2.8	-2.8	-2.7	-2.7	-2.7	-2.7	-2.7	-2.7	-2.7	-2.7	-2.7	-2.6	-2.6	20
18	-1.9	-2.6	-2.9	-3.0	-3.1	-3.1	-3.1	-3.1	-3.1	-3.1	-3.1	-3.1	-3.1	-3.1	-3.0	-3.0	-3.0	-3.0	-3.0	-3.0	-3.0	18
16	-2.3	-3.0	-3.3	-3.5	-3.5	-3.6	-3.6	-3.6	-3.6	-3.5	-3.5	-3.5	-3.5	-3.5	-3.5	-3.5	-3.5	-3.5	-3.4	-3.4	-3.4	16
14	-2.8	-3.5	-3.9	-4.3	-4.1	-4.1	-4.1	-4.1	-4.1	-4.1	-4.1	-4.1	-4.1	-4.1	-4.1	-4.1	-4.0	-4.0	-4.0	-4.0	-4.0	14
12	-3.5	-4.3	-4.6	-4.9	-4.8	-4.9	-4.9	-4.9	-4.9	-4.9	-4.9	-4.9	-4.9	-4.8	-4.8	-4.8	-4.8	-4.8	-4.8	-4.8	-4.8	12
10	-4.4	-5.2	-5.6	-5.7	-5.8	-5.9	-5.9	-5.9	-5.9	-5.9	-5.9	-5.8	-5.8	-5.8	-5.8	-5.8	-5.8	-5.8	-5.8	-5.8	-5.8	10
8	-5.6	-6.5	-6.9	-7.1	-7.2	-7.2	-7.2	-7.2	-7.2	-7.2	-7.2	-7.2	-7.2	-7.2	-7.1	-7.1	-7.1	-7.1	-7.1	-7.1	-7.1	8
6	-7.4	-8.3	-8.7	-8.9	-9.0	-9.1	-9.1	-9.1	-9.1	-9.1	-9.1	-9.1	-9.1	-9.0	-9.0	-9.0	-9.0	-9.0	-9.0	-9.0	-9.0	6
4	-10.0	-10.0	-10.0	-10.0	-10.0	-10.0	-10.0	-10.0	-10.0	-10.0	-10.0	-10.0	-10.0	-10.0	-10.0	-10.0	-10.0	-10.0	-10.0	-10.0	-10.0	4
2	-10.0	-10.0	-10.0	-10.0	-10.0	-10.0	-10.0	-10.0	-10.0	-10.0	-10.0	-10.0	-10.0	-10.0	-10.0	-10.0	-10.0	-10.0	-10.0	-10.0	-10.0	2
0	-10.0	-10.0	-10.0	-10.0	-10.0	-10.0	-10.0	-10.0	-10.0	-10.0	-10.0	-10.0	-10.0	-10.0	-10.0	-10.0	-10.0	-10.0	-10.0	-10.0	-10.0	0
	2	3	4	5	6	7	8	9	10	11	12	13	14	16	18	20	22	24	26	28	30	

ELEVATION ANGLE IN DEGREES

FREQUENCY IN MEGAHERTZ

Fig. 3-47B. Example of monopole vertical antenna gains.

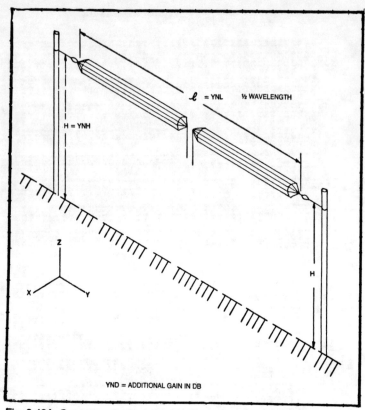

Fig. 3-48A. Geometry of a horizontal half-wavelength dipole antenna.

Item-D: Transmitter power. The power delivered to the antenna, after feeder and mismatch loss, etc. This power is expressed in kilowatts (kW) and decimal part of kilowatts.

Item-E: The required S/N ratio. The ratio in dB of the hourly median received signal power (dBw) in the signal occupied bandwidth relative to the hourly median noise in a 1.0-Hz bandwidth. This S/N ratio being required to provide the type and quality of service desired. The above noise is also in terms of dBw. See Fig. 3-20 for typical required S/N ratios. If the required S/N ratio is unknown, the required service may be given. Both the type and quality may be specified, e.g. broadcast quality voice and music or four channel RTTY (radio-teletype) at 60 words per minute, one error/thousand characters etc.

Item-F: Man-made noise at the receiver. Either the type of area(e.g. urban, sub-urban, or rural) or the expected noise in a 1.0-Hz bandwidth at 3.0 MHz may be specified. In remote unpopulous areas, it is interesting that cosmic (galactic) noise will "normally" dominate over man-made noise.

Item-G: Circuit reliability required. When a LUHF (LUF) or service probability computation is required, a minimum acceptable circuit reliability must be specified. Recall that circuit reliability is an estimate of the percen-

ANTENNA PATTERN

FREQUENCY RANGE 2.0 TO 30.0 — ANTENNA TYPE HORIZ HW DIPOLE — HEIGHT -0.250 — LENGTH -0.500 — ANGLE 0.000 — AZIMUTH 0.000 — EX(1) 0.000 — EX(2) 0.000 — EX(3) 0.000 — EX(4) 0.000 — CONDUCT. 0.001 — DIELECT. 4.000

ELEVATION ANGLE IN DEGREES

	2	3	4	5	6	7	8	9	10	11	12	13	14	16	18	20	22	24	26	28	30	
90	5.3	5.0	4.8	4.6	4.6	4.5	4.5	4.4	4.4	4.4	4.4	4.4	4.4	4.4	4.4	4.4	4.4	4.4	4.4	4.4	4.4	90
88	5.3	5.0	4.8	4.6	4.6	4.5	4.5	4.4	4.4	4.4	4.4	4.4	4.4	4.4	4.4	4.4	4.4	4.4	4.4	4.4	4.4	88
86	5.3	5.0	4.8	4.6	4.6	4.5	4.5	4.4	4.4	4.4	4.4	4.4	4.4	4.4	4.4	4.4	4.4	4.4	4.4	4.4	4.4	86
84	5.3	5.0	4.8	4.7	4.6	4.5	4.5	4.5	4.4	4.4	4.4	4.4	4.4	4.4	4.4	4.4	4.4	4.4	4.4	4.4	4.4	84
82	5.4	5.0	4.8	4.7	4.6	4.5	4.5	4.5	4.4	4.4	4.4	4.4	4.4	4.4	4.4	4.4	4.4	4.4	4.4	4.4	4.4	82
80	5.4	5.0	4.8	4.7	4.6	4.6	4.5	4.5	4.5	4.5	4.4	4.4	4.4	4.4	4.4	4.4	4.4	4.4	4.4	4.4	4.4	80
78	5.4	5.1	4.9	4.7	4.6	4.6	4.5	4.5	4.5	4.5	4.5	4.5	4.4	4.4	4.4	4.4	4.4	4.4	4.4	4.4	4.4	78
76	5.4	5.1	4.9	4.8	4.7	4.6	4.6	4.5	4.5	4.5	4.5	4.5	4.5	4.5	4.5	4.5	4.4	4.4	4.4	4.4	4.4	76
74	5.5	5.1	4.9	4.8	4.7	4.6	4.6	4.6	4.5	4.5	4.5	4.5	4.5	4.5	4.5	4.5	4.5	4.5	4.5	4.5	4.5	74
72	5.5	5.2	5.0	4.8	4.7	4.7	4.6	4.6	4.6	4.6	4.5	4.5	4.5	4.5	4.5	4.5	4.5	4.5	4.5	4.5	4.5	72
70	5.5	5.2	5.0	4.9	4.8	4.7	4.7	4.6	4.6	4.6	4.6	4.6	4.5	4.5	4.5	4.5	4.5	4.5	4.5	4.5	4.5	70
68	5.6	5.2	5.0	4.9	4.8	4.7	4.7	4.7	4.6	4.6	4.6	4.6	4.6	4.6	4.6	4.5	4.5	4.5	4.5	4.5	4.5	68
66	5.6	5.3	5.1	4.9	4.8	4.8	4.7	4.7	4.7	4.6	4.6	4.6	4.6	4.6	4.6	4.6	4.6	4.6	4.6	4.6	4.5	66
64	5.6	5.3	5.1	5.0	4.9	4.8	4.8	4.7	4.7	4.7	4.7	4.6	4.6	4.6	4.6	4.6	4.6	4.6	4.6	4.6	4.6	64
62	5.7	5.4	5.2	5.0	4.9	4.8	4.8	4.8	4.7	4.7	4.7	4.7	4.6	4.6	4.6	4.6	4.6	4.6	4.6	4.6	4.6	62
60	5.7	5.4	5.2	5.0	4.9	4.9	4.8	4.8	4.7	4.7	4.7	4.7	4.7	4.6	4.6	4.6	4.5	4.5	4.6	4.6	4.6	60
58	5.7	5.4	5.2	5.1	5.0	4.9	4.8	4.8	4.8	4.7	4.7	4.7	4.7	4.7	4.6	4.6	4.6	4.6	4.6	4.6	4.6	58
56	5.7	5.4	5.2	5.1	5.0	4.9	4.9	4.8	4.8	4.7	4.7	4.7	4.7	4.7	4.7	4.6	4.6	4.6	4.6	4.6	4.6	56
54	5.7	5.4	5.2	5.1	5.0	4.9	4.9	4.8	4.8	4.8	4.7	4.7	4.7	4.7	4.6	4.6	4.5	4.6	4.5	4.6	4.6	54
52	5.7	5.4	5.2	5.1	5.0	4.9	4.9	4.8	4.8	4.8	4.7	4.7	4.7	4.7	4.6	4.6	4.6	4.6	4.6	4.6	4.6	52
50	5.7	5.4	5.2	5.1	5.0	4.9	4.9	4.8	4.8	4.7	4.7	4.7	4.7	4.6	4.6	4.6	4.6	4.6	4.6	4.6	4.6	50
48	5.6	5.4	5.2	5.1	5.0	4.9	4.8	4.8	4.7	4.7	4.7	4.7	4.6	4.6	4.6	4.6	4.6	4.5	4.5	4.5	4.5	48
46	5.6	5.3	5.2	5.0	4.9	4.9	4.8	4.7	4.7	4.7	4.6	4.6	4.6	4.6	4.5	4.5	4.5	4.5	4.5	4.5	4.5	46
44	5.5	5.3	5.1	5.0	4.9	4.8	4.7	4.7	4.6	4.5	4.5	4.6	4.5	4.5	4.5	4.5	4.4	4.4	4.4	4.4	4.4	44
42	5.4	5.2	5.0	4.9	4.8	4.7	4.7	4.6	4.6	4.5	4.5	4.5	4.5	4.4	4.4	4.4	4.4	4.3	4.3	4.3	4.3	42
40	5.3	5.1	4.9	4.8	4.7	4.6	4.6	4.5	4.5	4.4	4.4	4.4	4.4	4.3	4.3	4.3	4.3	4.2	4.2	4.2	4.2	40
38	5.1	4.9	4.8	4.7	4.6	4.5	4.5	4.4	4.4	4.3	4.3	4.3	4.2	4.2	4.2	4.2	4.1	4.1	4.1	4.1	4.1	38
36	4.9	4.8	4.6	4.5	4.4	4.4	4.3	4.3	4.2	4.2	4.1	4.1	4.1	4.1	4.0	4.0	4.0	4.0	4.0	3.9	3.9	36
34	4.7	4.6	4.5	4.4	4.3	4.2	4.1	4.1	4.0	4.0	4.0	3.9	3.9	3.9	3.8	3.8	3.8	3.8	3.8	3.8	3.7	34
32	4.5	4.3	4.2	4.1	4.0	4.0	3.9	3.9	3.8	3.8	3.8	3.7	3.7	3.7	3.6	3.6	3.6	3.6	3.6	3.5	3.5	32
30	4.2	4.1	4.0	3.9	3.8	3.7	3.7	3.6	3.6	3.5	3.5	3.5	3.5	3.4	3.4	3.4	3.3	3.3	3.3	3.3	3.3	30
28	3.8	3.7	3.7	3.6	3.5	3.4	3.4	3.3	3.3	3.2	3.2	3.2	3.2	3.1	3.1	3.1	3.0	3.0	3.0	3.0	3.0	28
26	3.4	3.4	3.3	3.2	3.1	3.1	3.0	3.0	2.9	2.9	2.9	2.8	2.8	2.8	2.7	2.7	2.7	2.7	2.7	2.6	2.6	26
24	3.0	2.9	2.9	2.8	2.7	2.7	2.6	2.6	2.5	2.5	2.5	2.4	2.4	2.4	2.3	2.3	2.3	2.3	2.2	2.2	2.2	24
22	2.4	2.4	2.4	2.3	2.2	2.2	2.1	2.1	2.1	2.0	2.0	2.0	1.9	1.9	1.9	1.8	1.8	1.8	1.8	1.8	1.7	22
20	1.8	1.8	1.8	1.7	1.7	1.6	1.6	1.5	1.5	1.5	1.4	1.4	1.4	1.3	1.3	1.3	1.3	1.2	1.2	1.2	1.2	20
18	1.1	1.1	1.1	1.1	1.0	1.0	0.9	0.9	0.9	0.9	0.9	0.8	0.7	0.7	0.7	0.6	0.6	0.6	0.6	0.6	0.6	18
16	0.3	0.3	0.3	0.3	0.2	0.2	0.2	0.1	0.1	0.1	0.0	0.0	-0.0	-0.1	-0.1	-0.1	-0.1	-0.2	-0.2	-0.2	-0.2	16
14	-0.7	-0.6	-0.6	-0.7	-0.7	-0.7	-0.8	-0.8	-0.8	-0.9	-0.9	-0.9	-0.9	-1.0	-1.0	-1.0	-1.1	-1.1	-1.1	-1.1	-1.1	14
12	-1.9	-1.8	-1.8	-1.8	-1.8	-1.8	-1.9	-1.9	-1.9	-2.0	-2.0	-2.0	-2.0	-2.1	-2.1	-2.1	-2.1	-2.2	-2.2	-2.2	-2.2	12
10	-3.3	-3.2	-3.2	-3.2	-3.2	-3.2	-3.2	-3.3	-3.3	-3.3	-3.3	-3.4	-3.4	-3.4	-3.5	-3.5	-3.5	-3.5	-3.5	-3.6	-3.6	10
8	-5.1	-5.0	-4.9	-4.9	-4.9	-4.9	-5.0	-5.0	-5.0	-5.0	-5.1	-5.1	-5.1	-5.1	-5.2	-5.2	-5.2	-5.2	-5.3	-5.3	-5.3	8
6	-7.4	-7.3	-7.3	-7.2	-7.2	-7.2	-7.3	-7.3	-7.3	-7.3	-7.4	-7.4	-7.4	-7.4	-7.5	-7.5	-7.5	-7.5	-7.5	-7.6	-7.6	6
4	-10.0	-10.0	-10.0	-10.0	-10.0	-10.0	-10.0	-10.0	-10.0	-10.0	-10.0	-10.0	-10.0	-10.0	-10.0	-10.0	-10.0	-10.0	-10.0	-10.0	-10.0	4
2	-10.0	-10.0	-10.0	-10.0	-10.0	-10.0	-10.0	-10.0	-10.0	-10.0	-10.0	-10.0	-10.0	-10.0	-10.0	-10.0	-10.0	-10.0	-10.0	-10.0	-10.0	2
0	-10.0	-10.0	-10.0	-10.0	-10.0	-10.0	-10.0	-10.0	-10.0	-10.0	-10.0	-10.0	-10.0	-10.0	-10.0	-10.0	-10.0	-10.0	-10.0	-10.0	-10.0	0
	2	3	4	5	6	7	8	9	10	11	12	13	14	16	18	20	22	24	26	28	30	

FREQUENCY IN MEGAHERTZ

Fig. 3-48B. Example of horizontal half-wavelength dipole antenna gains.

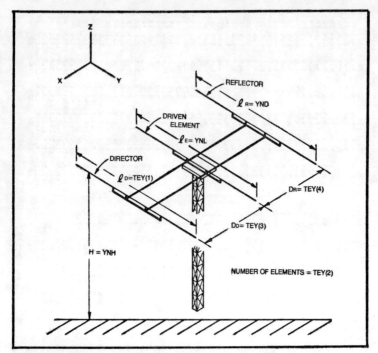

Fig. 3-49A. Geometry of a horizontal Yagi antenna.

tage of days of the month of calculation during which the signal quality will be acceptable and is a product of the fraction of days of ionospheric support by regular layers (i.e. E_s or sporadic-E expected) and the fraction of days during which the S/N (signal-to-noise ratio) is adequate.

Item-H: Is multipath computation requested? In some computerized methods this choice may be made (see options in Fig. 3-45). Multipath probability is an estimate of the liklihood that two or more sky-wave ionospheric propagation modes will exist within the constraints of the specified transmitter power (or more correctly the e.i.r.p.) and time delay tolerance.

Item-I: Minimum tolerable power ratio. The minimum tolerable power ratio between sky-wave ionospheric propagation modes at the receiver antenna input terminals to permit satisfactory system performance in the presence of multiple signals.

Item-J: Maximum tolerable time delays. This is the maximum tolerable difference in delay times between sky-wave ionospheric propagation modes, with received signal power levels within the item-I criterion above, the permit satisfactory system performance in the presence of multipath (multimode, e.g. a 3E and 1F mix) signals.

Antenna Parameters (Transmitting Or Receiving) (BOX-5):

(See Fig. 3-27b). Antenna parameters, transmitting or receiving—recall that in HF work, the transmitting and receiving antenna are consi-

ELEVATION ANGLE IN DEGREES

FREQUENCY RANGE 2.0 TO 30.0	ANTENNA TYPE HORIZONTAL YAGI					HEIGHT 30.000	LENGTH -0.500	ANTENNA PATTERN ANGLE AZIMUTH -0.580 0.000				EX(1) -0.450		EX(2) 5.000		EX(3) -0.128		EX(4) -0.235		CONDUCT. 0.001	DIELECT. 4.000	
90	1.0	-0.8	-4.9	-7.6	-2.7	-0.5	-1.0	-4.5	-6.9	-3.0	-1.1	-2.0	-3.1	-1.1								90
88	0.1	-0.8	-5.8	-8.5	-3.6	-1.4	-1.9	-5.3	-7.8	-3.9	-2.0	-2.0	-4.0	-2.0								88
86	-0.9	-1.9	-6.8	-9.6	-4.7	-2.5	-2.9	-6.3	-8.9	-5.0	-3.0	-3.0	-5.2	-3.0								86
84	-2.2	-3.1	-8.0	-10.0	-6.1	-3.8	-4.2	-7.5	-10.0	-6.4	-4.2	-4.2	-6.7	-4.2								84
82	-3.9	-4.7	-9.6	-10.0	-7.8	-5.4	-5.7	-8.9	-10.0	-8.3	-5.7	-5.7	-6.6	-5.7								82
80	-6.0	-6.8	-10.0	-10.0	-10.0	-7.6	-7.8	-10.0	-10.0	-10.0	-8.0	-8.0	-8.6	-8.0								80
78	-6.9	-10.0	-10.0	-10.0	-10.0	-10.0	-10.0	-10.0	-10.0	-10.0	-10.0	-10.0	-10.0	-10.0								78
76	-10.0	-10.0	-10.0	-10.0	-10.0	-10.0	-10.0	-10.0	-10.0	-10.0	-10.0	-10.0	-10.0	-10.0								76
74	-10.0	-10.0	-10.0	-10.0	-10.0	-10.0	-10.0	-10.0	-10.0	-10.0	-10.0	-10.0	-10.0	-10.0								74
72	-10.0	-10.0	-10.0	-10.0	-10.0	-10.0	-10.0	-10.0	-10.0	-10.0	-10.0	-10.0	-10.0	-10.0								72
70	-8.6	-9.5	-10.0	-10.0	-9.3	-7.5	-9.8	-10.0	-10.0	-10.0	-7.0	-7.0	-7.1	-7.3								70
68	-5.4	-6.7	-10.0	-10.0	-9.3	-5.3	-6.1	-9.8	-9.2	-8.9	-6.7	-6.7	-7.1	-6.5								68
66	-3.0	-3.1	-6.7	-10.0	-7.9	-3.6	-4.0	-7.1	-7.1	-6.5	-4.5	-4.5	-4.4	-4.5								66
64	-1.1	-1.1	-4.4	-10.0	-6.3	-1.4	-2.1	-5.3	-4.1	-3.3	-2.3	-2.1	-3.3	-2.1								64
62	0.5	0.6	-2.5	-8.6	-5.5	-0.8	-0.5	-3.0	-1.9	-1.1	-0.6	-0.6	-2.3	-0.8								62
60	1.8	2.1	-0.8	-6.6	-4.6	0.8	1.8	-1.0	0.0	0.6	-0.6	0.6	-3.3	-0.6								60
58	3.0	3.4	0.8	-4.6	-5.5	1.8	2.6	0.6	1.7	2.0	1.0	0.8	-4.5	2.0								58
56	4.0	4.4	2.2	-2.9	-4.6	2.6	3.3	2.1	3.2	3.0	2.3	1.8	-4.1	3.8								56
54	4.8	5.3	3.4	-1.3	-3.5	3.2	4.0	3.3	4.2	4.3	3.4	3.4	-4.1	4.6								54
52	5.5	5.5	4.5	0.3	-2.6	3.6	4.4	4.2	4.5	5.2	4.1	4.2	1.3	4.1								52
50	6.2	6.6	5.6	1.7	-1.3	3.7	5.2	5.6	5.4	6.4	5.5	4.7	1.3	6.0								50
48	6.7	7.2	6.6	3.1	-0.5	4.0	5.6	6.0	6.4	7.1	5.7	5.3	6.8	6.9								48
46	7.2	7.9	7.4	4.4	1.2	2.6	6.1	6.7	7.5	7.6	5.7	7.0	8.9	6.7								46
44	7.6	8.5	8.1	5.6	1.7	2.4	6.2	5.6	7.6	8.5	7.2	5.3	6.4	5.1								44
42	8.0	9.1	8.7	6.7	2.7	3.1	3.7	4.7	6.9	6.1	5.3	3.0	5.1	1.7								42
40	8.2	9.6	9.6	7.7	4.5	1.7	-0.1	3.4	7.6	6.2	3.5	-0.2	1.7	4.6								40
38	8.4	10.0	10.1	8.6	5.9	3.8	-0.8	1.6	6.9	8.4	7.2	-0.2	4.6	10.3								38
36	8.6	10.3	10.6	9.3	7.2	5.6	1.2	-0.4	4.2	8.6	9.6	1.6	1.6	8.1								36
34	8.6	10.6	10.9	10.0	8.3	7.3	3.8	-0.7	-0.8	6.0	9.9	7.3	1.1	8.0								34
32	8.6	10.9	11.3	10.6	9.3	8.7	6.1	1.6	-0.2	3.7	7.1	9.4	7.3	-0.2								32
30	8.6	11.0	11.5	11.1	10.2	9.8	8.0	4.6	-0.6	1.8	5.0	9.4	10.5	10.3								30
28	8.4	11.0	11.7	11.5	10.8	10.6	9.6	7.2	3.3	-0.2	2.3	6.3	10.5	10.5								28
26	8.2	10.9	11.8	11.7	11.4	11.2	10.8	9.2	6.5	3.3	-0.1	4.4	6.0	10.7								26
24	7.9	10.9	11.8	11.8	11.6	11.6	11.2	10.8	9.0	6.5	1.7	7.5	-0.1	6.0								24
22	7.6	10.4	11.6	11.6	11.8	12.0	12.5	12.0	10.8	9.0	6.7	11.0	11.5	-0.1								22
20	7.1	10.1	11.4	11.4	11.9	12.5	13.0	13.4	12.4	10.4	7.6	12.4	12.2	11.5								20
18	6.5	9.6	11.0	11.0	11.8	12.7	13.0	13.1	13.5	12.7	11.0	9.1	12.7	11.7								18
16	5.8	8.9	11.0	11.2	11.9	12.6	13.3	13.2	13.0	13.4	12.4	10.1	9.1	2.0								16
14	4.9	8.1	9.8	10.6	10.7	11.7	12.6	13.1	13.3	13.2	12.7	11.9	11.1	10.1								14
12	3.8	7.1	8.8	9.7	10.7	10.8	11.8	12.3	13.3	13.5	13.4	13.5	12.6	12.1								12
10	2.4	5.8	7.5	8.6	9.6	9.4	10.5	11.3	13.3	13.3	13.4	13.9	14.1	7.7								10
8	0.7	4.1	5.9	6.6	8.0	7.3	6.4	11.3	11.8	11.8	12.2	13.2	12.1	2.7								8
6	-1.7	1.8	3.7	4.9	6.0	4.2	5.5	6.4	7.1	7.7	2.9	8.4	7.6	3.3								6
4	-5.1	-1.6	0.4	1.6	2.8	-1.5	-0.2	0.8	1.5	2.1	1.4	2.1	5.9	3.3								4
2	-10.0	-7.4	-5.5	-4.2	-2.9	-10.0	-10.0	-10.0	-10.0	-10.0	-10.0	-10.0	-10.0	-4.6								2
0	-10.0	-10.0	-10.0	-10.0	-10.0	-10.0	-10.0	-10.0	-10.0	-10.0	-10.0	-10.0	-10.0	-10.0								0
	2	3	4	5	6	7	8	9	10	11	12	13	14	16	18	20	22	24	26	28	30	

FREQUENCY IN MEGAHERTZ

Fig. 3-49B. Example of horizontal Yagi antenna gains.

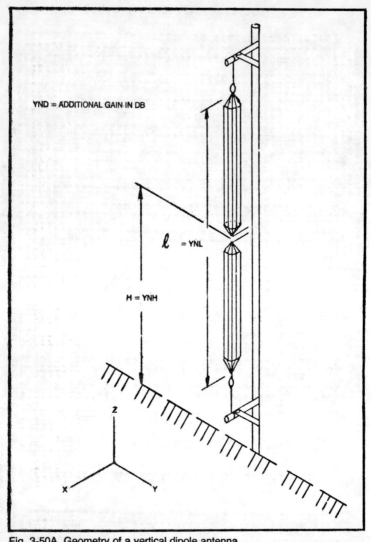

Fig. 3-50A. Geometry of a vertical dipole antenna.

dered to have reciprocal characteristics, following the laws of HF propagation and antenna reciprocity. This box indicates the data required for sixteen HF antenna types see Fig. 3-46 and 3-46B through Fig. 3-61A and 3-61B. These figures are to assist in the completion of BOX-5. Both geometrical configuration and tabularized antenna gain patterns are given for each example antenna in these figures. Should the customer utilize another antenna type, its performance may sometimes be approximated by selecting an antenna of similar pattern. Some of the simpler antennas, such as dipoles and verticals, permit the inclusion of additional gain to assist in

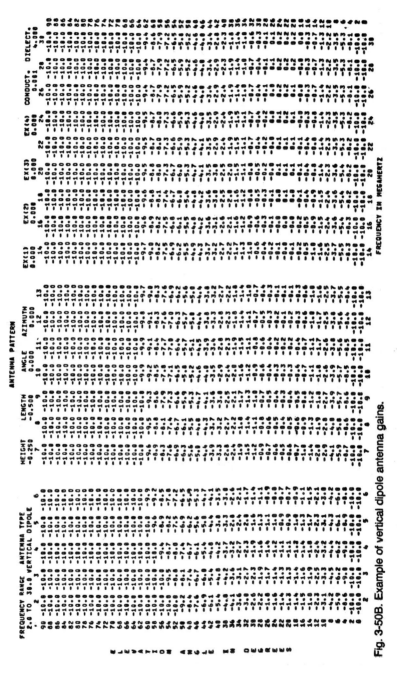

Fig. 3-50B. Example of vertical dipole antenna gains.

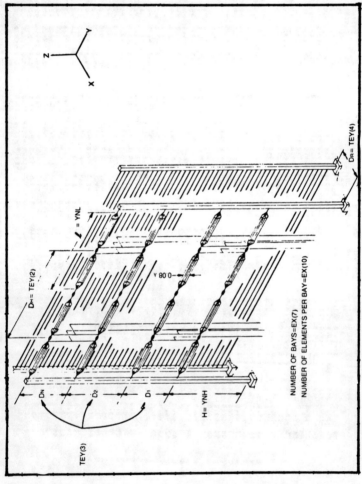

Fig. 3-51A. Geometry of a two-bay, four-stack curtain array antenna.

(MFMUFES3 74/12/30)

ANTENNA PATTERN

FREQUENCY RANGE 2.0 TO 30.0 — ANTENNA TYPE CURTAIN

HEIGHT 16.000 — LENGTH 22.000 — ANGLE 2.000 — AZIMUTH 0.000 — EX(1) 4.000 — EX(2) 26.000 — EX(3) 13.000 — EX(4) 7.000 — CONDUCT. 0.001 — DIELECT. 4.000

ELEVATION ANGLE IN DEGREES

Elev.	2	3	4	5	6
90	-10.0	-10.0	-10.0	-10.0	-10.0
88	-10.0	-10.0	-10.0	-10.0	-10.0
86	-10.0	-10.0	-10.0	-10.0	-10.0
84	-7.1	-10.0	-10.0	-10.0	-10.0
82	-4.5	-10.0	-10.0	-10.0	-10.0
80	-2.6	-8.2	-10.0	-10.0	-10.0
78	-1.0	-6.4	-10.0	-10.0	-10.0
76	0.4	-4.9	-10.0	-10.0	-10.0
74	1.6	-3.5	-10.0	-10.0	-10.0
72	2.6	-2.3	-10.0	-10.0	-10.0
70	3.5	-1.1	-10.0	-10.0	-10.0
68	4.4	0.0	-8.9	-10.0	-10.0
66	5.2	1.0	-7.5	-9.4	-10.0
64	5.9	2.0	-6.2	-8.9	-10.0
62	6.5	3.0	-4.8	-8.4	-10.0
60	7.1	3.8	-3.3	-8.1	-9.7
58	7.7	4.7	-1.9	-7.8	-8.1
56	8.2	5.5	-0.5	-7.5	-6.8
54	8.6	6.3	0.8	-7.0	-5.9
52	9.1	7.0	2.2	-6.1	-5.2
50	9.5	7.7	3.4	-4.7	-4.8
48	9.8	8.4	4.6	-2.9	-4.8
46	10.1	9.0	5.8	-0.9	-5.0
44	10.4	9.5	6.9	1.1	-5.1
42	10.6	10.0	7.9	3.0	-4.4
40	10.8	10.5	8.9	4.8	-2.6
38	10.9	10.9	9.7	6.5	-0.0
36	11.0	11.3	10.5	8.0	2.7
34	11.0	11.6	11.2	9.3	5.2
32	11.0	11.8	11.8	10.5	7.5
30	10.9	11.9	12.3	11.6	9.4
28	10.8	12.0	12.7	12.5	11.1
26	10.6	12.0	13.1	13.3	12.5
24	10.3	11.9	13.3	13.9	13.7
22	9.9	11.9	13.3	14.3	14.6
20	9.4	11.5	13.3	14.6	15.3
18	8.9	11.0	13.1	14.7	15.7
16	8.1	10.5	12.7	14.5	15.9
14	7.3	9.7	12.1	14.2	15.8
12	6.2	8.8	11.3	13.6	15.5
10	4.8	7.5	10.2	12.6	14.7
8	3.1	5.9	8.7	11.2	13.4
6	0.8	3.6	6.5	9.1	11.5
4	-2.6	0.3	3.3	6.0	8.4
2	-3.5	-5.5	-2.6	0.2	2.7
0	-10.0	-10.0	-10.0	-10.0	-10.0
	2	3	4	5	6

Elev.	7	8	9	10	11	12	13	14
90	-10.0	-10.0	-10.0	-10.0	-10.0	-10.0	-10.0	-10.0
88	-10.0	-10.0	-10.0	-10.0	-10.0	-10.0	-10.0	-10.0
86	-10.0	-10.0	-10.0	-10.0	-10.0	-10.0	-10.0	-8.5
84	-10.0	-10.0	-10.0	-10.0	-10.0	-10.0	-9.5	-5.2
82	-10.0	-10.0	-10.0	-9.5	-10.0	-10.0	-7.6	-3.0
80	-10.0	-10.0	-10.0	-7.5	-10.0	-10.0	-6.4	-1.6
78	-10.0	-10.0	-9.3	-5.9	-10.0	-10.0	-5.8	-0.6
76	-10.0	-10.0	-8.4	-4.5	-8.1	-10.0	-5.5	-0.1
74	-9.8	-8.7	-7.7	-3.4	-6.1	-10.0	-5.5	0.0
72	-9.2	-7.4	-7.3	-2.5	-4.2	-10.0	-5.6	-0.3
70	-8.8	-6.3	-6.9	-1.8	-2.5	-10.0	-5.9	-1.0
68	-8.8	-5.2	-6.5	-1.4	-1.0	-10.0	-6.7	-2.2
66	-9.0	-4.3	-6.0	-1.2	0.2	-6.5	-8.7	-3.7
64	-9.7	-3.6	-5.3	-1.3	1.2	-3.4	-10.0	-5.2
62	-10.0	-3.0	-4.3	-1.6	1.9	-0.8	-10.0	-6.4
60	-10.0	-2.7	-3.1	-2.2	2.4	1.3	-6.4	-7.7
58	-10.0	-2.7	-1.9	-2.7	2.4	2.9	-0.1	-10.0
56	-10.0	-3.0	-0.8	-2.9	2.0	4.0	4.1	-10.0
54	-10.0	-3.8	-0.0	-2.3	1.2	4.6	7.0	-10.0
52	-10.0	-5.4	0.4	-1.0	-0.0	4.7	9.1	-3.0
50	-8.9	-8.2	0.5	0.5	-1.2	4.2	10.3	1.8
48	-6.2	-10.0	-0.0	1.7	-1.1	2.9	10.8	4.8
46	-4.4	-10.0	-1.3	2.4	0.5	1.0	10.5	6.6
44	-3.3	-10.0	-3.8	2.5	2.3	-0.3	9.2	7.3
42	-2.8	-6.5	-8.9	1.8	3.6	0.9	6.8	6.8
40	-3.0	-3.5	-10.0	-0.1	4.0	3.2	4.9	5.1
38	-3.6	-1.9	-9.9	-4.0	3.5	4.8	6.5	2.0
36	-3.7	-1.4	-3.9	-10.0	1.5	5.2	9.1	0.8
34	-1.8	-1.9	-1.2	-10.0	-2.9	4.3	10.5	3.9
32	1.5	-3.0	-0.3	-3.0	-10.0	1.5	10.4	6.2
30	5.0	-1.9	-1.0	-0.1	-6.4	-6.1	8.3	6.6
28	7.9	2.2	-2.5	0.4	-0.5	-10.0	2.3	4.8
26	10.4	6.4	-0.3	-1.3	1.2	-0.8	-10.0	-1.1
24	12.4	9.8	5.0	-1.6	-0.2	1.9	4.0	-10.0
22	14.0	12.4	9.4	4.4	-1.9	0.4	6.8	0.3
20	15.3	14.5	12.8	9.7	4.8	-1.5	4.6	2.3
18	16.2	16.1	15.3	13.6	10.8	6.6	4.7	-1.5
16	16.8	17.2	17.1	16.3	14.9	12.7	13.9	4.1
14	17.1	17.9	19.3	18.1	17.6	16.6	19.5	12.4
12	17.0	18.1	18.9	19.3	19.3	19.1	23.0	17.2
10	16.5	17.9	19.0	19.7	20.2	20.5	25.0	20.0
8	15.4	17.0	18.4	19.4	20.2	20.9	25.8	21.2
6	13.6	15.4	16.9	18.2	19.2	20.2	25.4	21.1
4	10.6	12.6	14.2	15.6	16.9	17.9	23.3	19.3
2	5.0	7.0	8.7	10.2	11.5	12.7	18.2	14.3
0	-10.0	-10.0	-10.0	-10.0	-10.0	-10.0	-10.0	-10.0
	7	8	9	10	11	12	13	14

Elev.	16	18	20	22	24	26	28	30
90	-10.0	-10.0	-10.0	-10.0	-10.0	-10.0	-10.0	-10.0
88	-10.0	-10.0	-4.6	-6.0	-10.0	-10.0	-10.0	-10.0
86	-9.8	-8.5	1.4	-0.2	-7.8	-10.0	-10.0	-10.0
84	-4.9	-6.0	4.9	2.8	-4.2	-10.0	-10.0	-10.0
82	-1.9	-5.2	7.3	4.6	-1.5	-10.0	-10.0	-10.0
80	0.7	-5.6	9.1	5.7	0.6	-10.0	-10.0	-10.0
78	2.9	-7.7	10.3	6.1	2.2	-10.0	-10.0	-10.0
76	4.9	-10.0	11.2	6.1	3.5	-10.0	-10.0	-10.0
74	6.7	-10.0	11.6	6.0	4.4	-10.0	-10.0	-10.0
72	8.2	-10.0	11.5	6.2	4.8	-10.0	-10.0	-10.0
70	9.5	-6.5	10.8	7.0	4.7	-10.0	-10.0	-10.0
68	10.5	-2.4	9.3	8.2	4.0	-10.0	-10.0	-10.0
66	11.2	1.4	6.9	9.2	2.6	-10.0	-10.0	-9.9
64	11.4	5.1	2.9	9.6	0.4	-10.0	-10.0	-8.0
62	11.1	8.4	-4.1	9.4	-2.1	-10.0	-10.0	-7.7
60	10.3	11.0	-10.0	8.1	-2.7	-10.0	-10.0	-9.1
58	8.7	12.9	-9.3	5.6	-1.5	-10.0	-10.0	-10.0
56	6.3	14.0	-3.1	1.2	-0.8	-10.0	-10.0	-10.0
54	2.9	14.3	2.3	-6.5	-1.6	-10.0	-10.0	-8.0
52	-0.6	13.6	6.5	-10.0	-4.1	-10.0	-10.0	-7.4
50	-2.7	11.8	9.1	-10.0	-9.0	-10.0	-10.0	-9.5
48	-6.2	8.5	10.2	-7.2	-10.0	-10.0	-10.0	-10.0
46	-10.0	3.5	9.8	-1.6	-10.0	-10.0	-10.0	-10.0
44	-1.1	-1.2	7.8	1.4	-10.0	-10.0	-10.0	-10.0
42	5.4	-3.7	3.5	2.1	-10.0	-10.0	-10.0	-10.0
40	8.7	-10.0	-3.4	0.4	-10.0	-10.0	-10.0	-10.0
38	10.0	-0.2	-7.9	-4.0	-10.0	-10.0	-10.0	-10.0
36	9.5	6.7	-10.0	-10.0	-10.0	-10.0	-10.0	-10.0
34	6.8	9.4	-8.5	-10.0	-10.0	-10.0	-10.0	-10.0
32	2.5	9.4	0.4	-10.0	-10.0	-10.0	-10.0	-10.0
30	4.5	6.4	3.1	-10.0	-10.0	-10.0	-10.0	-10.0
28	8.0	0.7	2.1	-9.7	-10.0	-10.0	-10.0	-10.0
26	8.7	4.5	-3.4	-9.9	-10.0	-10.0	-10.0	-10.0
24	6.2	7.6	-5.7	-10.0	-10.0	-10.0	-10.0	-10.0
22	-3.4	6.7	-0.5	-10.0	-10.0	-10.0	-10.0	-10.0
20	-2.2	-0.1	-0.8	-10.0	-10.0	-10.0	-10.0	-10.0
18	3.7	-6.6	-8.0	-10.0	-10.0	-10.0	-10.0	-10.0
16	0.4	2.0	-10.0	-10.0	-10.0	-10.0	-10.0	-10.0
14	5.2	-2.9	-7.2	-10.0	-10.0	-10.0	-10.0	-10.0
12	14.6	6.2	-10.0	-10.0	-10.0	-10.0	-10.0	-10.0
10	19.4	14.8	1.0	-10.0	-10.0	-10.0	-10.0	-10.0
8	21.9	18.8	7.3	-10.0	-10.0	-10.0	-10.0	-10.0
6	22.6	20.4	9.7	-6.8	-3.3	-10.0	-10.0	-4.6
4	21.2	19.5	9.3	-4.9	-1.3	-10.0	-10.0	1.1
2	16.4	15.1	5.1	-7.8	-4.3	-10.0	-10.0	-0.4
0	-10.0	-10.0	-10.0	-10.0	-10.0	-10.0	-10.0	-10.0
	16	18	20	22	24	26	28	30

FREQUENCY IN MEGAHERTZ

Fig. 3-51B. Example of curtain antenna gains.

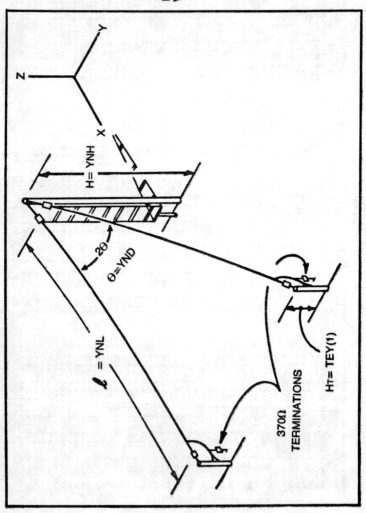

Fig. 3-52A. Geometry of a sloping Vee antenna.

ANTENNA PATTERN

FREQUENCY RANGE 2.0 TO 30.0 — ANTENNA TYPE TERM. SLOPING V — HEIGHT 15.240 — LENGTH 121.900 — ANGLE 22.500 — AZIMUTH 0.000 — EX(1) 1.829 — EX(2) 0.000 — EX(3) 0.000 — EX(4) 0.000 — CONDUCT. 0.001 — DIELECT. 4.000

ELEVATION ANGLE IN DEGREES	2	3	4	5	6	7	8	9	10	11	12	13	14	16	18	20	22	24	26	28	30	ELEVATION ANGLE IN DEGREES
90	-10.0	-10.0	-10.0	-10.0	-10.0	-10.0	-10.0	-10.0	-10.0	-10.0	-10.0	-10.0	-10.0	-10.0	-10.0	-10.0	-10.0	-10.0	-9.0	-9.8	-10.0	90
88	-10.0	-10.0	-10.0	-10.0	-9.6	-10.0	-10.0	-10.0	-10.0	-10.0	-10.0	-10.0	-10.0	-10.0	-10.0	-10.0	-9.8	-7.6	-9.6	-10.0	-10.0	88
86	-10.0	-10.6	-9.2	-10.3	-9.6	-10.0	-10.0	-10.0	-10.0	-10.0	-10.0	-8.7	-10.0	-10.0	-10.0	-10.0	-9.4	-10.0	-10.0	-10.0	-10.0	86
84	-10.0	-10.0	-8.6	-10.0	-10.0	-10.0	-10.0	-10.0	-10.0	-10.0	-10.0	-9.7	-9.8	-8.1	-10.0	-10.0	-10.0	-10.0	-7.9	-9.0	-10.0	84
82	-10.0	-10.0	-8.1	-10.0	-10.0	-10.0	-10.0	-10.0	-10.0	-10.0	-10.0	-10.0	-7.6	-10.0	-10.0	-10.0	-10.0	-6.8	-9.0	-10.0	-10.0	82
80	-10.0	-10.0	-7.9	-10.0	-10.0	-9.4	-10.0	-10.0	-10.0	-10.0	-9.7	-10.0	-8.3	-10.0	-10.0	-10.0	-8.0	-10.0	-10.0	-9.5	-9.3	80
78	-10.0	-10.0	-7.8	-9.9	-10.0	-9.1	-10.0	-10.0	-10.0	-10.0	-8.8	-10.0	-10.0	-10.0	-8.8	-10.0	-10.0	-10.0	-6.6	-8.8	-10.0	78
76	-10.0	-10.0	-8.0	-8.5	-10.0	-9.5	-10.0	-10.0	-10.0	-10.0	-10.0	-8.5	-10.0	-7.6	-9.4	-10.0	-10.0	-6.5	-8.0	-10.0	-10.0	76
74	-9.8	-10.0	-8.3	-7.4	-10.0	-10.0	-10.0	-10.0	-10.0	-10.0	-10.0	-7.2	-10.0	-6.3	-10.0	-10.0	-8.6	-7.4	-10.0	-6.7	-9.4	74
72	-9.2	-10.0	-8.8	-6.5	-10.0	-10.0	-9.1	-10.0	-10.0	-10.0	-10.0	-8.4	-6.9	-8.9	-10.0	-9.3	-8.5	-10.0	-5.4	-9.1	-10.0	72
70	-8.7	-10.0	-9.5	-6.1	-9.8	-10.0	-9.2	-10.0	-10.0	-10.0	-10.0	-10.0	-6.0	-10.0	-7.2	-10.0	-10.0	-7.2	-6.5	-10.0	-6.9	70
68	-8.3	-10.0	-10.0	-6.9	-7.7	-10.0	-10.0	-10.0	-10.0	-10.0	-10.0	-10.0	-7.8	-10.0	-6.3	-10.0	-10.0	-5.6	-10.0	-4.5	-10.0	68
66	-7.9	-9.8	-10.0	-6.3	-6.3	-10.0	-10.0	-9.1	-10.0	-10.0	-8.2	-10.0	-10.0	-5.4	-10.0	-10.0	-7.8	-10.0	-4.8	-8.9	-7.9	66
64	-7.5	-8.6	-10.0	-6.4	-5.3	-10.0	-10.0	-9.4	-10.0	-10.0	-8.6	-7.9	-10.0	-4.3	-10.0	-6.5	-10.0	-10.0	-4.7	-10.0	-5.1	64
62	-7.2	-7.6	-10.0	-7.0	-4.8	-7.5	-10.0	-10.0	-9.5	-10.0	-10.0	-6.5	-8.7	-6.4	-6.8	-8.7	-10.0	-5.4	-10.0	-2.7	-10.0	62
60	-6.9	-6.6	-10.0	-7.7	-4.6	-5.6	-10.0	-10.0	-9.3	-10.0	-10.0	-7.3	-5.5	-10.0	-3.5	-10.0	-7.3	-8.9	-5.7	-7.2	-3.7	60
58	-6.7	-5.8	-8.9	-8.5	-4.7	-4.4	-8.6	-10.0	-10.0	-9.7	-10.0	-10.0	-5.1	-10.0	-4.2	-7.5	-8.5	-10.0	-3.5	-9.9	-3.5	58
56	-6.5	-5.1	-7.5	-9.0	-5.2	-3.8	-5.9	-10.0	-10.0	-9.2	-10.0	-10.0	-7.1	-4.8	-9.7	-3.2	-10.0	-6.7	-10.0	-1.9	-10.0	56
54	-6.3	-4.5	-6.1	-8.8	-5.9	-3.6	-4.2	-9.1	-10.0	-10.0	-9.1	-10.0	-10.0	-2.7	-10.0	-3.8	-6.4	-8.0	-9.5	-4.6	-1.7	54
52	-6.2	-3.9	-4.9	-7.9	-6.8	-3.8	-3.2	-5.7	-10.0	-10.0	-8.9	-9.9	-10.0	-3.4	-3.6	-10.0	-2.9	-10.0	-4.5	-10.0	-1.8	52
50	-6.1	-3.5	-3.8	-6.3	-7.3	-4.4	-2.8	-3.7	-8.2	-10.0	-10.0	-7.9	-10.0	-7.1	-1.0	-8.7	-5.1	-4.6	-7.8	-3.2	-10.0	50
48	-6.1	-3.1	-2.8	-4.7	-7.0	-5.2	-2.9	-2.6	-4.8	-10.0	-10.0	-8.6	-7.2	-10.0	-1.5	-1.6	-10.0	-3.5	-7.7	-3.6	-2.4	48
46	-6.1	-2.7	-2.0	-3.3	-5.8	-5.8	-3.4	-2.1	-2.9	-6.3	-10.0	-10.0	-6.8	-10.0	-5.3	0.4	-3.8	-10.0	-3.6	-9.8	-1.1	46
44	-6.1	-2.5	-1.3	-2.9	-4.1	-5.8	-4.2	-2.3	-1.9	-3.4	-7.8	-10.0	-9.0	-5.3	-10.0	-0.9	0.8	-7.4	-7.2	-4.5	-8.7	44
42	-6.2	-2.3	-0.8	-0.9	-2.4	-4.6	-4.8	-3.0	-1.7	-1.9	-4.0	-8.9	-10.0	-4.8	-6.8	-6.4	1.1	0.5	-10.0	-5.4	-5.4	42
40	-6.3	-2.2	-0.3	0.0	-0.9	-2.8	-4.5	-3.8	-2.1	-1.3	-1.9	-4.4	-9.4	-5.3	-2.5	-9.0	-2.3	2.1	0.1	-10.0	-4.8	40
38	-6.4	-2.1	0.0	0.9	0.4	-1.6	-3.0	-4.0	-2.9	-1.5	-1.0	-1.9	-4.5	-8.7	-1.7	-2.1	-8.7	-0.2	2.7	-0.8	-9.3	38
36	-6.6	-2.2	0.3	1.4	1.4	0.6	-1.0	-2.9	-3.4	-2.2	-1.1	-0.8	-1.8	-8.2	-3.4	0.3	-2.0	-6.8	0.9	3.1	0.4	36
34	-6.9	-2.2	0.4	1.8	2.3	1.9	0.8	-0.9	-2.5	-2.8	-1.8	-0.8	-0.6	-3.8	-6.9	-0.2	1.6	-1.7	-5.6	1.4	3.5	34
32	-7.1	-2.4	0.5	2.2	2.9	3.0	2.4	1.2	-0.5	-2.1	-2.4	-1.5	-0.6	-1.1	-6.0	-3.1	1.9	2.7	-0.8	-5.2	1.4	32
30	-7.5	-2.5	0.5	2.4	3.4	3.9	3.7	3.0	1.7	0.0	-1.5	-2.0	-1.3	-0.1	-2.1	-6.2	-0.3	3.4	3.8	0.6	-5.0	30
28	-7.8	-2.8	0.4	2.5	3.8	4.5	4.7	4.4	3.6	2.4	0.8	-0.8	-1.5	-0.3	-0.8	-2.9	-4.5	1.5	4.6	5.0	2.4	28
26	-8.3	-3.1	0.2	2.4	4.0	4.9	5.4	5.5	5.2	4.4	3.2	1.7	0.1	-0.9	0.4	0.0	-3.2	-2.9	2.6	5.6	6.1	26
24	-8.8	-3.5	-0.1	2.3	4.0	5.2	6.0	5.3	6.3	6.0	5.3	4.2	2.8	-0.0	-0.0	1.0	0.3	-2.0	-1.0	3.3	6.2	24
22	-9.4	-4.0	-0.5	2.1	3.9	5.3	6.2	6.9	7.2	7.2	6.9	6.2	5.3	2.7	0.4	0.9	1.8	1.0	-1.6	-1.3	3.4	22
20	-10.0	-4.6	-1.0	1.7	3.7	5.2	6.3	7.2	7.7	8.0	8.0	7.7	7.2	5.4	2.9	1.3	2.0	2.8	2.2	0.1	-0.8	20
18	-10.0	-5.3	-1.6	1.2	3.3	4.9	6.2	7.2	7.9	8.4	8.7	8.7	8.6	7.6	5.8	3.6	2.5	3.3	4.1	3.0	2.3	18
16	-10.0	-6.1	-2.3	0.5	2.7	4.5	5.9	7.0	7.9	8.6	9.0	9.3	9.4	9.0	8.0	6.3	4.6	4.1	4.0	5.6	5.7	16
14	-10.0	-7.0	-3.2	-0.3	2.0	3.8	5.3	6.6	7.6	8.4	9.0	9.5	9.0	9.9	9.4	8.4	7.0	5.0	5.7	6.5	7.3	14
12	-10.0	-8.2	-4.3	-1.3	1.0	2.9	4.5	5.8	6.9	7.9	8.6	9.2	9.7	10.2	10.2	9.7	8.8	7.8	7.1	7.3	8.1	12
10	-10.0	-9.6	-5.6	-2.5	-0.3	1.7	3.4	4.8	6.0	7.0	7.8	8.6	9.1	9.9	10.2	10.1	9.7	9.0	8.4	8.2	8.7	10
8	-10.0	-10.6	-7.4	-4.3	-1.9	0.1	1.8	3.3	4.5	5.6	6.6	7.4	8.0	9.0	9.6	9.8	9.7	9.4	8.9	8.7	8.9	8
6	-10.0	-10.0	-9.7	-6.6	-4.2	-2.1	-0.3	1.2	2.5	3.6	4.6	5.5	6.2	7.3	8.1	8.5	8.7	8.6	8.3	8.2	8.3	6
4	-10.0	-10.0	-10.0	-10.0	-7.5	-5.4	-3.6	-2.0	-0.7	0.5	1.5	2.4	3.2	4.4	5.3	5.9	6.2	6.3	6.2	6.2	6.3	4
2	-10.0	-10.0	-10.0	-10.0	-10.0	-10.0	-9.4	-7.8	-6.5	-5.3	-4.2	-3.3	-2.5	-1.2	-0.2	0.5	0.8	1.0	1.1	1.1	1.2	2
0	-10.0	-10.0	-10.0	-10.0	-10.0	-10.0	-10.0	-10.0	-10.0	-10.0	-10.0	-10.0	-10.0	-10.0	-10.0	-10.0	-10.0	-10.0	-10.0	-10.0	-10.0	0
	2	3	4	5	6	7	8	9	10	11	12	13	14	16	18	20	22	24	26	28	30	

FREQUENCY IN MEGAHERTZ

Fig. 3-52B. Example of terminated sloping Vee antenna gains.

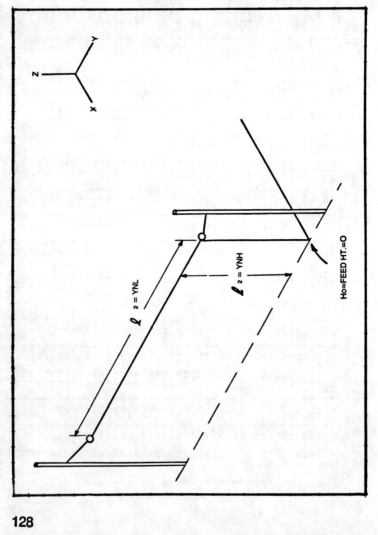

Fig. 3-53A. Geometry of an inverted L antenna.

128

ANTENNA PATTERN

FREQUENCY RANGE	ANTENNA TYPE	HEIGHT	LENGTH	ANGLE	AZIMUTH	EX(1)	EX(2)	EX(3)	EX(4)	CONDUCT.	DIELECT.
2.0 TO 30.0	INVERTED L	10.000	21.340	-0.000	0.000	0.000	0.000	0.000	0.000	0.001	4.000

ELEVATION ANGLE IN DEGREES

	2	3	4	5	6	7	8	9	10	11	12	13	14	16	18	20	22	24	26	28	30	
90	-4.4	-2.0	-0.0	3.3	5.6	4.1	1.6	-0.7	-3.1	-7.6	-10.0	-10.0	-10.0	-10.0	-6.2	0.7	0.1	-3.5	-10.0	-10.0	-10.0	90
88	-4.4	-2.0	-0.0	3.3	5.6	4.2	1.6	-0.7	-3.1	-7.5	-10.0	-10.0	-10.0	-10.0	-6.1	0.7	0.1	-3.5	-10.0	-10.0	-10.0	88
86	-4.3	-2.0	-0.0	3.3	5.6	4.2	1.6	-0.6	-3.1	-7.5	-10.0	-10.0	-10.0	-10.0	-6.1	0.7	0.1	-3.4	-10.0	-10.0	-10.0	86
84	-4.3	-2.0	-0.0	3.3	5.6	4.2	1.6	-0.6	-3.0	-7.4	-10.0	-10.0	-10.0	-10.0	-6.1	0.7	0.1	-3.4	-10.0	-10.0	-10.0	84
82	-4.2	-2.0	-0.1	3.3	5.6	4.2	1.7	-0.6	-3.0	-7.3	-10.0	-10.0	-10.0	-10.0	-6.1	0.7	0.2	-3.3	-10.0	-10.0	-10.0	82
80	-4.2	-2.0	-0.1	3.3	5.6	4.2	1.7	-0.5	-2.9	-7.2	-10.0	-10.0	-10.0	-10.0	-6.0	0.6	0.3	-3.1	-10.0	-10.0	-10.0	80
78	-4.1	-2.0	-0.1	3.3	5.6	4.2	1.7	-0.4	-2.8	-7.0	-10.0	-10.0	-10.0	-10.0	-6.0	0.6	0.3	-3.0	-9.8	-10.0	-10.0	78
76	-4.0	-1.9	-0.1	3.2	5.6	4.3	1.8	-0.4	-2.7	-6.9	-10.0	-10.0	-10.0	-10.0	-5.9	0.5	0.4	-2.8	-8.9	-10.0	-9.2	76
74	-3.9	-1.9	-0.1	3.2	5.6	4.3	1.9	-0.3	-2.6	-6.7	-10.0	-10.0	-10.0	-10.0	-5.9	0.4	0.5	-2.7	-8.1	-10.0	-8.3	74
72	-3.7	-1.9	-0.1	3.2	5.6	4.3	1.9	-0.2	-2.5	-6.5	-10.0	-10.0	-10.0	-9.9	-5.8	0.2	0.6	-2.5	-7.3	-9.7	-7.6	72
70	-3.6	-1.9	-0.1	3.2	5.6	4.3	2.0	-0.1	-2.3	-6.3	-10.0	-10.0	-10.0	-9.2	-5.7	0.1	0.7	-2.3	-6.7	-8.6	-6.9	70
68	-3.5	-1.6	-0.1	3.2	5.6	4.4	2.1	0.0	-2.2	-6.0	-10.0	-10.0	-10.0	-8.7	-5.6	-0.1	0.7	-2.1	-6.1	-7.5	-6.4	68
66	-3.4	-1.8	-0.1	3.2	5.6	4.4	2.1	0.2	-2.0	-5.8	-10.0	-10.0	-10.0	-8.1	-5.5	-0.4	0.8	-1.9	-5.6	-6.6	-6.0	66
64	-3.2	-1.8	-0.2	3.1	5.6	4.4	2.2	0.3	-1.9	-5.6	-9.6	-10.0	-10.0	-7.7	-5.3	-0.7	0.8	-1.8	-5.2	-5.7	-5.7	64
62	-3.1	-1.8	-0.2	3.1	5.6	4.5	2.3	0.4	-1.7	-5.4	-9.2	-10.0	-10.0	-7.3	-5.2	-1.0	0.8	-1.6	-4.8	-4.9	-5.4	62
60	-3.0	-1.8	-0.2	3.0	5.5	4.5	2.4	0.5	-1.5	-5.2	-8.8	-9.9	-10.0	-7.0	-5.0	-1.4	0.7	-1.5	-4.6	-4.1	-5.2	60
58	-2.9	-1.8	-0.3	3.0	5.5	4.5	2.5	0.7	-1.4	-5.0	-8.5	-9.4	-10.0	-6.7	-4.8	-1.9	0.6	-1.4	-4.4	-3.4	-5.1	58
56	-2.8	-1.7	-0.3	2.9	5.5	4.5	2.5	0.8	-1.2	-4.8	-8.2	-8.8	-9.8	-6.5	-4.5	-2.4	0.5	-1.3	-4.2	-2.7	-4.9	56
54	-2.7	-1.7	-0.4	2.9	5.4	4.5	2.6	0.9	-1.0	-4.6	-7.9	-8.4	-9.1	-6.3	-4.3	-2.9	0.4	-1.2	-4.1	-2.1	-4.7	54
52	-2.6	-1.8	-0.4	2.9	5.4	4.5	2.6	1.0	-0.9	-4.4	-7.6	-7.9	-8.5	-6.1	-4.0	-3.4	0.2	-1.2	-4.1	-1.6	-4.5	52
50	-2.5	-1.8	-0.5	2.7	5.3	4.5	2.7	1.1	-0.7	-4.2	-7.4	-7.5	-7.9	-6.0	-3.7	-3.7	-0.1	-1.2	-4.2	-1.1	-4.3	50
48	-2.4	-1.8	-0.6	2.5	5.2	4.5	2.7	1.2	-0.5	-4.1	-7.2	-7.1	-7.3	-5.9	-3.4	-3.9	-0.3	-1.2	-4.3	-0.7	-4.0	48
46	-2.3	-1.8	-0.7	2.5	5.1	4.5	2.8	1.3	-0.4	-3.9	-7.0	-6.7	-6.7	-5.8	-3.0	-3.7	-0.6	-1.3	-4.5	-0.4	-3.7	46
44	-2.3	-1.9	-0.8	2.3	5.0	4.4	2.8	1.4	-0.3	-3.8	-6.9	-6.4	-6.2	-5.7	-2.7	-3.3	-0.8	-1.3	-4.7	-0.2	-3.3	44
42	-2.2	-2.0	-1.0	2.2	4.9	4.3	2.8	1.5	-0.2	-3.7	-6.8	-6.1	-5.7	-5.6	-2.4	-2.5	-0.9	-1.4	-4.9	-0.0	-3.0	42
40	-2.2	-2.0	-1.1	2.0	4.8	4.2	2.7	1.5	-0.1	-3.6	-6.8	-5.8	-5.2	-5.5	-2.0	-1.7	-1.0	-1.4	-5.2	0.1	-2.7	40
38	-2.2	-2.1	-1.3	1.8	4.6	4.1	2.7	1.5	-0.0	-3.5	-6.7	-5.6	-4.8	-5.4	-1.7	-0.7	-0.8	-1.4	-5.4	0.1	-2.5	38
36	-2.3	-2.2	-1.5	1.6	4.4	4.0	2.6	1.5	0.0	-3.5	-6.8	-5.4	-4.4	-5.2	-1.4	0.2	-0.5	-1.3	-5.7	-0.0	-2.4	36
34	-2.3	-2.4	-1.7	1.3	4.2	3.8	2.5	1.5	0.0	-3.5	-6.8	-5.3	-4.1	-5.1	-1.1	1.1	-0.1	-1.2	-5.8	-0.2	-2.3	34
32	-2.4	-2.5	-1.9	1.1	3.9	3.6	2.4	1.4	-0.0	-3.5	-6.9	-5.1	-3.8	-4.9	-0.9	1.9	0.4	-1.0	-5.8	-0.5	-2.5	32
30	-2.5	-2.7	-2.2	0.7	3.6	3.4	2.2	1.3	-0.1	-3.6	-7.0	-5.1	-3.5	-4.7	-0.7	2.6	1.0	-0.8	-5.8	-0.8	-2.7	30
28	-2.6	-3.0	-2.5	0.4	3.3	3.1	2.0	1.1	-0.2	-3.7	-7.2	-5.1	-3.3	-4.5	-0.6	3.2	1.5	-0.5	-5.7	-1.3	-3.1	28
26	-2.8	-3.2	-2.9	-0.0	2.9	2.7	1.7	0.9	-0.3	-3.9	-7.4	-5.1	-3.2	-4.3	-0.5	3.7	2.0	-0.3	-5.6	-1.7	-3.6	26
24	-3.0	-3.5	-3.3	-0.5	2.4	2.3	1.4	0.6	-0.6	-4.1	-7.7	-5.2	-3.1	-4.2	-0.5	4.0	2.5	-0.0	-5.4	-2.2	-4.2	24
22	-3.2	-3.9	-3.7	-1.0	1.9	1.9	1.0	0.3	-0.9	-4.4	-8.1	-5.3	-3.1	-4.1	-0.6	4.2	2.8	0.2	-5.3	-2.7	-4.9	22
20	-3.5	-4.3	-4.3	-1.5	1.3	1.4	0.5	-0.1	-1.2	-4.8	-8.5	-5.6	-3.2	-4.1	-0.8	4.3	3.0	0.3	-5.2	-3.1	-5.6	20
18	-3.9	-4.8	-4.9	-2.3	0.6	0.7	-0.0	-0.6	-1.7	-5.3	-9.0	-5.9	-3.4	-4.1	-1.1	4.2	3.0	0.2	-5.3	-3.5	-6.2	18
16	-4.4	-5.4	-5.6	-3.1	-0.2	-0.0	-0.7	-1.2	-2.3	-5.9	-9.7	-6.3	-3.7	-4.4	-1.6	3.9	2.8	0.1	-5.4	-3.9	-6.6	16
14	-5.0	-6.1	-6.4	-4.1	-1.1	-0.9	-1.5	-2.0	-3.0	-6.7	-10.0	-6.9	-4.2	-4.7	-2.2	3.5	2.5	-0.3	-5.8	-4.3	-7.0	14
12	-5.8	-7.0	-7.4	-5.2	-2.3	-2.0	-2.5	-2.9	-3.9	-7.6	-10.0	-7.6	-4.8	-5.3	-3.1	2.8	1.9	-0.9	-6.4	-4.9	-7.3	12
10	-6.7	-8.1	-8.7	-6.6	-3.7	-3.3	-3.7	-4.1	-5.1	-8.8	-10.0	-8.6	-5.7	-6.1	-4.2	1.8	0.9	-1.9	-7.2	-5.6	-7.8	10
8	-8.0	-9.5	-10.0	-8.3	-5.4	-4.9	-5.3	-5.6	-6.6	-10.0	-10.0	-9.9	-7.0	-7.3	-5.8	0.4	-0.4	-3.2	-8.5	-6.6	-8.7	8
6	-9.9	-10.0	-10.0	-10.0	-7.7	-7.2	-7.5	-7.7	-8.6	-10.0	-10.0	-10.0	-8.8	-9.1	-7.9	-1.7	-2.4	-5.2	-10.0	-8.3	-10.0	6
4	-10.0	-10.0	-10.0	-10.0	-10.0	-10.0	-10.0	-10.0	-10.0	-10.0	-10.0	-10.0	-10.0	-10.0	-10.0	-4.7	-5.4	-8.3	-10.0	-10.0	-10.0	4
2	-10.0	-10.0	-10.0	-10.0	-10.0	-10.0	-10.0	-10.0	-10.0	-10.0	-10.0	-10.0	-10.0	-10.0	-10.0	-10.0	-10.0	-10.0	-10.0	-10.0	-10.0	2
0	-10.0	-10.0	-10.0	-10.0	-10.0	-10.0	-10.0	-10.0	-10.0	-10.0	-10.0	-10.0	-10.0	-10.0	-10.0	-10.0	-10.0	-10.0	-10.0	-10.0	-10.0	0
	2	3	4	5	6	7	8	9	10	11	12	13	14	16	18	20	22	24	26	28	30	

FREQUENCY IN MEGAHERTZ

ELEVATION ANGLE IN DEGREES

Fig. 3-53B. Example of inverted L antenna gains.

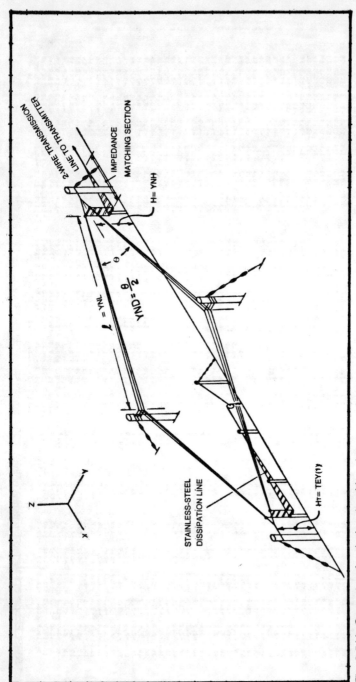

Fig. 3-54A. Geometry of a sloping rhombic antenna.

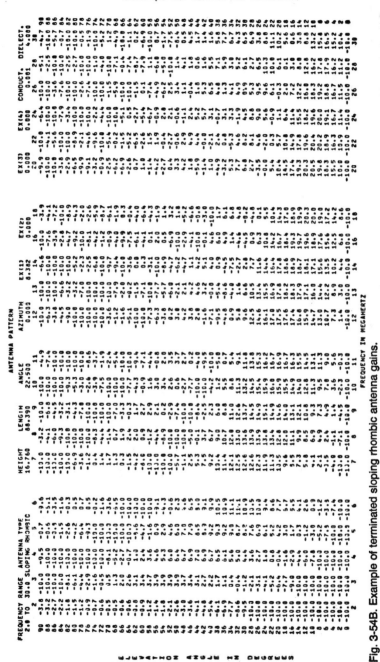

Fig. 3-54B. Example of terminated sloping rhombic antenna gains.

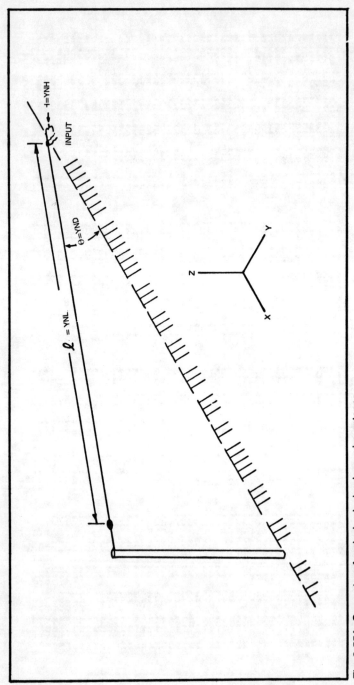

Fig. 3-55A. Geometry of a single sloping long-wire antenna.

ANTENNA PATTERN

FREQUENCY RANGE 2.0 TO 30.0 | ANTENNA TYPE SLOPING WIRE | HEIGHT 0.000 | LENGTH 91.440 | ANGLE 10.000 | AZIMUTH 0.000 | EX(1) 0.000 | EX(2) 0.000 | EX(3) 0.000 | EX(4) 0.000 | CONDUCT. 0.001 | DIELECT. 4.000

ELEVATION ANGLE IN DEGREES

	2	3	4	5	6	7	8	9	10	11	12	13	14	16	18	20	22	24	26	28	30	
90	1.8	-3.9	-1.0	-1.8	-2.4	-4.4	-10.0	-7.1	-4.3	-7.4	-6.0	-3.9	-0.5	-7.7	-6.0	-10.0	-0.6	-5.0	-5.4	-10.0	-3.1	90
88	1.8	-3.2	-0.9	-2.6	-1.8	-2.8	-10.0	-8.9	-5.0	-5.0	-3.3	-7.2	-0.5	-10.0	-5.9	-6.2	-2.4	-5.6	-4.7	-9.0	-8.8	88
86	1.7	-2.6	-0.6	-3.8	-1.5	-1.6	-10.0	-10.0	-6.8	-3.8	-1.2	-10.0	-1.5	-7.2	-9.5	-3.7	-5.1	-2.1	-10.0	-5.6	-7.1	86
84	1.6	-2.1	-0.1	-5.1	-1.4	-0.9	-10.0	-10.0	-10.0	-3.6	-0.2	-8.6	-3.2	-3.7	-10.0	-4.8	-2.5	-1.9	-9.4	-10.0	-2.7	84
82	1.5	-1.6	0.4	-6.7	-1.6	-0.5	-8.6	-10.0	-10.0	-4.5	-0.1	-4.5	-4.0	-2.7	-10.0	-10.0	-0.2	-5.3	-3.8	-10.0	-6.2	82
80	1.4	-1.1	1.0	-8.2	-2.0	-0.4	-6.2	-7.4	-10.0	-6.6	-1.1	-2.4	-2.5	-3.8	-6.0	-10.0	-0.9	-5.3	-4.9	-5.7	-10.0	80
78	1.3	-0.7	1.7	-8.9	-2.7	-0.8	-4.7	-5.0	-10.0	-10.0	-3.3	-1.8	-0.4	-7.3	-4.4	-6.2	-4.6	-1.4	-10.0	-5.4	-3.5	78
76	1.1	-0.4	2.3	-8.2	-3.8	-1.5	-3.8	-3.3	-7.3	-10.0	-6.9	-2.6	0.7	-10.0	-5.6	-3.5	-5.8	-1.1	-6.2	-10.0	-3.7	76
74	1.0	-0.1	2.8	-6.5	-5.0	-2.7	-3.4	-2.2	-4.9	-10.0	-8.9	-5.0	0.7	-6.1	-10.0	-3.9	-1.4	-4.8	-2.4	-9.5	-10.0	74
72	0.8	0.1	3.3	-4.7	-6.2	-4.4	-3.6	-1.6	-3.5	-6.5	-5.6	-9.6	-0.6	-2.4	-10.0	-7.8	0.1	-6.2	-4.3	-3.5	-7.2	72
70	0.6	0.3	3.8	-3.1	-6.8	-6.7	-4.3	-1.5	-2.9	-4.1	-2.3	-10.0	-3.1	-1.1	-7.6	-10.0	-1.6	-1.4	-10.0	-4.4	-2.5	70
68	0.4	0.4	4.2	-1.7	-6.2	-9.3	-5.7	-2.0	-2.9	-2.8	-0.3	-7.5	-5.4	-1.5	-3.3	-9.4	-7.7	-0.2	-4.6	-10.0	-4.1	68
66	0.1	0.4	4.5	-0.6	-4.6	-9.4	-8.0	-3.1	-3.5	-2.3	0.7	-3.7	-3.5	-4.0	-2.0	-3.7	-8.2	-3.0	-1.2	-5.9	-10.0	66
64	-0.1	0.5	4.8	0.3	-2.9	-6.6	-10.0	-5.0	-4.8	-2.6	0.8	-1.8	-0.4	-8.8	-3.0	-1.7	-1.9	-9.6	-2.7	-1.2	-4.5	64
62	-0.4	0.4	5.0	1.1	-1.2	-3.6	-10.0	-8.2	-7.1	-3.6	0.1	-1.1	1.5	-7.8	-7.0	-2.0	0.3	-4.1	-10.0	-2.3	-0.0	62
60	-0.7	0.3	5.1	1.6	0.1	-1.1	-10.0	-10.0	-10.0	-5.5	-1.5	-1.4	2.3	-2.8	-10.0	-5.2	-0.1	-0.4	-6.0	-10.0	-1.2	60
58	-1.1	0.2	5.2	2.1	1.3	0.7	-8.4	-10.0	-10.0	-8.9	-4.3	-2.8	2.0	-0.1	-6.5	-10.0	-3.1	-0.3	-1.1	-5.0	-9.9	58
56	-1.4	0.0	5.2	2.4	2.2	2.2	-4.8	-5.2	-10.0	-10.0	-9.6	-5.5	0.6	1.0	-1.6	-7.8	-10.0	-3.5	-0.7	0.1	-2.0	56
54	-1.8	-0.2	5.1	2.6	2.9	3.3	-2.3	-1.6	-8.8	-10.0	-10.0	-10.0	-2.2	0.6	0.5	-1.1	-7.3	-10.0	-4.4	0.1	2.6	54
52	-2.1	-0.4	5.0	2.7	3.4	4.2	-0.5	1.0	-4.5	-9.5	-9.4	-10.0	-7.4	-1.2	0.9	1.7	0.2	-8.7	-10.0	-4.7	2.2	52
50	-2.5	-0.7	4.8	2.7	3.8	4.8	0.9	3.0	-1.6	-4.6	-3.6	-10.0	-10.0	-5.0	-0.5	2.4	3.4	-1.4	-6.9	-10.0	-3.7	50
48	-3.0	-1.0	4.6	2.6	4.0	5.3	1.9	4.4	0.4	-1.5	-0.2	-5.9	-5.2	-10.0	-3.9	1.4	4.4	1.6	-1.5	-2.9	-5.8	48
46	-3.4	-1.4	4.3	2.4	4.2	5.6	2.6	5.5	1.9	0.7	2.2	-2.5	-0.6	-9.3	-10.0	-2.0	3.5	2.5	0.5	0.7	1.9	46
44	-3.9	-1.8	4.0	2.2	4.1	5.7	3.1	6.3	3.0	2.3	4.0	-0.1	2.2	-3.1	-9.7	-9.9	0.2	1.4	0.5	1.4	4.2	44
42	-4.4	-2.2	3.6	1.9	4.0	5.7	3.3	6.8	3.7	3.5	5.3	1.5	4.2	0.3	-2.7	-6.2	-7.6	-2.6	-1.6	0.1	3.7	42
40	-4.9	-2.7	3.2	1.5	3.8	5.5	3.4	7.2	4.2	4.3	6.2	2.7	5.6	2.4	0.9	0.4	-2.1	-10.0	-8.2	-4.1	0.8	40
38	-5.4	-3.2	2.7	1.0	3.5	5.3	3.3	7.3	4.4	4.9	6.9	3.5	6.6	3.8	2.9	3.7	4.0	-2.1	-10.0	-10.0	-7.8	38
36	-5.9	-3.7	2.2	0.5	3.1	4.9	3.1	7.2	4.5	5.1	7.2	4.1	7.3	4.8	4.2	5.6	7.1	3.5	-1.0	-4.0	-6.6	36
34	-6.5	-4.2	1.7	-0.1	2.6	4.4	2.7	6.9	4.3	5.1	7.3	4.3	7.7	5.4	4.9	6.6	8.9	6.4	3.4	2.3	2.9	34
32	-7.0	-4.8	1.1	-0.7	2.1	3.9	2.3	6.5	3.9	4.9	7.2	4.3	7.8	5.7	5.3	7.2	9.8	7.9	5.6	5.4	7.0	32
30	-7.6	-5.4	0.5	-1.3	1.4	3.3	1.7	6.0	3.4	4.5	6.8	4.0	7.7	5.7	5.4	7.4	10.2	8.6	6.7	7.0	9.2	30
28	-8.3	-6.0	-0.2	-2.0	0.8	2.6	1.0	5.3	2.8	4.0	6.3	3.5	7.3	5.4	5.2	7.3	10.2	8.8	7.1	7.6	10.1	28
26	-8.9	-6.6	-0.8	-2.7	0.1	1.8	0.3	4.6	2.0	3.2	5.5	2.8	6.6	4.8	4.6	6.8	9.9	8.7	7.1	7.7	10.4	26
24	-9.6	-7.3	-1.5	-3.5	-0.7	1.0	-0.5	3.8	1.1	2.4	4.6	1.9	5.7	3.9	3.7	6.0	9.2	8.1	6.6	7.3	10.1	24
22	-10.0	-8.0	-2.3	-4.2	-1.4	0.2	-1.3	3.0	0.3	1.5	3.6	0.9	4.6	2.7	2.5	4.8	8.1	7.1	5.7	6.5	9.5	22
20	-10.0	-8.7	-3.0	-5.0	-2.2	-0.6	-2.1	2.1	-0.6	0.6	2.7	-0.1	3.5	1.4	1.0	3.1	6.4	5.5	4.2	5.2	8.3	20
18	-10.0	-9.5	-3.9	-5.8	-3.0	-1.4	-2.9	1.3	-1.4	-0.3	1.8	-1.1	2.5	0.2	-0.5	1.2	4.2	3.2	1.9	3.0	6.3	18
16	-10.0	-10.0	-4.7	-6.7	-3.8	-2.2	-3.7	0.5	-2.2	-1.0	1.0	-1.8	1.7	-0.7	-1.7	-0.5	1.8	0.3	-1.4	-0.6	2.9	16
14	-10.0	-10.0	-5.6	-7.6	-4.7	-3.1	-4.5	-0.3	-2.9	-1.7	0.4	-2.3	1.2	-1.2	-2.0	-0.9	0.9	-1.5	-4.2	-4.7	-2.5	14
12	-10.0	-10.0	-6.7	-8.6	-5.7	-4.0	-5.4	-1.1	-3.6	-2.4	-0.1	-2.8	0.8	-1.3	-1.9	-0.4	1.8	-0.4	-3.0	-3.6	-2.6	12
10	-10.0	-10.0	-7.9	-9.8	-6.9	-5.2	-6.5	-2.1	-4.5	-3.2	-0.8	-3.4	0.4	-1.5	-1.7	0.3	3.0	1.3	-0.6	-0.4	1.9	10
8	-10.0	-10.0	-9.4	-10.0	-8.3	-6.6	-7.8	-3.3	-5.7	-4.3	-1.8	-4.3	-0.4	-2.0	-2.0	0.3	3.4	2.2	0.6	1.4	4.2	8
6	-10.0	-10.0	-10.0	-10.0	-10.0	-8.5	-9.7	-5.2	-7.5	-6.0	-3.4	-5.9	-1.8	-3.3	-3.0	-0.5	2.9	1.8	0.5	1.5	4.7	6
4	-10.0	-10.0	-10.0	-10.0	-10.0	-10.0	-10.0	-8.0	-10.0	-8.7	-6.1	-8.5	-4.4	-5.7	-5.3	-2.6	0.9	-0.0	-1.2	-0.1	3.3	4
2	-10.0	-10.0	-10.0	-10.0	-10.0	-10.0	-10.0	-10.0	-10.0	-10.0	-10.0	-10.0	-9.5	-10.0	-10.0	-7.5	-4.0	-4.8	-5.9	-4.7	-1.3	2
0	-10.0	-10.0	-10.0	-10.0	-10.0	-10.0	-10.0	-10.0	-10.0	-10.0	-10.0	-10.0	-10.0	-10.0	-10.0	-10.0	-10.0	-10.0	-10.0	-10.0	-10.0	0
	2	3	4	5	6	7	8	9	10	11	12	13	14	16	18	20	22	24	26	28	30	

FREQUENCY IN MEGAHERTZ

ELEVATION ANGLE IN DEGREES

Fig. 3-55B. Example of sloping long-wire antenna gains.

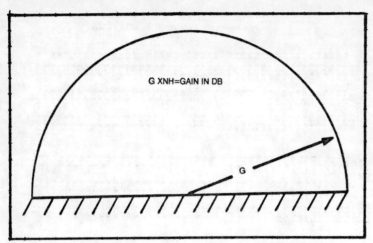

Fig. 3-56A. Illustration of the gain pattern of a constant gain antenna.

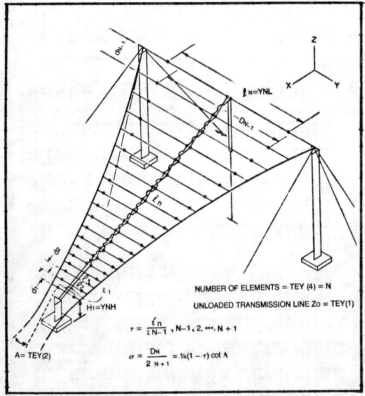

Fig. 3-57A. Geometry of a horizontally polarized crossed-dipole, log periodic antenna.

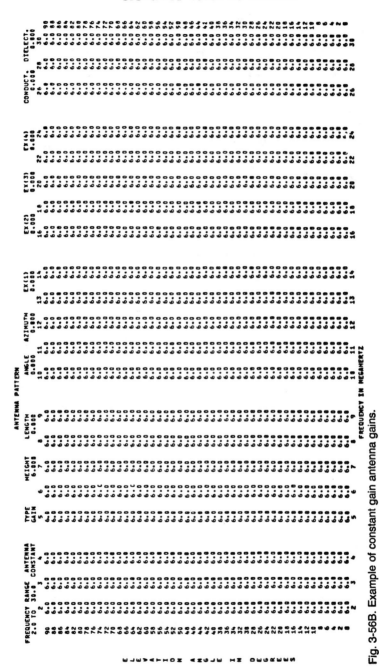

Fig. 3-56B. Example of constant gain antenna gains.

ANTENNA PATTERN

FREQUENCY RANGE 2.0 TO 30.0   ANTENNA TYPE HORIZONTAL LPA   HEIGHT 2.957   LENGTH 33.468   ANGLE 54.000   AZIMUTH 0.000   EX(1) 450.000   EX(2) 25.000   EX(3) 0.800   EX(4) 12.000   CONDUCT. 0.001   DIELECT. 4.000

ELEVATION ANGLE IN DEGREES	2	3	4	5	6	7	8	9	10	11	12	13	14	16	18	20	22	24	26	28	30	ELEVATION ANGLE IN DEGREES
90	5.3	5.1	3.4	1.4	0.3	-1.0	-5.3	-4.2	-6.8	-7.2	-7.9	-10.0	-10.0	-10.0	-10.0	-10.0	-10.0	-7.2	-4.6	-3.7	-2.7	90
88	5.3	5.2	3.5	1.6	0.5	-0.8	-5.5	-3.9	-6.3	-6.5	-7.6	-10.0	-10.0	-10.0	-10.0	-10.0	-10.0	-6.9	-4.2	-3.4	-2.4	88
86	5.3	5.2	3.5	1.8	0.6	-0.6	-5.5	-3.6	-5.6	-5.7	-7.0	-9.8	-10.0	-10.0	-10.0	-10.0	-10.0	-6.6	-3.8	-3.2	-2.1	86
84	5.3	5.2	3.7	2.0	0.8	-0.4	-5.2	-3.2	-4.8	-4.8	-6.4	-8.6	-9.8	-10.0	-10.0	-10.0	-10.0	-6.5	-3.6	-3.1	-1.9	84
82	5.3	5.3	3.8	2.2	1.0	-0.1	-4.7	-2.7	-4.0	-3.9	-5.6	-7.3	-8.4	-10.0	-10.0	-10.0	-10.0	-6.4	-3.5	-3.1	-1.7	82
80	5.3	5.3	3.9	2.4	1.3	0.2	-4.0	-2.1	-3.1	-3.0	-4.7	-6.0	-7.1	-10.0	-10.0	-10.0	-9.8	-6.4	-3.4	-3.1	-1.6	80
78	5.3	5.3	4.1	2.7	1.6	0.6	-3.1	-1.6	-2.2	-2.1	-3.8	-4.8	-5.8	-9.1	-10.0	-10.0	-9.4	-6.5	-3.5	-3.2	-1.6	78
76	5.3	5.4	4.2	3.0	1.9	0.9	-2.2	-0.9	-1.4	-1.3	-2.8	-3.6	-4.7	-7.4	-9.5	-9.6	-9.1	-6.6	-3.6	-3.4	-1.6	76
74	5.3	5.5	4.4	3.2	2.2	1.3	-1.3	-0.3	-0.5	-0.5	-1.9	-2.5	-3.6	-5.9	-7.9	-8.2	-8.7	-6.6	-3.8	-3.6	-1.7	74
72	5.3	5.5	4.6	3.5	2.5	1.8	-0.4	0.3	0.3	0.3	-1.0	-1.4	-2.5	-4.5	-6.5	-6.8	-8.1	-6.7	-4.1	-4.0	-2.0	72
70	5.3	5.6	4.8	3.8	2.9	2.2	0.5	1.0	1.0	1.0	-0.2	-0.4	-1.5	-3.2	-5.1	-5.5	-7.4	-6.6	-4.4	-4.4	-2.3	70
68	5.3	5.7	5.0	4.1	3.2	2.6	1.3	1.6	1.8	1.7	0.7	0.5	-0.6	-2.0	-3.8	-4.2	-6.5	-6.4	-4.8	-5.0	-2.8	68
66	5.2	5.7	5.2	4.5	3.6	3.1	2.2	2.2	2.5	2.4	1.5	1.4	0.2	-0.9	-2.6	-2.9	-5.6	-5.9	-5.1	-5.7	-3.4	66
64	5.2	5.8	5.4	4.8	4.0	3.5	2.9	2.8	3.1	3.0	2.2	2.2	1.1	0.2	-1.5	-1.7	-4.5	-5.2	-5.2	-6.4	-4.0	64
62	5.2	5.8	5.6	5.1	4.3	4.0	3.6	3.4	3.7	3.5	2.9	2.9	1.8	1.1	-0.4	-0.6	-3.4	-4.3	-5.0	-7.0	-4.6	62
60	5.1	5.9	5.7	5.4	4.7	4.4	4.3	4.0	4.3	4.1	3.6	3.6	2.6	2.0	0.6	0.4	-2.2	-3.2	-4.4	-7.1	-5.0	60
58	5.0	6.0	5.9	5.7	5.1	4.9	4.9	4.5	4.9	4.6	4.2	4.2	3.2	2.8	1.5	1.4	-1.0	-2.0	-3.4	-6.5	-5.0	58
56	5.0	6.0	6.1	6.0	5.5	5.3	5.5	5.0	5.3	5.0	4.8	4.8	3.9	3.6	2.4	2.3	0.1	-0.9	-2.3	-5.3	-4.4	56
54	4.9	6.0	6.3	6.2	5.8	5.7	6.0	5.4	5.8	5.4	5.3	5.4	4.5	4.3	3.2	3.1	1.2	0.2	-1.1	-3.7	-3.3	54
52	4.8	6.1	6.4	6.5	6.2	6.1	6.5	5.9	6.2	5.8	5.8	5.8	5.0	4.9	4.0	3.9	2.2	1.2	0.1	-2.0	-2.0	52
50	4.7	6.1	6.5	6.8	6.5	6.4	6.9	6.3	6.6	6.2	6.3	6.2	5.5	5.5	4.7	4.6	3.1	2.2	1.2	-0.4	-0.6	50
48	4.6	6.1	6.7	7.0	6.8	6.8	7.3	6.7	6.9	6.5	6.7	6.6	6.0	6.0	5.3	5.2	3.9	3.1	2.3	1.0	0.7	48
46	4.5	6.1	6.8	7.2	7.1	7.1	7.6	7.0	7.2	6.8	7.0	6.9	6.4	6.4	5.9	5.8	4.7	3.9	3.3	2.3	1.9	46
44	4.3	6.0	6.8	7.4	7.3	7.4	7.8	7.3	7.4	7.1	7.3	7.2	6.8	6.8	6.4	6.3	5.3	4.7	4.2	3.5	3.1	44
42	4.1	5.9	6.9	7.5	7.5	7.6	8.0	7.5	7.6	7.3	7.6	7.5	7.1	7.2	6.9	6.7	5.9	5.4	5.0	4.5	4.2	42
40	3.9	5.9	6.9	7.6	7.7	7.8	8.2	7.7	7.8	7.5	7.8	7.6	7.4	7.5	7.3	7.1	6.4	6.1	5.8	5.3	5.2	40
38	3.7	5.8	6.9	7.7	7.9	8.0	8.3	7.9	7.9	7.7	8.0	7.8	7.6	7.7	7.6	7.5	6.9	6.6	6.4	6.0	6.1	38
36	3.5	5.6	6.9	7.8	8.0	8.1	8.3	8.0	8.0	7.9	8.1	7.9	7.8	7.9	7.9	7.8	7.3	7.1	6.9	6.7	6.8	36
34	3.2	5.4	6.8	7.8	8.0	8.2	8.3	8.1	8.0	8.0	8.2	8.0	8.0	8.0	8.2	8.0	7.6	7.5	7.4	7.3	7.5	34
32	2.9	5.2	6.7	7.7	8.1	8.3	8.2	8.1	8.0	8.1	8.2	8.0	8.1	8.1	8.4	8.2	7.9	7.8	7.7	7.8	8.0	32
30	2.5	5.0	6.5	7.6	8.0	8.2	8.1	8.1	8.0	8.2	8.2	8.0	8.1	8.2	8.5	8.3	8.2	8.1	8.0	8.3	8.4	30
28	2.1	4.6	6.3	7.5	7.9	8.2	7.9	8.0	7.9	8.1	8.1	7.9	8.1	8.1	8.6	8.4	8.3	8.2	8.2	8.7	8.6	28
26	1.7	4.3	6.0	7.3	7.8	8.0	7.6	7.8	7.7	8.1	8.0	7.8	8.0	8.1	8.5	8.4	8.5	8.3	8.4	8.9	8.8	26
24	1.2	3.9	5.7	7.0	7.5	7.8	7.2	7.6	7.4	7.9	7.8	7.6	7.9	7.9	8.5	8.3	8.5	8.3	8.5	9.1	9.0	24
22	0.6	3.4	5.2	6.6	7.2	7.5	6.8	7.3	7.1	7.7	7.5	7.4	7.6	7.7	8.3	8.2	8.4	8.2	8.5	9.1	9.0	22
20	-0.1	2.8	4.7	6.2	6.8	7.1	6.3	6.9	6.7	7.4	7.1	7.1	7.3	7.4	8.0	7.9	8.2	8.0	8.4	8.9	8.9	20
18	-0.8	2.1	4.1	5.6	6.3	6.5	5.6	6.3	6.2	7.0	6.6	6.6	6.8	6.9	7.7	7.5	7.8	7.7	8.2	8.7	8.7	18
16	-1.7	1.3	3.4	4.9	5.6	5.9	4.8	5.7	5.5	6.4	6.0	6.0	6.3	6.4	7.1	7.0	7.3	7.3	7.8	8.2	8.4	16
14	-2.7	0.4	2.5	4.1	4.8	5.1	3.9	4.8	4.7	5.7	5.2	5.3	5.5	5.7	6.5	6.3	6.6	6.7	7.2	7.6	7.8	14
12	-3.9	-0.8	1.4	3.0	3.8	4.1	2.8	3.8	3.7	4.8	4.2	4.4	4.5	4.7	5.6	5.4	5.7	5.9	6.4	6.8	7.0	12
10	-5.3	-2.2	0.0	1.7	2.5	2.8	1.4	2.5	2.5	3.5	3.0	3.1	3.3	3.5	4.4	4.2	4.5	4.7	5.2	5.7	5.9	10
8	-7.2	-4.0	-1.7	0.0	0.8	1.1	-0.3	0.9	0.8	1.9	1.3	1.5	1.7	1.9	2.8	2.6	2.9	3.2	3.7	4.2	4.4	8
6	-9.5	-6.3	-4.0	-2.3	-1.4	-1.2	-2.6	-1.4	-1.4	-0.3	-0.9	-0.7	-0.5	-0.3	0.6	0.4	0.7	1.1	1.5	2.1	2.2	6
4	-10.0	-9.7	-7.4	-5.6	-4.0	-4.5	-6.0	-4.7	-4.7	-3.5	-4.2	-3.9	-3.8	-3.5	-2.6	-2.8	-2.6	-2.2	-1.7	-1.1	-1.0	4
2	-10.0	-10.0	-10.0	-10.0	-10.0	-10.0	-10.0	-10.0	-10.0	-9.3	-10.0	-9.7	-9.6	-9.3	-8.4	-8.6	-8.4	-7.9	-7.5	-6.9	-6.8	2
0	-10.0	-10.0	-10.0	-10.0	-10.0	-10.0	-10.0	-10.0	-10.0	-10.0	-10.0	-10.0	-10.0	-10.0	-10.0	-10.0	-10.0	-10.0	-10.0	-10.0	-10.0	0
	2	3	4	5	6	7	8	9	10	11	12	13	14	16	18	20	22	24	26	28	30	

FREQUENCY IN MEGAHERTZ

Fig. 3-57B. Example of horizontal crossedipole, log periodic antenna.

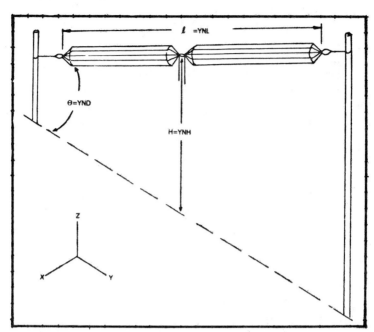

Fig. 3-58A. Geometry of an arbitrary tilted dipole antenna.

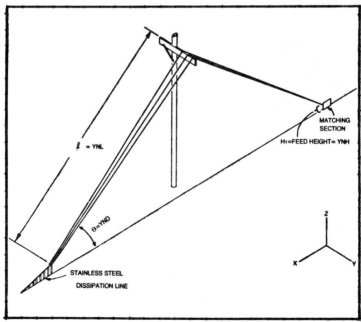

Fig. 3-59A. Geometry of a vertical half-rhombic antenna.

137

(HFMUFES3 74/12/30)

ANTENNA PATTERN

FREQUENCY RANGE	ANTENNA TYPE	HEIGHT	LENGTH	ANGLE	AZIMUTH	EX(1)	EX(2)	EX(3)	EX(4)	CONDUCT.	DIELECT.
2.6 TO 30.0	ARBITARY DIPOLE	-0.250	-0.500	45.000	0.000	0.000	0.000	0.000	0.000	0.001	4.000

ELEVATION ANGLE IN DEGREES

	2	3	4	5	6	7	8	9	10	11	12	13	14	16	18	20	22	24	26	28	30	
90	1.8	1.4	1.1	1.0	0.8	0.9	0.7	0.6	0.6	0.6	0.5	0.5	0.5	0.5	0.5	0.4	0.4	0.4	0.4	0.4	0.4	90
88	1.8	1.4	1.2	1.0	0.9	0.8	0.7	0.6	0.6	0.6	0.5	0.5	0.5	0.5	0.5	0.5	0.4	0.4	0.4	0.4	0.4	88
86	1.8	1.4	1.2	1.0	0.9	0.9	0.7	0.7	0.6	0.6	0.6	0.5	0.5	0.5	0.5	0.5	0.5	0.4	0.4	0.4	0.4	86
84	1.8	1.4	1.2	1.0	0.9	0.8	0.7	0.7	0.6	0.6	0.6	0.6	0.5	0.5	0.5	0.5	0.5	0.5	0.5	0.5	0.4	84
82	1.8	1.5	1.2	1.0	0.9	0.9	0.8	0.7	0.7	0.6	0.6	0.6	0.6	0.6	0.5	0.5	0.5	0.5	0.5	0.5	0.5	82
80	1.9	1.5	1.3	1.1	1.0	0.9	0.8	0.8	0.7	0.7	0.7	0.6	0.6	0.6	0.6	0.6	0.5	0.5	0.5	0.5	0.5	80
78	1.9	1.5	1.3	1.1	1.0	0.9	0.9	0.8	0.8	0.7	0.7	0.7	0.7	0.6	0.6	0.6	0.6	0.6	0.6	0.6	0.6	78
76	2.0	1.5	1.4	1.2	1.1	1.0	0.9	0.9	0.8	0.8	0.9	0.7	0.7	0.7	0.7	0.7	0.6	0.6	0.6	0.6	0.6	76
74	2.0	1.7	1.4	1.3	1.1	1.0	1.0	0.9	0.9	0.8	0.8	0.8	0.8	0.7	0.7	0.7	0.7	0.7	0.7	0.7	0.7	74
72	2.1	1.8	1.5	1.3	1.2	1.1	1.0	1.0	0.9	0.9	0.9	0.9	0.8	0.8	0.8	0.8	0.8	0.8	0.7	0.7	0.7	72
70	2.2	1.8	1.6	1.4	1.3	1.2	1.1	1.1	1.0	1.0	1.0	0.9	0.9	0.9	0.9	0.8	0.8	0.8	0.8	0.8	0.8	70
68	2.2	1.9	1.7	1.5	1.4	1.3	1.2	1.1	1.1	1.1	1.0	1.0	1.0	1.0	0.9	0.9	0.9	0.9	0.9	0.9	0.9	68
66	2.3	2.0	1.7	1.6	1.4	1.4	1.3	1.2	1.2	1.1	1.1	1.1	1.1	1.0	1.0	1.0	1.0	1.0	1.0	1.0	0.9	66
64	2.4	2.0	1.8	1.7	1.5	1.4	1.4	1.3	1.3	1.2	1.2	1.2	1.2	1.1	1.1	1.1	1.1	1.1	1.0	1.0	1.0	64
62	2.4	2.1	1.9	1.7	1.6	1.5	1.5	1.4	1.4	1.3	1.3	1.3	1.2	1.2	1.2	1.2	1.1	1.1	1.1	1.1	1.1	62
60	2.5	2.2	2.0	1.8	1.7	1.6	1.5	1.5	1.4	1.4	1.4	1.3	1.3	1.3	1.3	1.2	1.2	1.2	1.2	1.2	1.2	60
58	2.5	2.3	2.1	1.9	1.8	1.7	1.6	1.6	1.5	1.5	1.5	1.4	1.4	1.4	1.4	1.3	1.3	1.3	1.3	1.3	1.3	58
56	2.6	2.3	2.1	2.0	1.9	1.8	1.7	1.7	1.6	1.6	1.6	1.5	1.5	1.5	1.4	1.4	1.4	1.4	1.4	1.4	1.4	56
54	2.6	2.4	2.2	2.1	2.0	1.9	1.8	1.7	1.7	1.7	1.5	1.6	1.6	1.6	1.5	1.5	1.5	1.5	1.5	1.5	1.5	54
52	2.7	2.5	2.3	2.1	2.0	2.0	1.9	1.8	1.8	1.8	1.7	1.7	1.7	1.6	1.6	1.6	1.6	1.6	1.6	1.5	1.5	52
50	2.7	2.5	2.3	2.2	2.1	2.0	2.0	1.9	1.9	1.8	1.8	1.8	1.8	1.7	1.7	1.7	1.7	1.6	1.6	1.6	1.6	50
48	2.8	2.5	2.4	2.3	2.2	2.1	2.0	2.0	1.9	1.9	1.9	1.9	1.8	1.8	1.8	1.8	1.7	1.7	1.7	1.7	1.7	48
46	2.8	2.5	2.5	2.3	2.2	2.2	2.1	2.1	2.0	2.0	2.0	1.9	1.9	1.9	1.8	1.8	1.8	1.8	1.8	1.8	1.8	46
44	2.8	2.6	2.5	2.4	2.3	2.2	2.2	2.1	2.1	2.0	2.0	2.0	2.0	1.9	1.9	1.9	1.9	1.9	1.9	1.8	1.8	44
42	2.8	2.5	2.5	2.4	2.3	2.3	2.2	2.2	2.1	2.1	2.1	2.0	2.0	2.0	2.0	2.0	1.9	1.9	1.9	1.9	1.9	42
40	2.8	2.5	2.5	2.4	2.4	2.3	2.2	2.2	2.2	2.1	2.1	2.1	2.1	2.0	2.0	2.0	2.0	2.0	2.0	1.9	1.9	40
38	2.7	2.6	2.5	2.4	2.4	2.3	2.3	2.2	2.2	2.2	2.1	2.1	2.1	2.1	2.0	2.0	2.0	2.0	2.0	2.0	2.0	38
36	2.7	2.5	2.5	2.4	2.3	2.3	2.3	2.2	2.2	2.2	2.1	2.1	2.1	2.1	2.1	2.0	2.0	2.0	2.0	2.0	2.0	36
34	2.6	2.5	2.4	2.4	2.3	2.3	2.2	2.2	2.2	2.1	2.1	2.1	2.1	2.1	2.0	2.0	2.0	2.0	2.0	2.0	2.0	34
32	2.5	2.4	2.3	2.3	2.2	2.2	2.2	2.1	2.1	2.1	2.1	2.1	2.0	2.0	2.0	2.0	2.0	2.0	2.0	2.0	1.9	32
30	2.3	2.3	2.2	2.2	2.2	2.1	2.1	2.1	2.0	2.0	2.0	2.0	2.0	2.0	1.9	1.9	1.9	1.9	1.9	1.9	1.9	30
28	2.2	2.1	2.1	2.1	2.0	2.0	2.0	1.9	1.9	1.9	1.9	1.9	1.9	1.8	1.8	1.8	1.8	1.8	1.8	1.8	1.8	28
26	2.0	1.9	1.9	1.9	1.8	1.8	1.8	1.8	1.8	1.8	1.7	1.7	1.7	1.7	1.7	1.7	1.7	1.7	1.7	1.6	1.6	26
24	1.7	1.7	1.7	1.6	1.6	1.6	1.6	1.6	1.6	1.5	1.5	1.5	1.5	1.5	1.5	1.5	1.5	1.5	1.5	1.5	1.5	24
22	1.4	1.4	1.4	1.3	1.3	1.3	1.3	1.3	1.3	1.3	1.3	1.3	1.3	1.2	1.2	1.2	1.2	1.2	1.2	1.2	1.2	22
20	1.0	1.0	1.0	1.0	1.0	1.0	1.0	1.0	0.9	0.9	0.9	0.9	0.9	0.9	0.9	0.9	0.9	0.9	0.9	0.9	0.9	20
18	0.6	0.5	0.5	0.5	0.5	0.5	0.5	0.5	0.5	0.5	0.5	0.5	0.5	0.5	0.5	0.5	0.5	0.5	0.5	0.5	0.5	18
16	-0.0	-0.1	-0.1	-0.0	-0.0	-0.0	-0.0	-0.0	-0.0	-0.0	-0.0	-0.0	-0.0	-0.0	-0.0	-0.0	-0.0	-0.1	-0.1	-0.1	-0.1	16
14	-0.7	-0.8	-0.8	-0.7	-0.7	-0.7	-0.7	-0.7	-0.7	-0.7	-0.7	-0.7	-0.7	-0.7	-0.7	-0.7	-0.7	-0.7	-0.7	-0.7	-0.7	14
12	-1.6	-1.5	-1.6	-1.5	-1.6	-1.5	-1.6	-1.6	-1.6	-1.6	-1.5	-1.5	-1.5	-1.5	-1.5	-1.6	-1.6	-1.6	-1.6	-1.6	-1.6	12
10	-2.6	-2.8	-2.8	-2.7	-2.7	-2.7	-2.7	-2.7	-2.7	-2.7	-2.6	-2.6	-2.6	-2.6	-2.6	-2.6	-2.6	-2.6	-2.6	-2.6	-2.6	10
8	-4.1	-4.2	-4.2	-4.2	-4.2	-4.2	-4.1	-4.1	-4.1	-4.1	-4.1	-4.1	-4.1	-4.1	-4.1	-4.1	-4.1	-4.1	-4.1	-4.1	-4.1	8
6	-6.0	-6.2	-6.3	-6.2	-6.2	-6.2	-6.2	-6.1	-6.1	-6.1	-6.1	-6.1	-6.1	-6.1	-6.1	-6.1	-6.1	-6.1	-6.1	-6.1	-6.1	6
4	-9.0	-9.3	-9.3	-9.3	-9.2	-9.2	-9.2	-9.2	-9.1	-9.1	-9.1	-9.1	-9.1	-9.1	-9.1	-9.1	-9.1	-9.1	-9.1	-9.1	-9.1	4
2	-10.0	-10.0	-10.0	-10.0	-10.0	-10.0	-10.0	-10.0	-10.0	-10.0	-10.0	-10.0	-10.0	-10.0	-10.0	-10.0	-10.0	-10.0	-10.0	-10.0	-10.0	2
0	-10.0	-10.0	-10.0	-10.0	-10.0	-10.0	-10.0	-10.0	-10.0	-10.0	-10.0	-10.0	-10.0	-10.8	-10.0	-10.0	-10.0	-10.0	-10.0	-10.0	-10.0	0
	2	3	4	5	6	7	8	9	10	11	12	13	14	16	18	20	22	24	26	28	30	

FREQUENCY IN MEGAHERTZ

Fig. 3-58B. Example of arbitrary tilted dipole antenna gains.

ANTENNA PATTERN

FREQUENCY RANGE	ANTENNA TYPE	HEIGHT	LENGTH	ANGLE	AZIMUTH	EX(1)	EX(2)	EX(3)	EX(4)	CONDUCT.	DIELECT.
2.0 TO 30.0	HALF RHOMBIC	0.000	80.000	28.000	0.000	0.000	0.000	0.000	0.000	0.001	4.000

ELEVATION ANGLE IN DEGREES

	2	3	4	5	6	7	8	9	10	11	12	13	14	16	18	20	22	24	26	28	30	
90	-7.3	-1.1	-1.2	-6.0	-8.7	-10.0	-7.8	-2.8	-4.2	-1.1	-3.4	-10.0	-10.0	-2.7	-3.0	-7.0	-9.7	-1.1	-4.2	-10.0	-6.5	90
88	-6.7	-1.3	-0.3	-5.2	-5.1	-10.0	-10.0	-3.5	-5.1	-0.9	-1.4	-5.4	-9.1	-7.4	-3.6	-2.2	-10.0	-4.3	-2.9	-3.4	-10.0	88
86	-6.1	-1.6	0.5	-4.4	-4.2	-7.8	-10.0	-5.2	-5.9	-1.7	-0.1	-3.0	-4.9	-10.0	-5.5	-0.2	-5.2	-10.0	-5.3	-0.2	-5.4	86
84	-5.5	-2.0	1.1	-3.5	-2.8	-5.3	-10.0	-8.6	-6.5	-3.6	0.3	-1.6	-2.5	-8.5	-7.9	-0.3	-2.2	-5.2	-9.6	-1.4	-2.0	84
82	-5.0	-2.5	1.6	-2.5	-1.9	-3.1	-8.6	-10.0	-7.6	-7.1	-0.2	-0.6	-1.7	-3.0	-10.0	-3.7	-2.2	-0.4	-8.5	-10.0	-3.3	82
80	-4.5	-3.1	2.0	-1.5	-1.3	-1.3	-5.2	-10.0	-9.6	-10.0	-2.0	-0.2	-1.9	-0.1	-7.6	-10.0	-3.4	1.0	-2.5	-6.0	-7.4	80
78	-4.0	-3.8	2.2	-0.5	-1.1	0.2	-2.8	-9.0	-10.0	-10.0	-6.1	-0.7	-2.2	1.4	-3.0	-7.6	-6.1	-1.0	-0.7	0.1	-9.3	78
76	-3.6	-4.7	2.3	0.5	-1.0	1.2	-0.8	-4.9	-8.7	-9.7	-10.0	-2.9	-2.5	1.6	-0.8	-1.4	-10.0	-10.0	-0.8	1.4	-1.1	76
74	-3.2	-5.6	2.3	1.4	-1.1	1.9	0.9	-2.4	-5.0	-6.0	-10.0	-8.2	-3.5	0.1	0.2	1.4	-4.8	-8.8	-3.8	-1.9	1.3	74
72	-2.9	-6.5	2.2	2.2	-1.0	2.3	2.3	-0.5	-2.3	-3.0	-5.8	-10.0	-6.7	-4.0	0.9	2.3	-0.0	-1.5	-10.0	-10.0	0.5	72
70	-2.6	-7.3	1.9	2.9	-0.6	2.2	3.4	1.1	-0.6	-0.6	-1.9	-7.6	-10.0	-10.0	0.5	1.0	2.3	1.4	-2.0	-6.0	-9.3	70
68	-2.3	-7.9	1.4	3.3	0.1	1.9	4.2	2.7	0.6	1.1	0.7	-2.0	-8.2	-10.0	-2.2	-2.4	3.2	1.9	2.7	-0.9	-4.0	68
66	-2.1	-8.0	0.8	3.6	1.0	1.3	4.5	4.0	1.7	2.2	2.7	1.1	-2.3	-4.3	-10.0	-5.7	1.8	0.9	4.3	1.8	2.6	66
64	-2.0	-7.6	-0.0	3.7	1.9	0.6	4.5	5.0	2.9	2.6	4.0	3.4	1.1	-0.7	-8.3	-7.1	-4.2	0.2	3.0	3.2	4.2	64
62	-1.8	-6.9	-1.0	3.7	2.8	0.3	4.0	5.7	4.2	2.9	4.6	5.0	3.5	1.7	-0.3	-6.8	-10.0	-2.2	-2.7	3.5	2.6	62
60	-1.7	-6.1	-2.3	3.4	3.4	0.4	3.2	5.9	5.4	3.4	4.6	6.0	5.4	3.3	3.5	-1.6	-1.7	-10.0	-8.7	-0.4	-0.0	60
58	-1.7	-5.2	-3.7	2.9	3.9	1.1	2.1	5.7	6.2	4.4	4.2	6.3	6.8	4.3	5.7	2.6	2.5	-2.1	-4.4	-10.0	-2.6	58
56	-1.6	-4.4	-5.4	2.3	4.1	2.0	0.9	5.0	6.6	5.5	4.0	5.9	7.4	5.4	6.8	5.5	4.7	4.1	0.1	0.4	-10.0	56
54	-1.6	-3.6	-7.1	1.4	4.1	2.9	0.4	3.8	6.5	6.4	4.5	5.0	7.3	6.9	6.9	7.7	5.9	7.1	4.3	4.7	2.2	54
52	-1.7	-3.0	-8.5	0.2	3.8	3.5	0.6	2.2	5.9	6.9	5.6	4.3	6.4	8.1	6.5	9.0	7.1	8.3	7.6	6.5	7.2	52
50	-1.7	-2.4	-6.9	-1.3	3.3	3.9	1.5	0.6	4.7	6.8	6.5	4.6	5.0	8.7	6.7	8.9	8.8	8.2	9.7	7.8	9.1	50
48	-1.8	-2.0	-8.2	-3.3	2.5	4.0	2.4	-0.3	2.8	6.1	6.9	5.6	4.1	8.2	7.8	7.8	10.0	8.1	10.1	9.9	9.1	48
45	-1.9	-1.6	-6.8	-5.9	1.3	3.7	3.1	0.1	0.6	4.7	6.7	6.4	4.6	6.7	8.7	6.6	9.7	9.3	9.0	11.1	9.6	46
44	-2.0	-1.2	-5.4	-9.9	-0.2	3.1	3.4	1.1	-1.2	2.5	5.7	6.6	5.6	4.4	8.6	7.0	8.0	10.2	8.3	10.3	11.1	44
42	-2.1	-1.0	-4.1	-10.0	-2.4	2.2	3.4	2.1	-1.3	-0.5	3.8	6.0	6.1	2.8	7.2	8.0	5.9	9.4	9.5	8.1	11.0	42
40	-2.3	-0.8	-2.9	-10.0	-5.4	0.7	2.9	2.7	-0.0	-3.0	0.8	4.4	5.8	3.4	4.3	8.0	6.0	6.7	9.6	8.1	8.4	40
38	-2.5	-0.6	-2.0	-9.1	-10.0	-1.4	2.0	2.7	1.2	-2.7	-3.4	1.5	4.4	4.5	0.8	6.2	7.0	4.0	7.5	9.0	6.5	38
36	-2.6	-0.5	-1.2	-5.4	-10.0	-4.5	0.4	2.2	1.9	-0.6	-5.6	-3.4	1.6	4.6	0.7	2.1	6.4	4.9	2.9	7.4	7.6	36
34	-2.8	-0.4	-0.5	-3.3	-10.0	-10.0	-2.0	1.1	1.9	0.8	-2.7	-9.0	-4.0	3.3	2.5	-3.7	3.2	5.4	1.8	2.1	6.6	34
32	-3.1	-0.4	-0.0	-1.7	-5.6	-10.0	-6.2	-1.0	1.1	1.4	-0.1	-4.3	-10.0	-0.3	2.7	-1.2	-5.3	2.9	3.6	-2.1	0.3	32
30	-3.3	-0.4	0.4	-0.4	-3.2	-10.0	-10.0	-4.7	-0.7	0.9	1.0	-0.4	-4.1	-10.0	0.4	1.3	-5.0	-6.8	1.4	1.7	-5.5	30
28	-3.5	-0.4	0.8	0.5	-0.9	-4.4	-10.0	-10.0	-4.4	-0.7	0.9	1.2	0.3	-7.5	-7.6	0.3	1.0	-4.3	-10.0	-0.7	1.3	28
26	-3.8	-0.5	1.0	1.5	0.8	-1.0	-4.7	-10.0	-10.0	-4.5	-0.7	1.3	2.1	0.6	-7.7	-6.5	1.2	2.7	-0.2	-10.0	-1.9	26
24	-4.1	-0.7	1.2	2.1	2.1	1.4	-0.4	-3.8	-10.0	-10.0	-4.6	-0.3	2.0	3.7	2.0	-4.9	-3.2	3.4	5.2	3.8	-1.3	24
22	-4.4	-0.8	1.3	2.5	3.1	3.1	2.3	0.8	-2.0	-6.9	-10.0	-4.1	0.3	4.6	5.6	4.0	-0.7	0.7	5.7	7.7	7.3	22
20	-4.8	-1.1	1.3	2.9	3.8	4.3	4.3	3.7	2.4	0.3	-2.8	-5.5	-2.9	3.8	6.9	7.7	6.3	3.3	3.8	7.6	9.8	20
18	-5.3	-1.4	1.2	3.0	4.3	5.2	5.6	5.7	5.3	4.4	2.9	0.8	-1.1	1.7	6.5	9.0	9.7	8.8	6.8	6.3	8.9	18
16	-5.8	-1.8	0.9	3.0	4.6	5.7	6.6	7.0	7.2	7.0	6.4	5.5	4.1	1.9	4.7	8.4	10.7	11.5	11.2	9.9	8.8	16
14	-6.4	-2.3	0.6	2.9	4.6	6.0	7.1	7.9	8.4	8.7	8.7	8.5	7.9	6.0	4.7	6.6	9.7	11.9	13.0	13.2	12.5	14
12	-7.2	-3.0	-0.0	2.4	4.3	5.9	7.2	8.3	9.1	9.8	10.2	10.3	10.3	9.6	8.2	7.0	7.7	10.1	12.2	13.7	14.4	12
10	-8.1	-4.0	-0.9	1.6	3.7	5.5	7.0	8.3	9.3	10.2	10.8	11.3	11.7	11.9	11.4	10.4	9.1	8.5	9.5	11.3	13.1	10
8	-9.4	-5.3	-2.1	0.5	2.7	4.6	6.2	7.7	8.9	9.9	10.8	11.5	12.1	12.9	13.2	13.1	12.5	11.4	10.0	8.8	8.7	8
6	-10.0	-7.1	-3.8	-1.2	1.1	3.1	4.8	6.3	7.7	8.8	9.9	10.8	11.5	12.8	13.6	14.0	14.2	13.9	13.3	12.2	10.7	6
4	-10.0	-9.9	-6.6	-3.9	-1.6	0.4	2.2	3.8	5.2	6.5	7.6	8.6	9.5	11.0	12.2	13.0	13.6	13.9	14.0	13.9	13.5	4
2	-10.0	-10.0	-10.0	-9.2	-6.9	-4.8	-2.9	-1.3	0.2	1.5	2.7	3.7	4.7	6.3	7.7	8.7	9.6	10.2	10.6	10.9	11.1	2
0	-10.0	-10.0	-10.0	-10.0	-10.0	-10.0	-10.0	-10.0	-10.0	-10.0	-10.0	-10.0	-10.0	-10.0	-10.0	-10.0	-10.0	-10.0	-10.0	-10.0	-10.0	0
	2	3	4	5	6	7	8	9	10	11	12	13	14	16	18	20	22	24	26	28	30	

FREQUENCY IN MEGAHERTZ

ELEVATION ANGLE IN DEGREES

Fig. 3-59B. Example of side-loaded vertical, half-rhombic antenna gains.

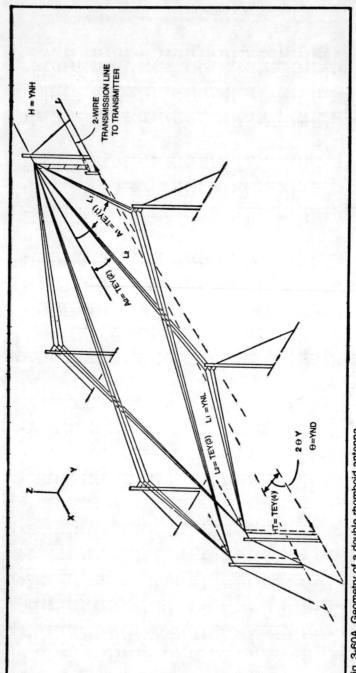

Fig. 3-60A. Geometry of a double rhomboid antenna.

ANTENNA PATTERN

FREQUENCY RANGE	ANTENNA TYPE	HEIGHT	LENGTH	ANGLE	AZIMUTH	EX(1)	EX(2)	EX(3)	EX(4)	CONDUCT.	DIELECT.
2.0 TO 30.0	DOUBLE RHOMBOID	62.000	150.000	0.000	0.000	25.500	17.500	214.750	62.000	0.001	4.000

ELEVATION ANGLE IN DEGREES / FREQUENCY IN MEGAHERTZ

	2	3	4	5	6	7	8	9	10	11	12	13	14	16	18	20	22	24	26	28	30	
90	-10.0	-4.8	-10.0	-2.6	-10.0	-10.0	-10.0	0.2	-10.0	-6.2	-10.0	0.8	-10.0	-10.0	-10.0	-10.0	-10.0	-10.0	-10.0	-10.0	-10.0	90
88	-10.0	-10.0	-10.0	-2.4	-10.0	-10.0	-6.1	-3.7	-10.0	0.9	-10.0	-10.0	-6.3	-10.0	0.5	-7.1	-6.5	-4.2	-10.0	1.9	-1.2	88
86	-10.0	-10.0	-7.5	-4.1	-10.0	-6.8	-1.4	-10.0	-3.7	-4.3	-10.0	-9.4	-1.0	0.9	-10.0	0.1	-5.5	-10.0	-4.4	-7.5	-10.0	86
84	-10.0	-10.0	-2.7	-9.2	-5.3	-8.4	-4.2	-10.0	-0.7	-10.0	-5.7	-6.9	-10.0	-1.3	-3.7	-10.0	-10.0	-9.8	-5.5	-10.0	1.4	84
82	-10.0	-6.1	0.1	-10.0	1.2	-10.0	-10.0	-10.0	-5.5	-7.2	-5.4	-10.0	-8.6	-10.0	-10.0	-2.1	-4.2	-10.0	-4.5	0.3	-6.7	82
80	-10.0	-3.7	1.4	-10.0	4.3	-10.0	-1.6	-2.3	-10.0	3.2	-10.0	-8.4	-9.2	0.9	-0.1	-10.0	-10.0	-8.1	-10.0	-3.9	-10.0	80
78	-10.0	-2.8	1.5	-10.0	4.7	-10.0	-0.3	-3.5	-10.0	3.1	-10.0	0.9	-10.0	-0.5	0.6	-8.4	-10.0	-4.4	-9.2	-9.8	-8.8	78
76	-10.0	-3.1	-0.8	-10.0	2.5	-9.6	-5.5	-10.0	-10.0	-10.0	-1.6	-10.0	2.0	-3.3	-10.0	-10.0	-8.2	-0.2	-2.5	-10.0	-3.7	76
74	-10.0	-4.7	-4.1	-10.0	-3.3	-1.1	-10.0	1.3	-7.1	-7.6	0.9	-10.0	1.8	-10.0	-1.7	-1.3	-0.3	-10.0	-1.6	-3.8	-9.4	74
72	-10.0	-8.0	-10.0	-7.0	-10.0	2.5	-10.0	2.4	-10.0	-10.0	-10.0	-0.5	-10.0	-10.0	-10.0	-10.0	-10.0	-3.9	-4.9	-10.0	-9.9	72
70	-10.0	-10.0	-7.6	-4.9	-10.0	2.6	-9.2	-4.1	-5.8	-1.5	-6.4	-2.1	-3.6	1.5	-5.9	-9.9	4.7	3.1	-10.0	-0.8	4.3	70
68	-10.0	-10.0	-0.9	-4.3	-10.0	-0.5	-1.4	-10.0	-0.5	-10.0	-10.0	-10.0	5.2	-3.2	-10.0	0.2	-3.8	-10.0	-10.0	-7.7	-10.0	68
66	-5.4	-10.0	1.8	-5.7	-4.9	-10.0	0.2	-10.0	-3.2	-2.0	-0.5	-10.0	-1.9	-10.0	-8.2	1.9	-10.0	0.7	-1.1	-3.6	-0.6	66
64	-1.8	-10.0	2.5	-10.0	0.8	-10.0	-1.7	1.8	-10.0	0.6	-10.0	-10.0	-4.8	-7.1	5.6	-10.0	-2.8	-10.0	-7.6	5.8	-9.7	64
62	1.0	-9.7	1.6	-10.0	2.4	-10.0	-7.4	6.4	-10.0	-5.8	4.7	-5.1	-10.0	1.1	-6.1	-8.8	-1.7	-7.1	5.7	-10.0	-7.9	62
60	3.3	-10.0	-1.3	-5.0	1.0	-4.4	-10.0	5.7	2.1	-10.0	5.4	-1.2	-8.3	-10.0	-9.4	4.7	-10.0	-10.0	-3.1	-0.1	-0.8	60
58	5.3	-10.0	-8.0	0.1	-4.3	3.4	-10.0	-0.7	8.3	-10.0	-4.4	7.6	-10.0	1.3	5.4	-10.0	-0.4	6.0	-4.4	-9.2	2.1	58
56	7.0	-10.0	-10.0	2.4	-10.0	5.8	-10.0	-10.0	8.3	2.1	-10.0	4.5	3.8	-10.0	-10.0	-10.0	8.2	-10.0	0.9	-6.4	-10.0	56
54	8.4	-10.0	-7.1	2.6	-7.8	4.3	-0.3	-10.0	1.5	7.9	-6.8	-10.0	5.8	-10.0	-2.3	4.5	-9.2	1.7	-6.7	4.0	-10.0	54
52	9.6	-10.0	-4.3	0.5	-3.6	-3.8	5.2	-10.0	-10.0	7.0	1.7	-7.6	-6.4	5.2	-10.0	-1.3	-2.8	-5.6	4.9	-0.6	-2.2	52
50	10.5	-6.8	-5.5	-5.7	-1.2	-5.6	5.6	-4.5	-6.2	-6.8	5.4	3.1	-10.0	3.4	-10.0	5.6	-8.0	-8.3	4.3	-6.9	7.6	50
48	11.3	-3.4	-10.0	-10.0	0.1	2.2	-1.2	2.8	-10.0	-0.9	-0.1	2.2	6.1	-8.8	10.1	-10.0	5.8	4.1	-6.1	8.2	-10.0	48
46	12.0	0.2	-10.0	-5.2	-0.8	2.7	-1.5	3.6	-4.8	-5.9	-1.6	-1.6	7.2	-10.0	9.8	-4.7	0.9	2.7	0.7	4.0	1.3	46
44	12.5	3.6	-2.8	-3.5	-5.3	3.1	7.1	-10.0	-0.2	-4.3	1.4	-4.5	-2.5	7.1	-10.0	10.1	-10.0	1.3	6.0	-8.4	6.7	44
42	12.8	6.6	1.1	-9.1	-10.0	-3.0	8.4	5.2	-2.4	-1.5	-10.0	3.7	-7.1	11.5	-10.0	5.8	5.8	-10.0	7.6	3.8	-5.4	42
40	13.0	9.1	2.5	-8.8	-5.3	-5.9	5.7	11.0	-0.3	-5.0	2.4	-10.0	3.7	0.9	6.5	-5.5	4.9	11.1	-10.0	12.3	-0.0	40
38	13.1	11.2	2.5	2.3	-6.1	-10.0	-0.6	11.6	10.9	-7.5	-3.4	3.8	-8.2	-0.9	9.7	-1.4	-2.7	6.2	10.4	-10.0	11.7	38
36	13.0	12.9	2.1	6.7	-10.0	-8.7	-8.6	8.2	14.0	9.6	-10.0	-0.8	4.1	1.1	-0.3	5.8	-5.9	-5.8	11.7	7.1	-10.0	36
34	12.8	14.2	4.3	8.7	3.2	-9.1	-10.0	0.3	12.8	14.8	8.1	-10.0	0.4	-10.0	7.5	-10.0	3.5	-7.0	-6.2	11.8	3.2	34
32	12.4	15.1	8.3	8.7	9.2	-5.9	-10.0	-10.0	7.3	15.2	15.2	6.9	-3.8	3.8	-2.2	13.1	-10.0	7.7	-0.6	3.2	7.5	32
30	12.0	15.8	11.8	6.9	12.1	6.7	-10.0	-10.0	-4.4	11.1	16.5	15.4	6.6	-10.0	-7.8	7.0	14.7	-10.0	7.9	-0.7	9.3	30
28	11.4	16.2	14.4	4.9	12.7	12.2	3.1	-10.0	-10.0	1.2	13.2	17.4	16.0	6.1	-5.8	-10.0	13.5	15.4	-6.8	6.2	-3.3	28
26	10.7	16.4	16.3	8.9	11.2	14.8	11.4	-0.1	-10.0	-10.0	3.9	14.2	18.1	9.8	9.4	-7.2	-8.8	16.3	16.4	2.1	0.3	26
24	9.8	16.3	17.7	13.6	7.1	15.6	15.5	10.6	-1.8	-10.0	-10.0	4.5	14.3	18.4	7.3	10.3	-10.0	-4.7	17.4	18.2	5.3	24
22	8.9	16.0	18.5	16.9	8.8	12.5	16.8	15.9	18.4	-1.8	-10.0	-10.0	3.1	18.5	17.3	11.4	11.2	-10.0	-6.1	16.9	20.4	22
20	7.8	15.5	18.9	19.1	15.1	6.8	15.3	17.8	16.2	10.9	-0.3	-10.0	-10.0	11.1	20.4	16.7	12.5	12.4	-10.0	-10.0	13.7	20
18	6.6	14.7	19.8	20.4	13.8	13.3	8.9	16.4	18.3	16.7	11.9	2.2	-10.0	-10.0	14.7	21.6	18.1	10.2	13.1	-10.0	-10.0	18
16	5.2	13.8	18.7	21.1	21.4	19.1	12.8	9.2	16.4	18.3	17.1	13.1	4.9	-10.0	-5.9	16.8	22.6	21.8	7.0	11.4	-8.8	16
14	3.7	12.6	18.1	21.2	22.7	22.4	20.1	14.6	6.8	14.5	17.3	17.0	13.9	-10.0	-6.8	-8.7	14.6	22.6	23.6	17.7	-0.2	14
12	2.8	11.2	17.1	20.9	23.2	24.1	23.8	22.0	18.1	18.2	8.4	14.1	15.2	7.7	2.9	-0.4	-0.3	6.7	19.8	23.8	22.4	12
10	0.8	9.6	15.8	20.1	22.9	24.7	25.5	25.4	24.3	21.9	17.9	10.8	2.6	9.3	-10.0	9.4	5.2	6.2	11.8	3.5	17.6	10
8	-2.2	7.5	14.8	18.7	22.8	24.4	26.0	26.8	26.9	26.4	25.1	22.9	19.7	7.7	-8.5	2.5	10.9	7.6	7.8	10.8	20.8	8
6	-5.0	4.9	11.7	16.6	20.3	23.1	25.1	26.6	27.4	27.8	27.6	26.9	25.6	20.9	10.9	-4.5	-1.1	-1.5	1.8	4.3	16.8	6
4	-8.7	1.3	8.2	13.3	17.3	20.3	22.7	24.6	25.9	26.8	27.2	27.2	26.7	24.2	18.2	-10.0	14.1	16.3	13.3	1.4	4.8	4
2	-10.0	-4.7	2.3	7.5	11.6	14.8	17.4	19.5	21.0	22.2	22.9	23.3	23.2	21.7	17.3	1.7	12.8	17.9	18.0	13.7	-4.0	2
0	-10.0	-10.0	-10.0	-10.0	-10.0	-10.0	-10.0	-10.0	-10.0	-10.0	-10.0	-10.0	-10.0	-10.0	-10.0	-10.0	-10.0	-10.0	-10.0	-10.0	-10.0	0
	2	3	4	5	6	7	8	9	10	11	12	13	14	16	18	20	22	24	26	28	30	

FREQUENCY IN MEGAHERTZ

Fig. 3-60B. Example of sloping double rhomboid antenna gains.

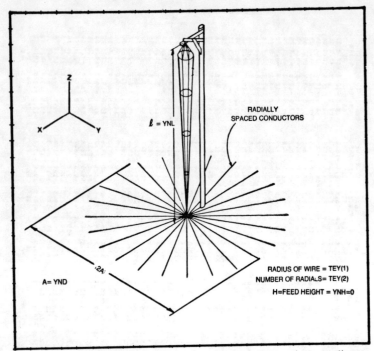

Fig. 3-61A. Geometry of a vertical antenna with a radial-conductor earth system.

this type of approximation. Should this not be adequate the computer program will allow input of externally computed (or measured) patterns via "antenna type-18".

Transmitting Antennas (BOX-6):

Transmitting antennas. Tabulate the type of transmitting or receiving antennas to be used. The beginning and ending frequency for each antenna must be included in all cases, and these frequencies must not overlap. Up to three antennas may be used for the transmitter and/or receiver.

Receiving Antennas (BOX-7):

Receiving antennas. Tabulate similar information, as in box-6 above, for the receiving antennas.

ANTENNA PATTERN

FREQUENCY RANGE	ANTENNA TYPE	HEIGHT	LENGTH	ANGLE	AZIMUTH	EX(1)	EX(2)	EX(3)	EX(4)	CONDUCT.	DIELECT.
2.0 TO 30.0	VERTICAL W.G.S.	-0.000	-0.250	76.200	0.000	2.057	120.000	-0.000	-0.000	0.001	4.000

ELEVATION ANGLE IN DEGREES	2	3	4	5	6	7	8	9	10	11	12	13	14	16	18	20	22	24	26	28	30	ELEVATION ANGLE IN DEGREES
90	-10.0	-10.0	-10.0	-10.0	-10.0	-10.0	-10.0	-10.0	-10.0	-10.0	-10.0	-10.0	-10.0	-10.0	-10.0	-10.0	-10.0	-10.0	-10.0	-10.0	-10.0	90
88	-10.0	-10.0	-10.0	-10.0	-10.0	-10.0	-10.0	-10.0	-10.0	-10.0	-10.0	-10.0	-10.0	-10.0	-10.0	-10.0	-10.0	-10.0	-10.0	-10.0	-10.0	88
86	-10.0	-10.0	-10.0	-10.0	-10.0	-10.0	-10.0	-10.0	-10.0	-10.0	-10.0	-10.0	-10.0	-10.0	-10.0	-10.0	-10.0	-10.0	-10.0	-10.0	-10.0	86
84	-10.0	-10.0	-10.0	-10.0	-10.0	-10.0	-10.0	-10.0	-10.0	-10.0	-10.0	-10.0	-10.0	-10.0	-10.0	-10.0	-10.0	-10.0	-10.0	-10.0	-10.0	84
82	-10.0	-10.0	-10.0	-10.0	-10.0	-10.0	-10.0	-10.0	-10.0	-10.0	-10.0	-10.0	-10.0	-10.0	-10.0	-10.0	-10.0	-10.0	-10.0	-10.0	-10.0	82
80	-10.0	-10.0	-10.0	-10.0	-10.0	-10.0	-10.0	-10.0	-10.0	-10.0	-10.0	-10.0	-10.0	-9.6	-10.0	-10.0	-10.0	-10.0	-10.0	-10.0	-10.0	80
78	-9.7	-10.0	-10.0	-10.0	-10.0	-10.0	-10.0	-10.0	-10.0	-9.1	-9.0	-10.0	-10.0	-7.9	-10.0	-10.0	-8.8	-10.0	-10.0	-10.0	-10.0	76
76	-8.4	-10.0	-10.0	-10.0	-10.0	-10.0	-7.0	-10.0	-10.0	-10.0	-7.3	-10.0	-10.0	-8.4	-7.9	-10.0	-6.9	-8.8	-9.4	-7.1	-10.0	76
74	-7.3	-10.0	-10.0	-10.0	-7.9	-10.0	-6.5	-7.0	-8.5	-10.0	-6.5	-10.0	-6.8	-8.9	-6.2	-7.7	-7.7	-6.4	-8.7	-6.5	-7.0	74
72	-6.4	-10.0	-10.0	-10.0	-8.0	-5.4	-5.7	-10.0	-6.5	-10.0	-6.5	-4.7	-5.7	-7.4	-5.8	-5.9	-7.8	-5.3	-6.3	-7.8	-5.0	72
70	-5.6	-10.0	-4.0	-10.0	-7.7	-6.1	-4.9	-10.0	-5.1	-9.4	-6.7	-5.9	-4.6	-5.6	-6.3	-4.5	-5.9	-6.2	-4.4	-6.1	-6.1	70
68	-4.8	-10.0	-3.4	-8.3	-7.0	-6.7	-4.3	-10.0	-4.0	-6.3	-6.2	-7.0	-4.1	-4.2	-5.9	-4.1	-4.1	-5.8	-4.3	-3.9	-5.7	68
66	-4.2	-10.0	-2.8	-6.1	-6.0	-7.1	-3.9	-10.0	-3.2	-3.7	-4.9	-7.2	-4.1	-3.1	-4.3	-4.6	-3.0	-4.0	-5.0	-3.1	-3.5	66
64	-3.6	-10.0	-2.2	-4.3	-4.8	-7.0	-3.7	-1.9	-2.5	-1.9	-3.5	-5.8	-4.3	-2.5	-2.9	-4.3	-3.0	-2.4	-3.8	-3.9	-2.3	64
62	-3.1	-10.0	-1.6	-2.8	-3.7	-6.4	-3.6	-2.2	-1.9	-0.8	-2.3	-3.7	-3.8	-2.5	-1.8	-2.9	-3.5	-1.9	-2.0	-3.4	-2.9	62
60	-2.6	-10.0	-1.1	-1.6	-2.6	-5.4	-3.5	-2.7	-1.5	-10.0	-1.4	-1.7	-2.7	-2.7	-1.3	-1.5	-2.8	-2.4	-1.1	-1.6	-2.9	60
58	-2.2	0.4	-0.6	-0.5	-1.6	-4.1	-3.2	-3.1	-1.4	-0.1	-0.7	-0.2	-1.5	-2.6	-1.3	-0.6	-1.4	-2.4	-1.4	-0.5	-1.2	58
56	-1.8	0.6	-0.1	0.3	-0.7	-2.7	-2.6	-3.3	-1.4	-0.4	-0.2	0.6	-0.4	-1.7	-1.6	-0.3	-0.2	-1.1	-1.8	-0.7	0.0	56
54	-1.5	0.8	0.3	1.1	0.0	-1.4	-1.7	-3.0	-1.5	-0.9	0.1	1.0	0.4	-0.5	-1.4	-0.6	0.4	0.2	-0.8	-1.2	-0.2	54
52	-1.2	1.0	0.7	-10.0	0.7	-0.2	-0.8	-2.3	-1.4	-1.2	0.1	0.9	0.9	0.6	-0.5	-0.8	0.2	0.9	0.6	-0.4	-0.7	52
50	-0.9	1.1	1.0	-10.0	1.3	0.9	0.1	-1.2	-1.0	-1.4	-0.1	0.5	1.1	1.3	0.7	-0.3	-0.2	0.9	1.4	1.0	0.2	50
48	-0.7	1.2	1.3	-10.0	1.8	1.8	1.0	-0.1	-0.3	-1.1	-0.2	0.1	1.1	1.8	1.6	0.8	0.1	0.4	1.3	1.8	1.6	48
46	-0.4	1.3	1.6	-10.0	2.2	2.5	1.7	1.0	0.5	-0.4	-0.1	-0.2	0.8	1.9	2.3	1.9	1.1	0.5	0.8	1.7	2.2	46
44	-0.3	1.3	1.8	-10.0	2.6	-10.0	2.4	2.1	1.4	0.6	0.4	-0.1	0.6	1.7	2.5	2.6	2.2	1.5	0.9	1.1	1.9	44
42	-0.1	1.4	2.0	-10.0	-10.0	-10.0	3.0	2.9	2.3	1.7	1.2	0.5	0.8	1.3	2.3	2.9	3.0	2.6	1.9	1.3	1.3	42
40	0.0	1.4	2.1	3.3	-10.0	-10.0	-10.0	-10.0	3.0	2.7	2.1	1.5	1.3	1.2	1.9	2.8	3.3	3.4	3.1	2.5	1.8	40
38	0.1	1.4	2.2	3.4	-10.0	-10.0	-10.0	-10.0	-10.0	3.6	3.0	2.6	2.1	1.5	1.6	2.3	3.1	-10.0	-10.0	3.6	3.1	38
36	0.2	1.4	2.3	3.4	-10.0	-10.0	-10.0	-10.0	-10.0	-10.0	-10.0	3.6	3.1	2.3	1.8	2.0	2.6	3.3	-10.0	-10.0	-10.0	36
34	0.3	1.3	2.3	3.3	3.7	-10.0	-10.0	-10.0	-10.0	-10.0	-10.0	-10.0	3.9	3.2	2.6	2.2	2.2	2.7	3.4	-10.0	-10.0	34
32	0.3	1.2	2.2	3.2	3.7	-10.0	-10.0	-10.0	-10.0	-10.0	-10.0	-10.0	-10.0	-10.0	3.5	2.9	2.5	2.5	2.8	3.3	-10.0	32
30	0.3	1.1	2.2	3.1	3.6	-10.0	-10.0	-10.0	-10.0	-10.0	-10.0	-10.0	-10.0	-10.0	-10.0	3.9	3.4	2.9	2.7	2.8	3.2	30
28	0.2	1.0	2.0	2.9	3.5	-10.0	-10.0	-10.0	-10.0	-10.0	-10.0	-10.0	-10.0	-10.0	-10.0	-10.0	-10.0	3.9	3.4	3.1	2.9	28
26	0.1	0.8	1.8	2.7	3.4	4.0	-10.0	-10.0	-10.0	-10.0	-10.0	-10.0	-10.0	-10.0	-10.0	-10.0	-10.0	-10.0	-10.0	4.1	3.7	26
24	-0.0	0.5	1.6	2.4	3.1	3.8	-10.0	-10.0	-10.0	-10.0	-10.0	-10.0	-10.0	-10.0	-10.0	-10.0	-10.0	-10.0	-10.0	-10.0	-10.0	24
22	-0.2	0.3	1.3	2.0	2.8	3.4	-10.0	-10.0	-10.0	-10.0	-10.0	-10.0	-10.0	-10.0	-10.0	-10.0	-10.0	-10.0	-10.0	-10.0	-10.0	22
20	-0.5	-0.1	1.0	1.6	2.5	3.0	3.6	-10.0	-10.0	-10.0	-10.0	-10.0	-10.0	-10.0	-10.0	-10.0	-10.0	-10.0	-10.0	-10.0	-10.0	20
18	-0.8	-0.5	0.6	1.2	2.0	2.5	3.1	-10.0	-10.0	-10.0	-10.0	-10.0	-10.0	-10.0	-10.0	-10.0	-10.0	-10.0	-10.0	-10.0	-10.0	18
16	-1.2	-1.0	0.0	-0.6	1.4	1.9	2.6	3.0	-10.0	-10.0	-10.0	-10.0	-10.0	-10.0	-10.0	-10.0	-10.0	-10.0	-10.0	-10.0	-10.0	16
14	-1.8	-1.6	-0.6	-0.1	0.7	1.2	1.9	2.2	-10.0	-10.0	-10.0	-10.0	-10.0	-10.0	-10.0	-10.0	-10.0	-10.0	-10.0	-10.0	-10.0	14
12	-2.5	-2.4	-1.4	-0.9	-0.1	0.4	1.0	1.4	1.9	-10.0	-10.0	-10.0	-10.0	-10.0	-10.0	-10.0	-10.0	-10.0	-10.0	-10.0	-10.0	12
10	-3.4	-3.4	-2.5	-2.0	-1.2	-0.7	-0.1	0.3	-10.0	-10.0	-10.0	-10.0	-10.0	-10.0	-10.0	-10.0	-10.0	-10.0	-10.0	-10.0	-10.0	10
8	-4.6	-4.7	-3.8	-3.4	-2.6	-2.2	-1.5	-1.2	-10.0	-10.0	-10.0	-10.0	-10.0	-10.0	-10.0	-10.0	-10.0	-10.0	-10.0	-10.0	-10.0	8
6	-6.4	-6.5	-5.7	-5.3	-4.5	-4.1	-3.5	-3.1	-10.0	-10.0	-10.0	-10.0	-10.0	-10.0	-10.0	-10.0	-10.0	-10.0	-10.0	-10.0	-10.0	6
4	-9.2	-9.4	-8.6	-8.2	-7.4	-7.0	-6.4	-6.1	-10.0	-10.0	-10.0	-10.0	-10.0	-10.0	-10.0	-10.0	-10.0	-10.0	-10.0	-10.0	-10.0	4
2	-10.0	-10.0	-10.0	-10.0	-10.0	-10.0	-10.0	-10.0	-10.0	-10.0	-10.0	-10.0	-10.0	-10.0	-10.0	-10.0	-10.0	-10.0	-10.0	-10.0	-10.0	2
0	-10.0	-10.0	-10.0	-10.0	-10.0	-10.0	-10.0	-10.0	-10.0	-10.0	-10.0	-10.0	-10.0	-10.0	-10.0	-10.0	-10.0	-10.0	-10.0	-10.0	-10.0	0
	2	3	4	5	6	7	8	9	10	11	12	13	14	16	18	20	22	24	26	28	30	

FREQUENCY IN MEGAHERTZ

Fig. 3-61B. Example of vertical monopole with ground screeen antenna gains.

Chapter 4
Ionospheric Scatter Propagation

Within the approximate frequency range of 30.0 to 80.0 MHz, a minute portion of the electromagnetic radio-frequency energy beamed at the ionosphere's D-Layer (see Appendix 9 and Fig. 4-1) may be forward-scattered to distant receiver site via the common volume illuminated by the transmitting and receiving antennas. This ionoscatter (ionospheric forward scatter) transmission, most effective in great-circle distances of approximately 1000.0 to 2000.0 kilometers is thought to be due to dynamically changing electron density gradients (refraction index) in the ionosphere's D-Layer. This is conceptually not unlike the tropospheric turbulances (blobs) of the troposcatter (tropospheric scatter) propagation mode. The ionospheric forward scatter transmission mechanism also entails some "reflection" action from the ionospheric layer's electron gradients. Actually, this is refraction, albeit generally alluded to as ionospheric reflection.

Unlike the conventional HF ionospheric sky-wave propagation mechanism, the ionospheric scatter mode is essentially a brute-force single-frequency operation. Large transmission losses (losses between two isotropic antennas) make higher transmitter powers (e.g. 50.0 kilowatts), high-gain antennas (e.g. 20.0 dBi), and sensitive receivers (with up to date detection, error correction, and coding techniques) highly desirable. The transmission attenuation increases with frequency. A rough rule-of-thumb, over great-circle paths between approximately 1,000.0 and 2,000.0 kilometers, states that at about 30.0 MHz the scatter loss is of the order of some 80.0 dB above the so called "free-space" value, while at about 60.0 MHz the scatter loss may be in excess of 90.0 dB.

RAYLEIGH FADING

Although it does exhibit quiescent periods, the ionospheric-scatter signal may vary over many orders of magnitude. The rapid (within the minute) fading has been classified as Rayleigh distributed (see Appendix 13 entitled "Rayleigh Fading"). Fading rates might typically be from 0.1 to 5.0 Hz at a frequency of 45.0 MHz with a narrow-beam antenna (fading is exasperbated with broad-beam antennas) since these antennas accept more multipath components). Being relatively immune from polar absorp-

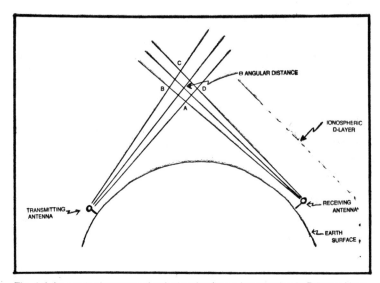

Fig. 4-1. Ionospheric scatter circuit showing beam intersection in D layer. Note the common volume and the scattering angle, 0.

tion effects resulting from solar flares (especially above about 40.0 MHz), an excellent use for the ionospheric-scatter mode of propagation is "thin-line" traffic through the auroral or trans-polar regions where conventional HF ionospheric "sky-wave" is susceptible to excessive absorption and interference. The "thin-line" traffic through the aurroral or trans-polar regions where conventional HF ionospheric "sky-wave" is susceptible to excessive absorption and interference. The "thin-line"traffic capacity (not unlike conventional skywave HF) is relatively low, several kilohertz of bandwidth. Ionospheric forward scatter has characteristically carried the equivalent of ten to fifteen 60-WPM(words per minute) multiplexed regular teleprinter channels at a character... error rate of 10^{-3}, or less, for approximately 99% of the time. Sophisticated terminal equipment and modulation/detection schemes should be able to better this error rate. At 50.0 MHz, a typical transmitter power of 5.0 kilowatts per 60-WPM teletypewriter channel might be required in temperature zones for 99% time circuit availability (reliability). Again, the use of sophisticated end-devices (being developed almost daily) may improve this figure. At about 30.0 MHz intelligible voice transmission requires approximately 10.0 kilowatts of transmitter power per voice channel and 90.0 kilowatts of transmitter power per voice power at about 50.0 MHz for circuit availability (reliability) of some 99%.

Atmospheric noise does not normally play an important role in the ionospheric scatter mode as it does in conventional HF transmission, especially in the polar regions of the earth. Diversity reception (see Appendix 12) is normally used in ionospheric forward-scatter transmission to combat short term (within-the minute) fading.

Many factors should be considered in the selection of operating frequencies, regulations, investments, etc. For example transmission loss increases with frequency, making the power requirement picture more attractive (lower transmitter power requirements) at the lower frequency portion of the band. At these lower frequencies, however, possible interference from conventional HF ionospheric types of propagation are more likely to become a threat, especially during the sunspot highs when (maximum-usuable frequencies) in the regular HF modes become high. In the arctic and arctic regions, polar disturbances may affect transmission at the lower frequencies (e.g. around 30.0 MHz). In temperate locations, absorption should not present such problems. Auroral "sputter", an abnormally rapid buzzing fade might occur around two days total (distributed) during a typical year. Meteor bursts into the ionosphere's might well account for 20 to 40 dB signals above the forward scatter mode signals. Solar flare which enhance D-region ionization for relatively brief time periods, provide for additional enhancement of signal levels, as does E_s (sporadic-E) which is comprised of dense patches of ionization, appearing sporadically at about the height of the E layer.

An important consideration in ionospheric forward-scatter propagation calculations is the fact that galactic (cosmic) noise is most generally the controlling noise, even though at relatively infrequent times other noise types (e.g. lightning or man-made) may temporarily rule. In temperate and polar regions, signal-to-galactic noise ratios will generally suffice for propagation computations. Receiver noise figures, unlike at HF should generally be kept down to a few decibels, even though galactic noise levels are most of the time higher than this by some 6.0 to 12.0 dB depending upon path location, antenna orientaton, and antenna beamwidth. When a signal, especially in the polar areas, is absorbed, so is the cosmic noise. This is because the cosmic (galactic) noise issuing from outer space must transit via the absorptive ionospheric D layer as must our desired signal. When this noise is sufficiently absorbed the receiver noise may become the controlling.

FORWARD-SCATTER PROBLEM

Let's use an example to illustrate the ionospheric forward-scatter propagation mode.

Example

The transmitter and receiver antenna gains are both 20.0 dBi (transmission-line attenuation and mismatch loss is ignored to concentrate on the propagation aspects of the problem). Intelligence to be transmitted on this ionospheric forward-scatter circuit is one voice channel, with a transmitted bandwidth of 3.0 kHz. Dual-diversity reception with a minimum combining ratio is used. The great-circle distance is 1,200.0 kilometers. Operating frequency is 50.0 MHz and grade of communications will be that required between experienced operators. The problem is to find the transmitter power necessary.

Solution

From Fig. 4-2, we see that at a frequency of 50.0 MHz, a great-circle path of 1,200.0 kilometers, transmitter power of 50.0 kilowatts, and antenna gains (at both ends of the path) of 20.0 dBi, we obtain a median S/N

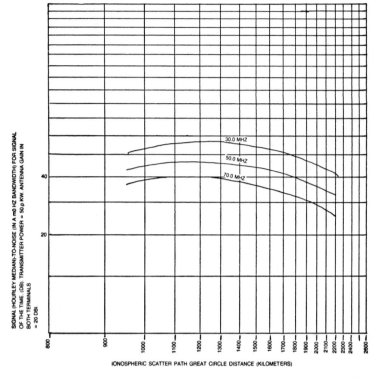

Fig. 4-2. Graph showing signal (hourly median) to noise (in a 1-Hz bandwidth) ratio for 99% of the time. Transmitter power = 50 KW. Transmitter and receiver antenna gain each 20 dBi. The signal is in the signal occuped bandwidth, while the noise is in a 1-Hz bandwidth.

(signal-to-noise) ratio of 46.0 dB. By referring to Fig. 3-20, we see that the required S/N ratio is 45.0 dB. Thus, we have an excess S/N availability of 1.0 dB. This means that we might reduce our transmitter power by this 1.0 dB amount, or from 50.0 kilowatts to 40.0 kilowatts (a considerable saving in energy expenditure). Getting back to the available /N ratios from Fig. 4-2, the terms may be confusing at first, but they *are* found in the propagation field and are included for their didactic value.

Chapter 5
Microwave and VHF/UHF Propagation

The term microwave appears to "conjure-up" different visions and concepts to different people. Some go along with *Webster's New World Dictionary of the American Langauge* definition: "Adj., designating or of that part of the electromagnetic spectrum lying between the far infra-red and some other lower frequency limit—commonly regarded as extending from 300,000 Mhz to 300.0 Mhz. Noun, any electromagnetic wave of microwave frequency". On the other hand, a popular electronics dictionary describes the term microwave as, "a term applied to radio waves in the range of frequencies between 1,000.0 MHz and upward—generally defines operation in the region where distributed constant circuits enclosed by conducting boundaries are used instead of the conventional lumped constant circuit components. And, at least one source fixed the microwave range as 150.0 MHz to 13.0 GHz. There appears to be no official telecommunications frequency deliniation for this so-called microwave band (see Appendix 17, entitled "Secret Spectrum"). To many persons in the telecommunications field the term microwave terrestrial WRH (Within Radio Horizon) radio circuits between parabolic antennas in a roughly defined radio frequency band lying between some 1,500.0 MHz and 20.0 GHz. We shall present this chapter based upon this latter precept. It should be kept in mind, nevertheless, that VHF/UHF radio circuits follow the same general rules in their propagational design aspects.

There are two general phases to microwave path selection. The first is the preliminary or desk phase. Obviously, if a given microwave path fails (by calculation or other judgements) to work at the desk, one would naturally be to spend valuable time and money in field work. The second phase is the field phase. To illustrate a typical computation, use Fig. 5-1 as a format. This worksheet model is entitled, "Preliminary Microwave Path Selection Considerations."

MICROWAVE PATH SELECTION (OFFICE PHASE)

The first task faced by the microwave propagation engineer is the procurement of any and all maps and charts of the area(s) in which mic-

rowave radio circuits might be installed. There are many types of maps and charts, and each its particular contribution to make. Some examples are:
1. Topographical maps published by the U.S. Geological Survey.
2. U.S. Army Corps of Engineers maps. These are particularly good.
3. Ordinary automobile maps.
4. USAF aeronautical maps.

The main point is to have on hand the most accurate topographical data available for the proper selection of your microwave path.

Now, let's go over Fig. 5-1 and consider our sample problem.

Item 1: After filling in the engineer's name, date and such obvious data, we fill in the path-considered information. In our case, for simplicity, this is a one hop microwave system and we desire a circuit from "Podunk" to "Kokomo". We list these names, and if pertinent, the coordinates. These are called the A end and C end, but can be called anything convenient. In this example, we have placed a passive repeater (reflector) at Mt. Gunnysack and have called this point B.

If the microwave system you are designing has only one hop or path (a microwave system can have from one to a multitude of hops), then fill in as indicated. Use one worksheet for each path or hop. Microwave hops are generally anywhere from "a couple-three" to some 160.0 kilometers in length, although there have been shorter and longer ones.

Items 1A and 1B: Arbitrarily designate the path ends, and any intermediate points, by letters. We have designated the Podunk end as A, the passive reflector as point B, and the Kokomo end as point C. From your maps/charts, obtain the geographical coordinates to the best possible accuracy.

Items 1C, 1D, and 1E: These items are for the inclusion, in the event they are necessary, of the location(s) of passive repeaters along the microwave path. Use names and/or accurate geographical coordinates. Geographical coordinates are also necessary sometimes because there may be no name for the location of interest.

It is at times, owing to topography and other considerations, not possible to obtain a clear WRH shot from point A to point C, and it might be advantageous to employ passive repeaters. These are simply large radio mirrors. This is quite a topic in itself, interesting, useful, and absorbing. The interested reader might do well to avail himself of a copy of PASSIVE REPEATER ENGINEERING (Manual No. 161A). This is free at this writing and is an excellent and clear presentation on this Passive Repeater topic. It is available from Chief Engineer, Microflect Co., 3575 25th Street S.E., Salem, Oregon-97302, U.S.A.

Items 1F and 1G: These sub-items are reserved for any other pertinent locations on the microwave path.

ITEM 2: It is helpful at this point of our microwave path planning to draw a sketch. This is shown in this worksheet example Fig. 5-1. This sketch should include as much useful data as has thus far been amassed, such as path length (including, legs), passive reflector sizes if known at this stage of the problem, passive reflector angles, antenna sizes or gains if known at this point of the problem, and any other pertinent factors such as topographical altitudes and so on.

MICROWAVE
"PRELIMINARY MICROWAVE PATH SELECTION CONSIDERATIONS"

ENGR. __John Jones.__   DATE   __Aug. 27/1978.__
ENGR. __Bill Smith.__

1. PATH CONSIDERED: (Part of Only one path (hop). __PODUNK__ to __KOKOMO__ system).

   a) "A" End (Name & Coordinates)   __PODUNK, ULSTER COUNTY, NY USA.__
   b) "C" End (Name & Coordinates)   __KOKOMO, GREENE COUNTY NY USA.__
   c) Passive Repeater #1 Location   __MT. GUNNYSACK, ULSTER COUNTY, NY. USA. (B)__
   d) Passive Repeater #2 Location   __NONE__
   e) Passive Repeater #3 Location   __NONE__
   f) Other   __NONE__
   g) Other   __NONE__

2. SKETCH PATH (Showing Path Lengths, Altitudes, Passive Repeater Sizes, Passive Repeater Reflection Angles, Antenna Sizes and Other Pertinent Factors):

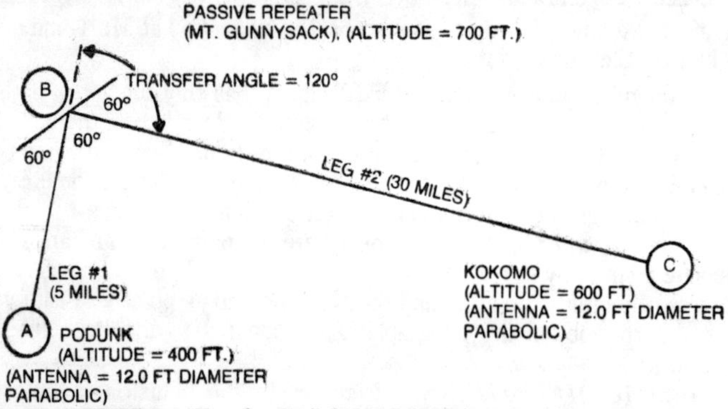

   PASSIVE REPEATER
   (MT. GUNNYSACK). (ALTITUDE = 700 FT.)

   TRANSFER ANGLE = 120°
   B   60°
   60°   60°
   60°

   LEG #2 (30 MILES)

   LEG #1
   (5 MILES)

   KOKOMO
   (ALTITUDE = 600 FT)
   (ANTENNA = 12.0 FT DIAMETER PARABOLIC)

   A   PODUNK
   (ALTITUDE = 400 FT.)
   (ANTENNA = 12.0 FT DIAMETER PARABOLIC)

   a) ALTITUDE A = __400__ ft., C = __600 ft.__ PASSIVE REPEATER B. = __700 ft.__
      ✓ ✓ ✓ ✓
   b) PATH LENGTHS (Miles) Leg #1 __5.0__, Leg #2 __30.0__, Leg #3 __✓__, Leg #4 __✓__
   c) PASSIVE REPEATER SIZES #1 __30.0 × 40.0 ft.__ #2 __✓__, #3 __✓__ Leg #4 __✓__
   d) ANTENNA SIZES (Ft) (A) __12.0 ft,__ (B) __12.0 ft.__
   e) PASSIVE REPEATER #1 Ref. ∠° __60°__ BILLBOARD #2 Ref. ∠° __✓__
      PASSIVE REPEATER #3 Ref. ∠° __✓__ BILLBOARD #4 Ref. ∠° __✓__

3. SOURCE OF PATH INFO (MAPS/SCALES, ETC.) __Maps "X", "Y", "Z" etc.__

4. PATH "MAP PROFILED" FROM ABOVE MAPS AND K-FACTORS CONSIDERED
   a) __0.5__   b) __4/3__   c) __× (Infinity).__

5. CLEARANCES RECOMMENDED (e.g. 0.6F at 4/3K and Corresponding Feet)
   __0.6 1st Fresnel Radius at 0.5K.__

6. ANTENNA/BILLBOARD NEAR FIELD: ☐ YES   ☒ NO
   DOUBLE PASSIVE REFLECTORS CLOSE COUPLED: ☐ YES ☐ NO N/A.

7. ESTIMATED ANT. HGTS. (A) __100 ft.__   (C) __50 Feet.__
   (Above Ground)

8. REQUISITE CLEARANCES OBTAINED: ☒ YES   ☐ NO
   (Worst Obstructions (s))

Fig. 5-1. Preliminary path selection form.

9. LIST ALL PATH GAINS/LOSSES AND CALCULATE MEDIAN RECEIVED SIGNAL

a) Freq. (Gc) __6.725 GHz__
b) Xmtr. Power Output __1.0 Watt (+ 30.0 dBm)__ dBm.
c) Xmtr R.F. Mux Loss __−1.0 dB__ dB.
d) Xmtr. WG Loss __−1.0 dB__ dB.
e) Xmtr. Ant. Gain __Parabola 12 ft.__ __(+45.6)__ (dBL)
f) F.S.L. Leg #1 __−127.2 dB__ dB
g) First Passive Repeater Gain __+115.7__ dB
h) F.S.L. Leg #2 __−142.7 dB__ dB.
i) Second Passive Repeater Gain __N/A__ dB.
j) Other __Rain Loss__ __0.0__ dB.
k) Other __None.__ dB.
l) Rcvr. Ant. Gain __Parabola 12 ft. + 45.6 dBi__ dB.
m) Rcvr. WG Loss __−1.0 dB__ dB.
n) Rcvr. RF Mux Loss __−1.0 dB__ dB.
o) Rcvr. Median Input Signal __−37.0 dBm__ dBm.
p) Rcvr. I.F. Bandwidth Assumed __30.0 MHz__ MHz.
q) Assumed Rcvr. (KTBF) __−89.1 dBm__ dBm.
r) Rcvr. FMI Threshold (KTBF + 10) __−79.1 dBm.__ dBm.
or digital microwave BER threshold
s) Estimated Path Reliability __99.999999% (dual diversity)__ (%).
t) Estimated Reflection Points:

	LEG # 1		LEG # 2		LEG. # 3		LEG. # 4	
	LOCATION	REF. COEFF.	LOCATION	REF. COEFF.	LOCATION	REF. COEFF.	LOCATION	REF. COEFF.
K-FACTOR (a) ∞	APPROX 2.2 MILES	BLOCKED	20.5 MILES	BLOCKED	N/A	N/A	N/A	N/A
K-FACTOR (b) 4/3	APPROX 2.2 MILES	BLOCKED	20.0 MILES	BLOCKED	N/A	N/A	N/A	N/A
K-FACTOR ½	APPROX 2.2 MILES	BLOCKED	19.5 MILES	BLOCKED	N/A	N/A	N/A	N/A

10. OTHER PERTINENT DATA, REMARKS AND "INPUTS"

11. PATH RECOMMENDED FOR FIELD PROFILING

YES ☒ NO ☐

(If no, state reasons therefore under REMARKS).

REMARKS AND CLARIFYING NOTATION:

ENGINEER SIGNATURE

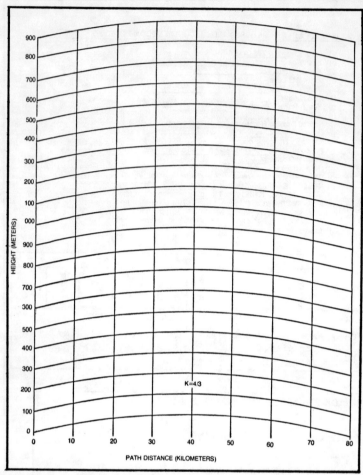

Fig. 5-2. Path profiling paper for k = 4/3.

Item 3: List the sources of your information for your tentative microwave path.

Item 4: Path map-profiled from above maps (and other information) and K-factors considered. Let's, at this point, consider some facets regarding K-factor and radio path terrain profiling. Although you won't necessarily be going through all of the following, it is quite important to the understanding of the microwave propagation problem. The K-factor in propagation work is the ratio of effective earth radius to the actual earth radius. This will become clear as we go along. Under so-called normal conditions on a typical microwave path, a microwave radio ray does not follow a geometrically straight line, but rather displays a trajectory best described as having a "toward-the-earth" bending radius somewhat greater than the real earth's radius of curvature. In brief, this condition is brought about by the fact that the troposphere's dielectric constant lapses (decreases in a certain manner with height or altitude) in a so-called "standard" atmosphere. Under these "standard" or "normal" meteorological conditions, the microwave radio ray may be represented conveniently by a straight line instead of a curved line if our ray is drawn upon a fictitious earth representation of radius 4/3 that of the earth's actual radius. This concept was the outgrowth of the excellent work of Shelling, Burrows, and Ferrell in what was very probably the first attempt (circa 1933) to utilize a model of atmospheric refractive index for the solution of propagational problems in microwave work. When weather conditions differ from the norm, microwave radio rays may assume curvature radii different from the curvature they would have under the above weather conditions. In microwave propagation problems, these curved lines are difficult to work with, especially with regard to reflected rays. So, to make the radio rays appear as straight lines under these radio-meteorological conditions, path profiling paper, based upon an earth's radius corresponding to these radio-meteorological conditions, is employed. See Fig. 5-2 for K = 4/3 and Fig. 5-3 for K = ½.

K-profile paper for topographical microwave path profiling is generally available from various microwave equipment manufacturing companies. However, you can construct your own in the following manner see Fig. 5-4. You can begin on any linear/linear graph paper or draw squares to make your own rectangular coordinate type graph paper. Next, assign an arbitrary distance scale along the x-axis and height scale along the y-axis. In our example, we have assigned the usually convenient distance of two miles to the inch along the x-axis. The sheet can be made as long as necessary to accomodate your particular microwave-path lengths. We have also assigned the usually convenient value of 100.0 feet to the inch on the y-axis.

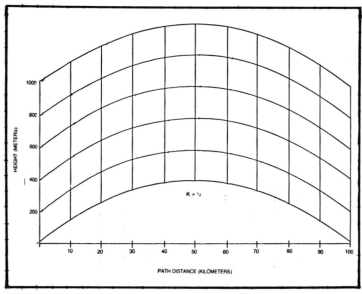

Fig. 5-3. Path profiling paper for k = ½.

153

This can also be made anything desirable and carried to any necessary height for your particular topography involved. Let's assume that twenty miles path distance and seven hundred feet of height or altitude will cover our particular needs for a given microwave path. As the K-paper stands, it will represent flat earth or what is known as an infinite K-factor; more on this in a little bit. To construct a K-paper for any other value of K-factor, use the following formula:

$$h = \frac{2d_1 \times d_2}{3K}$$

Where: h = Earth bulge in feet from the flat-earth reference.
d_1 = Distance in miles (statute) from a given end of the microwave path to an arbitrary point along the path.
d_2 = Distance in miles (statute) from the opposite end of the microwave path to the same arbitrary point along the path.
K = K-factor considered.

Let's assume a K-factor of 1.0 (true earth radius) for simplicity. This reduces our formula to $h = (2d_1 d_2)/3$. Now, let us compute h for a path distance of 0.0 (zero) miles. Obviously, this will yield $h = 0.0$ feet and we plot this point with an both at the 0-mile and 20-miles points. Next, consider the 2.0 and 18.0 statute-mile points which may be calculated using $d_1 = 2.0$ miles from either end of the path. Therefore,

$$h = \frac{2d_1 d_2}{3.0} = \frac{2(2.0 \times 18.0)}{3.0} = 24.0 \text{ feet.}$$

Mark an X at the 2.0 and 18.0 miles points for a height of 24.0 feet as shown in Fig. 5-4. Repeat this procedure for distances 4.0 and 16.0 miles and mark an X at the 42.6 foot point. Repeat the procedure for distances 6.0/14.0 miles, 8.0/12.0 miles, and 10.0/10.0 miles, and mark an x correspondingly. The x points are then all connected and the resulting curve then becomes the 0.0 height parabolic line for a K-factor of 1.0 (true earth). Next construct a parabolic line parallel to the x curved parabolic line as shown. Continue process until you run out of paper in the altitude direction or until you have enough altitude to accomodate your path profile topographical altitudes.

In case you'd prefer to work in the metric system, the basic principles are exactly the same, but the formula is slightly different.

$$h = \frac{d_1 d_2}{12.75\,K}$$

Where: d_1 and d_2 have the same meaning as previously but in kilometers instead of miles.
h is in meters instead of feet.

Remember that these curved lines are parabolic (see Appendix 18 entitled, "Parabolic World"). It is possible to expand the scale for other distances and altitudes by simply multiplying the distance by 2.0 and heights by 4.0. Thus, in Fig. 5-4, the distance scale would become 40.0 miles and the height scale would become 2,800.0 feet. Similarly, the scales may be shortened as desired for various microwave path plots. This is done by dividing the distance scale by 2.0 and the height scale by 4.0. It will be

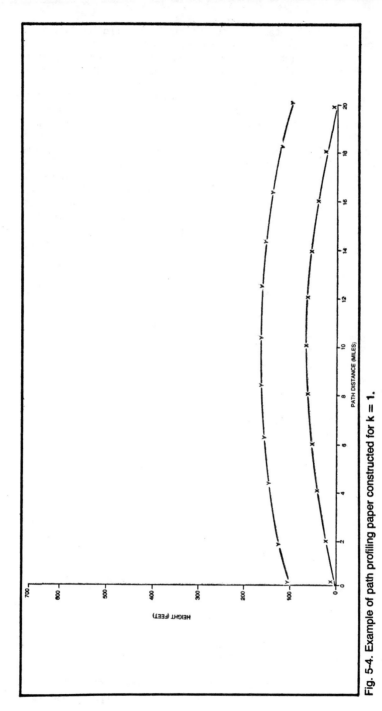

Fig. 5-4. Example of path profiling paper constructed for k = 1.

noted that the height and distance scales are different, and therefore angles cannot be measured directly on these plots without special devices. An ordinary protractor cannot be used without obtaining erroneous results. The different scales are employed to allow convenient distance versus height presentations (e.g. a height, say, of 100.00 feet couldn't even be seen if the height scale were the same as that of the distance scale (miles).

In the portion of the troposphere in which terrestrial microwaves are normally propagated, water vapor pressure and temperature "usually" decrease with altitude, and thus produce a quasi-linear diminution of index-of-refraction with height. The tropospheric index-of-refraction is responsible for the speed of microwave energy through it. Thus, in a "standard" atmosphere, the upper portion of our wave-front travels through the troposphere slightly faster than the lower wave-front portion, and the radio wave curves "downward" or earthward in varying degrees depending upon the rate with which the troposphere's dielectric constant changes with height. The converse of this is also true, an increasing refractive index with height will cause the rays to curve away from the earth. The effect of these radio-meteorological actions is to increase or decrease the radio horizon distance, making the use of different K-factor terrain profiling paper convenient. The "standard" 4/3 K radio-meteorological condition might typically exist at a given place on the earth for approximately 50.0 % to 80.0% of the time. To engineer a microwave path for this K-factor, would be a relatively straight-forward matter. This, however, would hardly be adequate for the higher circuit reliabilities (e.g. 99.99%). As previously pointed out, weather variations produce different K-factor conditions and these must be taken into account based upon area radio-meteorology observed over a sufficiently long time period, usually years.

Based upon the weather/climate patterns in the locality of a planned microwave path, experience, measurement, meteorological almanacs, and any other dependable sources, such as weather bureaus, a microwave propagationist judges and/or calculates the extremes of departure of the K-factor from the standard 4/3 k-factor condition and the percentage of time of such excursions (see Appendix 19 entitled "Refractivity and K-Factor Data").

Microwave-path engineering based upon a nominal 4/3 K-factor with extremes of ⅔ K and infinite K is not unusual. Even negative K-factors, concave earth, are sometimes taken into account. Basically, one should use K-factor extremes better than the desired microwave path propagation reliability. That is, if a path reliability of 99.9% is required, propagationally, one wouldn't use a K-factor allowing the microwave ray to be blocked for say, 2.0% of the time. Good engineering judgement must be used. At any rate, the infinity K-factor, of course, would be based upon the fact that under this condition a microwave ray follows the curvature of the earth. This is also known in propagational circles as the "flat-earth" condition. K-factors smaller and larger than 1.0 (true earth) will cause the microwave radio to bend away from the earth and toward the earth respectively, giving birth to the terms "earth bulging" and "earth flattening.".

As an example of judging K-factor conditions, a microwave engineer might note an early morning fog formed by nocturnal cooling of the surface of the earth. Almost immediately, this condition (barring the existence of other radio-meterological complications which must be analyzed) may be spotted as conducing to an earth-bulging condition. The engineer reasons that nothing has happened to the total amount of water in the air, but that the fog was brought about by condensation into droplets of some of the air's water vapor. In the form of vapor, water has much more effect upon the troposphere's radio refractive index than it does in its condensed form. The microwave radio engineer knows that this fog layer is frequently accompanied by a temperature inversion (temperature increasing with height), and that this combination is a "set-up" for a surface layer earth-bulging microwave propagation condition. As already pointed out, this militates toward K-factor conditions smaller than 4/3. Alternately, if a sharp rise in temperature occurs, or a drop in water vapor content, or both, with an increase in altitude or height, a decrease in refractive index with height can result causing the radio ray to be "down-ward bending" at a more than a k = 1.0 rate. If the refractive index rate of decrease as a function of height is sufficiently great, the radio may be bent sufficiently to strike the earth, be reflected upward, refracted to earth again, etc. This phenomenon is known as a surface duct or surface trapping and could give rise to microwave propagation well beyond the radio horizon, especially in ducts located over water. There are many values of K-factor occuring over the world, and the best radio meteorological information for a given location should always be used. See Appendix 19 entitled "Refractivity and K-Factor Data" for samples of cumulative distributions of K-factors in various parts of the world.

To continue, the earth-flattening condition is also called superrefraction, and if the microwave ray just follows the earth's curvature, the K-factor under these conditions would be infinity. Thus, the radio ray may be represented by a straight line on flat earth K-profile paper since the surface of a fictitious earth having an infinite radius would obviously be a plane. Similarly, earth-bulging is known as a sub-refractive condition causing the microwave ray to curve away from the earth more than usual. An earth-sagging radio-meteorological condition represents a concave earth (or negative K-factors) as the earth under these radio-meteorological conditions appears to a straight-line microwave ray. It is helpful to remember that the radio ray in varying degrees according to K-factor, bends toward the increasing atmospheric refraction index. Actually, as above implied the K factor, and thus the radio ray path through the troposphere, depends upon a change of radio refractive index with altitude or height and measurement of the quantities which affect radio refraction are frequently obtained by weather radiosonde (RAOB) balloons. This change of refractive index with height is called refractive index gradient, and the three measured quantities which combine to derive it are barometric pressure, temperature, and relative humidity. By measuring these three quantities both in the vicinity of the earth's surface and some altitude (several hundred meters is usually sufficient for terrestrial microwave), a K-factor may be computed. Additional measurements between the earth's surface and the final altitude of interest would, of course, render important information on the vertical refractive index which could be examined for the conditions giving rise to unusual microwave propagation conditions such as ducting or trapping. The refractive index at a particular altitude may be calculated from the formula:

$$N = (n - 1) \times 10^6 = \frac{77.6}{T}$$

$$P + \frac{4810\, e_s\, (RH)}{T}$$

Where: N = Refractivity
 n = Refractive Index
 T = Temperature in Kelvins
 P = Total atmospheric pressure
 e_s = Saturation vapor pressure in millibars at temperature T
 RH = Relative Humidity as a fractional part of 1.0 (e.g. 10% RH = 0.1)

N and n, is relatively simple when we consider that the earth surface index of refraction is generally about 1.0003 and that variations are often less than ten parts per million. Thus, it becomes easier to handle these quantities when used when the following equation: $N = (n - 1)\, 10^6$. We now see that the quantity n = 1.0003 becomes 300.0, a more convenient numerical range with which to work.

Now then, knowing the index of refraction and at some desired height, our k-factor may be computed, assuming the gradient is linear, as follows:

$$K = \left[\frac{1.0}{1.0 + \left(\frac{a}{n}\right)\left(\frac{dn}{dh}\right) \cos \Theta} \right]$$

Where: a = True earth radius (6370.0 kilometers)
 n = Radio refractive index

$\left(\frac{dn}{dh}\right)$ = Refractive index gradient relative to height, in the part of the troposphere in which the microwave propagates
 Θ = Angle of the microwave radio ray with respect to the local earth tangent.

An easier formula with which to work (although you won't actually be doing much of this type of calculation unless you're involved in radiometeorology) is as follows:

$$K = \left[\frac{157}{157 + \frac{dn}{dh} \cos \Theta} \right]$$

There are, of course, other means for obtaining the refractive index or refractivity value, although, as we have mentioned, you probably won't be obtaining your K-factor data this way, but by work already done (see

Appendix 19 entitled "Refractivity and K-Factor Data"). The methodology is presented here for a greater depth of understanding of the "mechanics" of microwave propagation and its difficulties or ramifications. Another method of obtaining tropospheric refractivity is through the employment of radio-free refractometers. Basically, a radio refractometer is a resonant cavity (or tuning capacitor) the troposphere for dielectric and is designed to directly measure the index of refraction as a function of the content of the cavity. The relationship between the change of resonant frequency of the cavity and the corresponding change of index of refraction error:

$$\frac{\Delta f}{f} = -\frac{\Delta n}{n}$$

Where: f = Original resonant frequency of the cavity, the tropospheric content of which has an index of refraction of n.

Since n is very close to 1.0, the following approximation may be employed with negligible error:

$$\frac{\Delta f}{f} = -\Delta n$$

Regardless of the method used, it must not be assumed that a single (or even several) measurement and computation of K-factor will yield the extremes and their percentages of occurrence in any given radio-meteorological location. Actually, it takes many measurements over a extended time period during all seasons, time of day, and radio meteorological conditions to establish realistic K-factors distributions. It should be pointed out that K-factors should be used which permit adequate path clearance for a period of time greater than the desired path propagation reliability.

It might be mentioned at this point that the K-factor values that are derived by radiosonde (RAOB) data has in some quarters been considered somewhat questionable due in part to the fact that the tropospheric values (pressure, temperature, etc.) are sampled sequentially, instead of simultaneously, during balloon ascent. Another objection has been that of possible sensor response lag, a sensor giving a somewhat lower altitude data at a somewhat higher altitude. Although these actions might tend to smooth out the data, some have questioned the resultant accuracy. Additionally, this upper air data is synoptically taken, for weather purposes the world over, generally only twice per day which in some radio-engineering quarters is considered an excessively small sampling to be representative. The refractometry method of measurement is, unfortunately, not frequently used. These objections, be as they may, the radiosonde balloon weather data is the most plentiful and in many cases the only such data available at all. The K-factor data in Appendix 19 is for the greatest part based upon radiosonde balloon measurement. As this subject is rather vast, the serious propagationist would do well to avail himself of two very excellent works:

RADIOMETEOROLOGY (Monograph 92)
U.S. Dept. Of Commerce, NBS
By B.R. Bean and E.J. Dutton

A WORLD ATLAS OF ATMOSPHERIC RADIO REFRACTIVITY
(ESSA Monograph 1) ITSA
By B.R. Bean, B.A. Cahoon, C.A. Samson and G.D. Thayer

These works are generally available from:

>The Superintendent Of Documents
>U.S. Government Printing Office
>Washington, D.C. 20402 U.S.A.

or

>NTIS (National Technical Information Service)
>5285 Port Royal Road
>Springfield, Virginia 22161

ITEM 5: This item has to do with recommended clearances. This item is of necessity tied in with the previous one concerning K-factors. This interpendence will become clear as we proceed. But first, some basic background. The subject of Fresnel zones, named after their discoverer Jean Augustin Fresnel, and the manner in which they apply to microwave path engineering, is indeed an interesting one. Fresnel zones as they apply to microwave path engineering, have to do with the required path clearance which is normally deliniated as a fraction of the first Fresnel zone radius. The direct radio ray, in the simple case, is a geometrically straight line from transmitting antenna to the receiving antenna down on appropriate K-paper. However, the infinite number of points from which geometry shows that reflected rays might reach the receiving antenna one half wavelength later than the direct ray, produces an ellipsoid of revolution, the antennas (transmitting and receiving) being located at the ellipsoid's foci (see Fig. 1 of Appendix 20) which shows the first Fresnel zone ellipsoid, the direct ray between antennas (solid line), and the reflection ray defining a point on the ellipsoid). This ellipsoid is defined as the first Fresnel zone and forms an intangible protective barrier which should not be violated by intruding obstructions except by specific design amounts. Thus, it follows that allowing clearance for the direct ray only would generally be (except in special cases) inadequate. With direct ray clearance only, there could be an excess path loss of 6.0 decibels for the case of a pure knife-edge obstruction of zero-reflection coefficient to some 20.0 decibels for other obstruction shapes of higher reflection coefficient. This could prove devastating on a microwave circuit unless a sufficient fade margin is allowed (decibels cost money).

The first Fresnal zone, or more accurately first Fresnel zone radius, is defined as the perpendicular distance from the direct ray line to the ellipsoidal surface at a given point along the microwave path. It is calculated by:

$$FR = 2280 \left[\frac{d_1 d_2}{f(d_1 + d_2)} \right]^{\frac{1}{2}} \text{ feet.}$$

Where: d_1 and d_2 are the distances in statute miles from a given point on a microwave path to the ends of the path (or path segment).

f = Frequency in MHz
FR = First Fresnel zone radius in feet.

In the metric system of measurement, the first Fresnel zone radius is computed by the following formula:

$$FR = 17.3 \times \left[\frac{d_1 d_2}{f(d_1 + d_2)} \right]^{1/2} \text{ meters.}$$

Where: d_1 and d_2 are the distances in kilometers from a given point on the microwave path to the ends of the path (or path segment).
f = Frequency in GHz
FR = First Fresnel zone radius in meters.

There are in addition, of course, the second, third, fourth etc. Fresnel zones, and these may be easily computed, at the same point along the microwave path, by multiplying the first Fresnel zone radius by the square root of the desired Fresnel zone number. That is, to compute the second Fresnel zone radius, one would multiply the first Fresnel zone radius by the square root of 2.0 or 1.414.

There are some applicable scientific guides, and much experience, involved in dictating the amount of Fresnel clearance to use and under which K-Factor conditions. It is a matter of design judgement and trade-offs (generally money rules). Frequently, 0.6 first Fresnel zone radius clearance is sufficient for WRH microwave propagation (see Appendix 8 entitled "Line of Sight"). This 0.6 first Fresnel zone radius clearance is stipulated because, regardless of the terrestrial reflection coefficient, the so called free-space path loss attenuation value obtains. Otherwise, the terrestrial reflection might add to or subtract from the direct ray field strength depending upon its phase relationship with the direct ray. For so-called "normal" microwave propagation, 0.6 first Fresnel zone radius clearance at 4/3 K-factor is generally considered. Please keep in mind though that the so-called "normal" condition or 4/3 K-factor generally does not obtain in any given location for more than approximately 50%—60% of the time. At other locations and times the K-factors may swing through certain other values for sufficient percentages of time to render a microwave system's propagational reliability something less than the high order of, 99.99%. For a tough coastal path, for example, where weather conditions play havoc with microwave propagation, a 0.3 first Fresnel zone radius clearance at ⅓ K-factor might be employed. When we say, for example, 0.6 or 0.3 first Fresnel zone clearance, we mean that any obstruction along the microwave path should not protrude into the first Fresnel ellipsoid more than 0.4 and 0.7 respectively of the first Fresnel zone radius at any point along the path.

ITEM 6: There are times in terrestrial microwave-path design when intervening topography between the desired path terminals, as well as the existence of other situations, need to locate a microwave terminal at the foot of a hill or mountain near a road passive repeater (also called reflector or radio-mirror) might be of great advantage. The use of passive repeaters can, in certain instances, eliminate active stations and access roads (and their cost and maintenance). These passive repeaters, however, must be applied correctly. For example, the relationship between the passive repeater and the near antenna, or between two passive repeaters in a dual

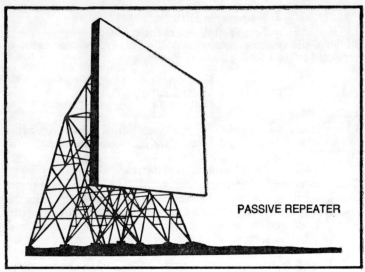

Fig. 5-5. Example of a passive repeater (courtesy Microflect Co., Inc.).

passive-repeater installation may have a near field (or a far field) and close coupled relationship depending upon the distance between them and their projected sizes (a passive repeater respectively may project anywhere from zero area to its full area depending upon the angle with which it is viewed by an incident radio wavefront. In this type of work, the terms near field/far field are generally reserved for antenna/passive reflector relationships, while the term, close coupled, is generally reserved for passive repeater/passive repeater relationships. A gain or a loss may result, depending upon the near-field/far-field antenna/passive repeater physical relationship. A check can be performed via the following formula:

$$-d = \left(\frac{2D^2}{L}\right)$$

Where: d = Near-field/Far-field boundary for an antenna/passive repeater relationship. This is otherwise known as the Fresnel-region/Fraunhofer-region boundary. (Do not confuse Fresnel Zone with Fresnel Region, they are different.)
L = Wavelength
D = Largest projected dimension of passive repeater, or antenna, whichever is the larger.

(Note: d, L, and D are all in the same units)

As for passive reflector installations, they may be employed singly with efficiency when the deflection (or transfer) angle is greater than 40° to 50°. When the deflection angle is less than this, double passive repeater installations are generally more efficient, other factors being equal. Coupling between the dual-passive repeater varies with frequency of operation, distance between passive repeaters, and their effective areas in a complex fashion. However, a simple calculation may be performed to determine whether or not the close-coupled conditions exists. Use the following mathematical expression:

$$2LD/A^2$$

Where: A_2 = Effective area of smaller passive repeater
L = Wavelength
D = Distance between passive repeaters

(Note: A, L, and D are all in the same units)

If the result of this calculation is less than 4.0, the two passive repeaters in the dual-passive repeater set-up are to be considered close coupled, and a special relationship exists making their combined gain quite different from that which might erroneously be derived by considering the individual passive repeater gains and the free-space loss between them as separate entities. Generally, it may be stated that the gain of close-coupled double-passive repeater configurations is the gain of the smaller of the two passive repeaters, less the coupling loss. Coupling loss is seldom more than 2.0 decibels, and frequently 1.0 decibel or less. See Fig. 5-5 and 5-6 respectively for illustrations of passive repeater configurations.

The gain of a single passive repeater is given by the formula:

$$\text{Gain} = 20 \, \text{Log}_{10} \frac{4\pi \, A \, \text{Cos} \, B}{L^2}$$

Where: A = Area of passive repeater when viewed perpendicular, i.e. its physical width multiplied by its physical height.

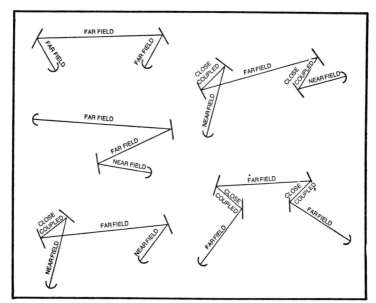

Fig. 5-6. Examples of different reflector arrangements (courtesy Microflect Co. Inc.).

B = One half the included angle formed by the incident and reflected radio rays.
L = Wavelength.
(Note: A and L^2 are in the same units. B is in degrees).

If antenna/passive repeater and passive repeater/passive repeater relationships are far-field and not close-coupled respectively, the individual antenna gains, free-space losses, and passive repeater gains in decibels are simply added algebracially.

Parabolic antenna gains may be calculated by the well known formula:

Gain (dBi) = $20 \log_{10} D + 20 \log_{10} F + 17.8$

Where: D = Parabolic antenna diameter in meters.
F = Operating frequency in GHz.

or

Gain (dBi) = $20 \log_{10} D + 20 \log_{10} F + 7.5$

Where: D = Parabolic antenna diameter in feet.
F = Frequency in GHz.

The basic free-space loss formula is:

FSL (dB) = $36.6 + 20 \log_{10} D + 20 \log_{10} F$

Where: FSL (dB = Free-space loss in decibels
D = Path distance in miles
F = Operating Frequency in MHz

The following "enigmatic-looking" versions of the free-space loss formula, however, may not be as familiar and are presented here as a bit of diversion.

1. FSL(dB) = $10 \log_{10} (4\pi D/L)$
2. FSL(dB) = $22 + 10 \log_{10} (D^2/L^2)$
3. FSL(dB) = $10 \log_{10} 4543 f^2 + 10 \log_{10} D^2$
4. FSL(dB) = $10 \log_{10} (4.56 \times 10.^3 f D^2)$

Where: In formulae 1 and 2, path distance, D, and wavelength, L, are in the same units, while formulae 3 and 4, D is path distance in miles while f is the operating frequency in MHz.

5. FSL(dB) = $96.6 + 20 \log_{10} F + 20 \log_{10} D$
6. FSL(dB) = $92.4 + 20 \log_{10} F + 20 \log_{10} D$

Where: In both formula 5 and 6 above, F is the operating frequency in GHz. In formula 5, D is the path distance in miles (statute). In formula 6, D is the path distance in kilometers.

As a point of interest, it might be emphasized here that free-space attentuation does not occur because space has some mysterious absorbing property or electromagnetic energy dissipating quality, but simply because radiated energy continuously spreads and "thins" out as we increase our distance from the radiating source (e.g. transmitting antenna), following the well-known inverse square law. This, of course, accounts for $10 \log_{10} D^2$ (or $20 \log_{10} D$), which is the same in our free-space loss formulae. An interesting way to envision free-space loss is to imagine transmitting isotropic source surrounded by a metal sphere concentrically juxtaposed about this source. Under these conditions, the metal sphere would capture all the radiated energy, and regardless of the distance between the isotrope and the sphere, there would be *no* free space loss since the sphere would absorb *all* the radiated energy.

ITEM 7: Now that we have map profiled our tentative microwave path on simultaneous K-factor paper, see Fig. 5-7, we can, using the principles and information set forth, estimate the required antenna heights which might be necessary in attaining our requisite clearance for the K-factor conditions under consideration in our geographical/radiometeorological location(s).

Consult Fig. 5-7. The first thing noted here is the fact that there are represented three super-imposed k-factor representations. The various sea-level "baselines" (parabolae) are shown, and represent earth-bulges corresponding to the mentioned K-factors. This is the way a smooth earth would appear to straight-line radio rays. Please keep in mind that the three profiles shown are the same one under the differing K-factor conditions. Only path end A is coincident for all three K-factors since it on the zero miles path distance of the representation. B_1, B_2, and B_3 are all the same point under the varying K-factors, as are C_1, C_2 and C_3. The B location represents a passive reflector, while A represents one microwave terminal and $C_1C_2C_3$ represents the other microwave terminal. Since our criterion in ITEM 4 specified a 0.5 K-factor and ITEM 5 required a 0.6 first Fresnel radius clearance at 0.5 K-factor, we see that this can be satisfied by tower heights at the terminal ends as shown (A end 100.0 foot tower and C end 50.0 foot tower). The 0.6 first Fresnel zone radii are partially shown by dotted lines. It is important at this point to call attention to the fact that the three profiles shown do not occur in "jumps" from one to the other, but most generally change from one to the other in a more continuous and smoothed fashion.

ITEM 8: In this format item, simply indicate whether or not the requisite design clearance is obtained under the worst obstruction conditions under consideration.

Item 9: With the foregoing background then, let us work out a preliminary problem. In this fictitious microwave path problem, we need a path, as called for in ITEM 1, from Podunk to Kokomo. Let's work out this problem to determine whether or not it's worth pursuing by field work.

ITEM 9A: Here we list the frequency of operation. This frequency depends upon a multitude of complex and inter-related factors such as availability and propagation characteristics. For example, the 2.0-GHz frequencies generally have more favorable fading characteristics, but the 11.0-GHz frequencies are sensitive to rain absorption and scattering, and this must be taken into account. For our purpose, in this problem, we arbitrarily chose an operating frequency of 6.725 GHz.

ITEM 9B: Here, we assume a tentative transmitter power of 1.0 watt (+ 30.0 dBm).

ITEM 9C: This is estimated RF multiplexer loss (circulators, diplexers, etc.). In our example, we assume a loss of 1.0 dB. This value may be obtained from manufacturers.

ITEM 9D: Transmitter waveguide loss is the attenuation of the waveguide from the RF multiplexer to the antenna. Herein we have assumed, for simplicity, 1.0 dB. In practice this value may be obtained from manufacturers literature. This value depends upon waveguide length contemplated and the antenna VSWR.

ITEM 9E: Parabolic (or other type) antenna gain. This may be computed or obtained for manufacturers' literature. If the gain is given as

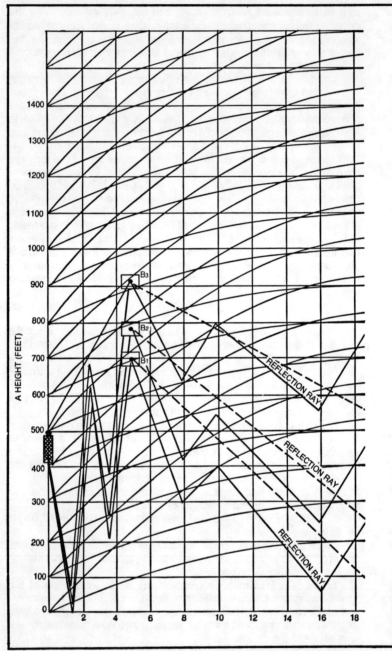

Fig. 5-7. Map profile of the tentative microwave path on simultaneous k-factor paper.

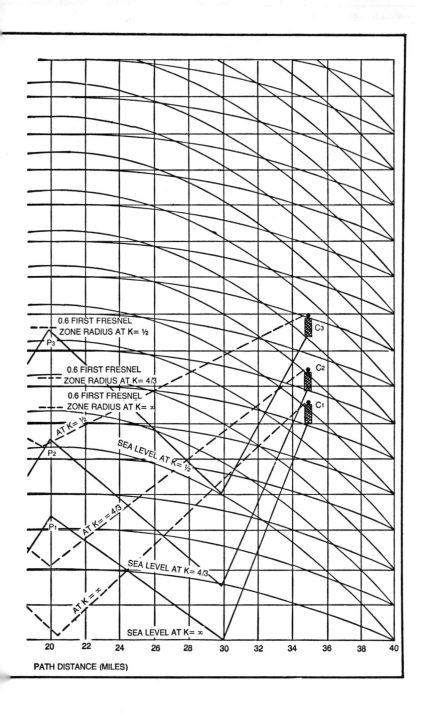

dBi, it is the gain over an isotropic source; if it is given in terms of dBd, it is the gain over a half wave dipole. Now, since a half-wave dipole is generally considered to have a gain of 2.1 dB over an isotrope, one can readily see that the comparison basis is of the utmost importance.

ITEM 9F: Free-space loss formulas were already given, and these may be used. Several companies also put out small cardboard calculators which determine such quantities. This is the FSL or free-space loss for leg l, i.e. from terminal A to the passive repeater at location B ($B_1B_2B_3$). This works out to be -127.2 dB. The minus sign is used by many propagation engineers to indicate loss. See terrain profiled of this path segment in Fig. 5-7. By the way, considering FSL for this segment is a valid move since the passive reflector, being 5.0 miles from the near antenna, is in a far-field condition. Recall from Item 6 that this turned out to be far-field from formula $d = (2D^2/L)$.

ITEM 9G: This item describes the gain of our passive repeater at Mt. Gunnysack. Recall that this subject of passive repeaters is a detailed one with an almost an infinite variety of configurations according to one's particular application. As such, it would be next to impossible to include all known passive repeater information herein. Accordingly, the writer again urges the serious propagationist to avail himself of the passive repeater literature previously mentioned. Returning now to ITEM 9G requiring the gain of our passive repeater at Mt. Gunnysack, we invoke the formula given previously-viz:

$$\text{Gain(dB)} = \frac{20 \text{ Log}_{10} 4\pi A \text{ Cos} B}{L^2}$$

Solving this equation gives a two-way passive repeater gain of 115.7 dBi. The gain is called two-way because it both receives and transmits (actually re-transmits) the signal. In the Microflect Co. literature already mentioned, there are all sorts of graphs and "goodies" which eliminate the drudgery (as well as error probability) from calculations and which make the design engineer's life more pleasant.

ITEM 9H: The second leg (Mt. Gunnysack to Terminal C) free space loss is computed by the same free space loss formula as in Sub Item 9F. Substituting and solving, we get a loss of 142.7 dB.

ITEM 9I: This is a possible second passive repeater gain. Since we do not employ a second passive repeater in our example., this is marked N/A or not-applicable.

ITEM 9J: For this sub-item, we go to any source giving the best and most accurate information on rain in a given area. Let's, for the sake of this particular problem, say that our rain rate most nearly matches that of Freiburg, Germany as given in Appendix 21. Here we see that the rainfall rate exceeds 40.0 millimeters per hour for 0.001% of the time. Converting our 40.0 millimeters per hour rain rate to terms of inches per hour (in order to use the loss data at the tail end of Appendix item 21), we have:

$$\text{Rain Rate in Inches/Hour} = \frac{\text{Rain Rate in Millimeters/Hour}}{25.4}$$

$$= \frac{40.0}{25.4} = 1.5 \text{ Inches/Hour.}$$

Interpolating between 0.6 Inches/Hour and 4.0 Inches/Hour rainfall rate and corresponding decibel additional path loss we obtain a figure of approximately 0.65 dB per mile (statute) at 7.0 GHz (near our operating frequency of 6.75 GHz). Let us assume that during an "average" rainfall that an "average" of three thunderheads (from the best information available) will cross our microwave path and that each such rain producing cell (thunderhead) has an effective diameter (rain producing length of 1.0 statute mile. This adds up to "three miles" of rain attenuation at the rate of 0.65 dB per mile or a total rain attenutation of 1.95 dB for the 30.0 mile portion of the path for 0.001% of the time.

As can be seen in Appendix 21, rain attenutation increases with frequency and not-infrequently rain attenuation is ignored below about 8.0 GHz. At any rate, herein we have undertaken to account for the rain loss by thus far allowing an additional 1.95 dB margin. There is however, another facet; during rain showers, the troposphere is usually thoroughly mixed (by turbulence of rain, winds, etc.). This breaks up air stratification and such other path loss-producing meteoroligical mechanisms and thus might be self-compensating. As for the short 5.0-mile path segment from microwave terminal A to passive repeater B, we assume herein that rain clouds will not cross it simultaneously with the long path segment from passive repeater B to microwave terminal C.

Additionally, we have assumed that our weather investigation of the area showed that this side of the path was on the leeward side of Mt. Gunnysack and therefore generally experiences much less precipitation. If this were not the case, we would proceed in a manner similar to the long path segment and come up with a rain attenuation estimate for the short segment. All in all, then, taking the above into consideration, we have allowed 0.0 dB for rain attenuation and have so indicated under item 9J. Each case is individual and should be reasoned out as well as possible. *As in many facets of radio propagation, there is considerable art and intelligent guess-work involved since we are at the state-of-the art in many cases. Precise and exact data are just not available and things are not as neat as in an Ohm's Law calculation.*

Additionally, we selected the 0.001% of the time value since we generally deal, in microwave work, with high orders of propagational reliability of the order of 99.999% and better. Some engineers even extrapolate available data to small percentage values indeed. At any rate, it can easily be appreciated that one would not , for example in such a high reliability microwave path, use a rain rate exceeded for say, 1.0% of the time or worse. Elegant engineering formulae are just not available at this stage of the science/art and much judgement is required.

ITEM 9K: This space is reserved for any other possible gains or losses.

ITEM 9L: This space is for indicating the receiving antenna gain. Let's asssume, as a reasonable start, the same gain as at the transmitting end. Note that actually although we call one end the transmitting end and the other end the receiving end, for convenience, both ends usually transmit *and* receive since usually a two-way circuit is involved. So, let's assume for an antenna gain of 45.6 dBi.

ITEM 9M: Here write in the waveguide loss between the receiver antenna and the receiver input terminals (do not include RF multiplexing loss is this figure). We assume a 1.0-dB loss for waveguide in our example. This data is actually available by measurement or from manufacturer's literature. In the planning stages, of course, use manufacturer's figures.

ITEM 9N: Receiver RF Multiplexer loss. This also may be measured or manufacturer's data as above. Herein we assume a loss of 1.0 dB.

ITEM 9O: This is the receiver median input signal or signal level the receiver sees for 50.0% of the time. Frequently this is confused with the so-called Free-Space value. While there is a 1.6 dB difference between the two, this difference is frequently ignored and the free-space value is taken as the median value. Rayleigh fading is assumed. At any rate, this received signal value may be obtained for item 9O by simply adding up algebraically all the system/gains and losses from transmitter output to receiver input. (Gains are plus and losses are minus.).

```
Transmitter Power Output ..........................+ 30.0 dBm
Transmitter RF Multiplex Loss ......................− 1.0 Db
Transmitter waveguide loss............................− 1.0 dB
Transmitter Antenna Gain .............................+ 45.6 dB
Free space loss (Leg-l) ...............................−127.2 dB
Passive Repeater Gain ................................+ 115.7 dB
Free Space Loss (Leg-2) .............................−142.7 dB
Receiving Antenna Gain ...............................+ 45.6 dB
Receiver Wave guide loss ..............................−1.0 dB
Receiver RF Multiplex Loss ...........................−1.0 dB
                                            Total .....−37.0 dBm
```

This value (−37.0 dBm) is the signal power input to the receiver. It represents the median signal (or free space value, if you will, or in some quarters it is called the unfaded value) input to the receiver.

ITEM 9P: This is the noise-bandwidth value (I.F. bandwidth) of the receiver contemplated for use on the microwave path being engineered, or in lieu of this an assumed representative I.F. bandwidth value. Such values are available from reputable manufacturers also for planning purposes. This noise bandwidth is actually a rectangular shaped ideal bandwidth of the same area under the bandpass curve as the usual 3.0-dB bandwidth of the receiver I.F. Generally, with but small error, we may consider the 3.0 dB bandwidth where a series of synchronous-tuned I.F. stages are employed the same as that for the rectangular one. We have herein assumed the value of 30.0 MHz. This value will be used in the next steps.

ITEM 9Q: Assumed receiver KTBF in dBm. This is simply the random noise (please see Appendix 14) in the receiver, against which the incoming signal of −37.0 dBm must compete. Most noise is generated, at these frequencies, in the receiver and not in the atmosphere as in HF or galaxy as in Ionospheric Scatter propagation. K is simply Boltzmann's Constant and is equal to −198.6 dBm/Kelvin/Hz. The value T is frequently, for terrestrial microwave purposes, taken at 300.0 Kelvins making it, in the KTBF expression, equal to 10 Log_{10} 300.0 = 10(2.477 = 24.77) dB. The value B, as we mentioned in the previous item, is 30.0 MHz so B in the expression KTBF would be equal to 10 Log_{10} 30,000,000 (Hz) = 10(7.477) = 74.77 dB. And, a not-atypical F (receiver noise figure) might be 10.0 dB.

So, adding up all the decibel values corresponding to KTBF, we have:

$$K = -198.60 \text{ dBm}$$
$$T = +24.77 \text{ dBm}$$
$$B = +74.77 \text{ dBm}$$
$$F = +10.00 \text{ dBm}$$
$$\text{Total} = -89.06 \equiv 89.1 \text{ dBm.}$$

This −89.10 dBm of noise power is then our assumed receiver KTBF value.

ITEM 9R: Here we calculate the value KTBF of item 9Q plus 10.0 dB. This value in dBm then is the FDM/FM (Frequency Division Multiplex/Frequency Modulation) information or TV Frequency-Modulated receiver threshold, or what is known as the Frequency Modulation Improvement Threshold. It might logically be questioned at this point as to why the receiver threshold isn't the same as the noise level or KTBF value previously calculated. Well, as we know, an FM (Frequency Modulation) detector is a device based upon peak signal or noise values. The peak factor (ratio of peak to rms value) of random noise is normally taken as 13.0 dB while the peak factor of the signal (a sinusoidal wave) is the well known 3.0 dB. In order, then, for the peak noise to equal the peak signal, the rms signal must obviously have an amplitude 10.0 dB greater than the random noise. Should the signal level be below this value, it would rapidly deteriorate with the signal's becoming useless. It is to be pointed out that the threshold for considering outage is not always the "KTBF + 10.0 dBm" in FDM/FM or Frequency-Modulated TV microwave systems. This value of threshold provides generally marginal service. Please see Appendix 23. The microwave path design engineer must decide his "outage" point, which, as can be seen, might be anything from a signal a few dB down to catastrophic failure of the microwave system. Sorry about that, but that's the way the cookie crumbles. Many manufacturers provide curves such as, shown in Appendix 27. But, let's go on with this capricious subject of propagation. We have covered the fade margin to the receiver FMI (Frequency Modulation Improvement) threshold. While applicable to FDM/FM and FM TV microwave, digital radio is quite a different story. In digital microwave, the threshold is based upon BER (Bit Error Rate). Typical digital microwave threshold values might range from some −80.0 dBm to −90.0 dBm for a BER of 10^{-6} (one bit error per a million bits) to some −65.0 dBm to −73.0 dBm for a BER of 10^{-7} (one bit error per ten million bits). These values are available, for planning purposes from reputable manufacturers. See Appendix 24 entitled, "Digital Microwave".

Item 9S: To estimate the microwave path Propagational reliability, we take the algebraic difference between item 9O (the receiver median input signal of −37.0 dBm) and item 9R (the receiver FMI threshold of −79.1 dBm). This gives a fade margin of 42.1 dB. This appendix item shows a short term (fast) fading propagational reliability in of 99.99% for WRH (within radio horizon) paths for single receiver (non-diversity) and in excess of 99.999999% propagational reliability for dual receiver (diversity). This kind of propagational reliability should be adequate for just about any type of service; however, we must make certain to attain the specification value. Please be aware that Appendix 25 shows generalized curves of propagational reliability as a function of fade margin versus path length. They should be adequate for many locations. A few locations might show worse and some might show better fading characteristics (please see Appendix No. 22 entitled "Fading Distributions"). The fade margins shown in Appendix 25 are of the Rayleigh type "quasi-worst" case. We reiterate, however, that the propagational "woods" are full of various fading curves and scientific reasoning therefore. Again, propagation, and especially fading, is not a neat phenomenon and frequently requires much judgement.

Item 9T: This is a very interesting and complex item also. Basically, we know that classically a radio wave may consist of a direct ray, as between antennas, and a terrestrially reflected ray. We also know that if these rays arrive at the receiving point in unfavorable phase/ampltiude relationships that the signal received may be seriously attenuated by the process of cancellation. It is therefore necessary to have the best available data as to location and nature of the reflection points (actually areas) along a given microwave path or path leg and the corresponding terrestrial reflection coefficients. Please see Fig. 5-7. In addition to the other data shown therein, as already mentioned, we note that the terrestrial reflection ray is shown for each of the three k-factors considered. The reflection point location varies as a function of k-factor. To satisfy the condition for a reflection ray, the terrestrial angle of incidence must equal the angle of reflection. The three reflection rays shown are, of course, the same one, only under differing k-factor conditions. At this office stage of planning, it is often difficult to estimate the terrestrial reflection coefficient since map topography may be in error and other information (such as, plant growth, land slides, flooding, new building construction, tree heights, whether or not trees are deciduous, etc.) may not be available, possibly leading to false conclusions and serious error. Suspect reflection areas may, however, be pointed out. In the obvious case of over-water paths, salt-flat paths, or some desert paths, reasonable estimates might be ventured and could be sufficient cause for office elimination of a path, thus saving considerable time and expenditure by removing from the field phase consideration of a poor-risk path.

Since this subject is actually more meaningfull when considered in content with an actual field-obtained mcrowave path profile and other field investigations, it is planned to treat it in greater depth in the following fieldphase of microwave path selection.

However, it might be mentioned at this point that from our office-phase microwave path map profile (please see Fig. 5-7) our reflection rays are effectively blocked for all three of our considered k-factors (infinity, 4/3 and ½), and so we eliminate this reflection loss from consideration at this phase of path planning, but defer the consideration to actual field-investigation. It might be stated however, that from the profile, it appears/ that there *might* be difficulty from a possible reflection from peak $P_1P_2P_3$ (all the same peak at different k-factors). For the sake of desk-planning, we assume that this peak is a zero reflection coefficient knife-edge. With more detailed field inspection and profiling, we will be in a better position to better judge it. Leg 1 (from terminal A to the passive reflector $B_1B_2B_3$) is treated similarly. The reflected rays are not shown for the sake of illustration clarity.

MICROWAVE PATH SELECTION (FIELD-PHASE)

Our office path profile may be checked by any of several methods, including aerial radar profiling employing an aircraft, civil engineering type measurements, ground and air altimetry and/or combinations of these or any other desired valid methods. The over-riding objective, regardless of methodology, is to obtain the absolutely best available and most accurate data for our terrain profile along the microwave path. This microwave path profile is plotted on the three k-factors as the desk version infinity, 4/3 and ½). A new profile may be drawn or the desk profile version corrected if it did not contain too many errors. Tower heights may have to be adjusted to retain the required Fresnal Zone radius clearance and keep the reflection rays blocked or reflection points (actually areas) at locations of low reflection coefficient. This latter point will be touched upon in more detail a little later on in this field-phase of microwave path selection. But, let's, as we did for the preliminary or office phase, use as a format the worksheet form, "Microwave Path Selection (Field Phase)". Please keep in mind that there is nothing sacrosanct about any of the forms proposed in this book and that you, the reader, may well, after a little practice, modify them to more conveniently fit your purpose and modus operandi. Let's go then to the first item.

Item 1: Here we again identify the path since the office-phase forms and the field-phase ones will form a package. The A end as before is Podunk, the B point is the passive repeater at Mt. Gunnysack and the C end is Kokomo. As before, geographical coordinates may also be recorded here as necessary and/or desired.

Item 2: Herein we record the rechecked path and path segment (leg) lengths. As shown in the field-phase worksheet form, Figure 5-8A, this is the same as for the desk or office phase, no new or better information having been obtained in the field.

Item 3: Here are recorded the profile data sources. As previously, we record the maps used and as additions we have employed aircraft radar profiling, ground altimetry, aircraft altimetry, and theodolite. Aircraft radar profiling is basically just that, an electronic package accurately measuring the altitude along the microwave path as flown in an aircraft. As can be seen, this renders data on the path topography and thus gives a profile. Ground altimetry is also just that, measuring by accurate altimeters, various heights or altitudes along the path to confirm results. Aircraft altimetry is generally of the "fly-by" type and while not as accurate as the ground altimetry, does render additional inputs and cross checks on the heights of topographical features along the microwave (or VHF or UHF) path. Theodolite type measurement is also not-infrequently employed for checking the heights and accurate locations of possible path obstructions and other path topography. Again, all possible methods at one's command are brought to bear to produce the most accurate profile possible.

Item 4: Here we list the k-factors used. These remain as on the preliminary office-phase worksheet form unless field-work turns up better information.

Item 5: This item is used as a check of the available "raw" clearance. That is without any antenna-elevating towers at the worst obstruction. Let'assume that Fig. 5-7 also represents final field-obtained profile. It will be obvious from the profile that the worst obstruction for the Mt.

```
                    MICROWAVE PATH SELECTION (Field-Phase)
ENGR Peter N. Saveskie              DATE _____
ENGR_____✓_____
```

1. MICROWAVE PATH

a) "A" END (NAME AND COORDINATES) Podunk.
b) "B" END (NAME AND COORDINATES) Mt. Gunnysack
c) "C" END (NAME AND COORDINATES) Kokomo
d) NAMES AND COORDINATES OF LEG ENDS (IF ANY) _____
   Mt. Gunnysack

2. PATH DISTANCE (MILES):
LEG #1 __5.0__ LEG #2 __30.0__ LEG #3 ✓ LEG #4 ✓

3. PROFILE DATA SOURCES:
MAPS (TYPE, SCALE, NUMBER, ETC.) MAPS, X, Y, Z. etc.

RADAR PROFILE __Yes__ GROUP ALTIMETRY __Yes__ AIRCRAFT ALTIMETRY __Yes__
OTHER THEODOLITE _____

4. K-FACTORS USED
a) __Infinity (∞)__ b) __4/3__ c) __½__

5. EXISTING RAW CLEARANCE CONSIDERING WORST OBSTRUCTION (ANTENNAS ON GROUND):

LEG. NO.	K-Factor a. = ∞			K-Factor b. = 4/3			K-Factor c. = ½		
	FEET	FRESNEL RAD. AT 6.725 GHz	FRESNEL RAD. AT N/A GHz	FEET	FRESNEL RAD. AT 6.725 GHz	FRESNEL RAD. AT N/A GHz	FEET	FRESNEL RAD. AT 6.725 GHz	FRESNEL RAD. AT N/A GHz
Leg #1	BLOCKED		N/A	BLOCKED		N/A	BLOCKED		N/A
Leg #2			N/A			N/A	≅ 30.0	> 0.6	N/A
Leg #3	N/A →		N/A →						
Leg #4	N/A →		N/A →						

6. RECOMMENDED CLEARANCE:
a) FREQ. __6.725__ GHz FRACTION OF FRESNEL RAD. __0.6 First__ AT K-FACTOR __½__.
   THIS GIVES __≅ 8__ th + FRESNEL ZONE RADIUS CLEARANCE AT __4/3__ K-FACTOR.
   AND GIVES __≅ 18__ th + FRESNEL ZONE RADIUS AT __∞__ K-FACTOR.
b) FREQ. __N/A__ GHz FRACTION OF FRESNEL RAD. __✓__ AT K-FACTOR __✓__
   THIS GIVES __✓__ FRESNEL ZONE RADIUS CLEARANCE AT __✓__ K-FACTOR.
   AND GIVES __✓__ FRESNEL ZONE RADIUS CLEARANCE AT __✓__ K-FACTOR.

7. ANTENNA HEIGHTS (OFF GROUND) CORRESPONDING TO STEPS 6 a) AND 6 b) ABOVE:

STEP 6 a)   STEP 6 b)

"A" END ANTENNA HFT. (FEET) __100.0__   "A" END ANTENNA HGT. (FEET) __N/A__
"C" END ANTENNA HFT. (FEET) __50.0__   "B" END ANTENNA HFT. (FEET) __N/A__

8. REFLECTION POINTS FOR K-FACTORS UNDER CONSIDERATION:

LEG #1

	K-FACTOR	REFLECTION POINT LOCATION	REFLECTIONS BLOCKED (Yes or No)
FREQ.	K-Factor a = ∞	20.5 Miles	YES
	K-Factor b = 4/3	20.0 Miles	YES
GHz	K-Factor c = ½	19.5 Miles	YES

	K-FACTOR	REFLECTION POINT LOCATION	REFLECTIONS BLOCKED (Yes or No)
FREQ.	K-Factor a = ✓	N/A	N/A
	K-Factor b = ✓	N/A	N/A
GHz	K-Factor c = ✓	N/A	N/A

LEG #2

	K-FACTOR	REFLECTION POINT LOCATION	REFLECTIONS BLOCKED (Yes or No)
FREQ.	K-Factor a = ∞		YES
	K-Factor b = 4/3		YES
GHz	K-Factor c = ½		YES

	K-FACTOR	REFLECTION POINT LOCATION	REFLECTIONS BLOCKED (Yes or No)
FREQ.	K-Factor a =	N/A	N/A
	K-Factor b =	N/A	N/A
GHz	K-Factor c =	N/A	N/A

Fig. 5-8. Path selection forms for the field phase.

LEG #3

FREQ.			
	K-Factor a = ___	N/A	N/A
GHz	K-Factor b = ___	N/A	N/A
	K-Factor c = ___	N/A	N/A

FREQ.			
	K-Factor a = ___	N/A	N/A
GHz	K-Factor b = ___	N/A	N/A
	K-Factor c = ___	N/A	N/A

LEG #4

FREQ.			
	K-Factor a = ___	N/A	N/A
GHz	K-Factor b = ___	N/A	N/A
	K-Factor c = ___	N/A	N/A

FREQ.			
	K-Factor a = ___	N/A	N/A
GHz	K-Factor b = ___	N/A	N/A
	K-Factor c = ___	N/A	N/A

9. FOR REFLECTIONS WHICH ARE NOT BLOCKED, INDICATE REFLECTION POINT LOCATION DESCRIPTION OF REFLECTION POINT TERRAIN AND REFLECTION COEFFICIENT REGARDING ALL PATH LEGS, FREQUENCIES, AND K-FACTORS SHOWN. ALSO INDICATE ADDITIONAL NECESSARY DECIBEL FADE MARGIN SHOULD UNBLOCKED REFLECTIONS EMINATE FROM EVEN FRESNEL ZONE WITHIN RANGE OF CONSIDERED K-FACTORS.

LEG #1

	K-FACTOR RANGE	REFLECTION POINT LOCATION	DESCRIPTION OF REFLECTION POINT TERRAIN	REFLECTION COEFFICIENT	ADDITIONAL NECESSARY FADE MARGIN (dB)
FREQ. GHz	K-Factor a = ___	BLOCKED	BLOCKED	BLOCKED	BLOCKED
	K-Factor b = ___	BLOCKED	BLOCKED	BLOCKED	BLOCKED
	K-Factor c = ___	BLOCKED	BLOCKED	BLOCKED	BLOCKED
FREQ. GHz	K-Factor a = ___	N/A	N/A	N/A	N/A
	K-Factor b = ___	N/A	N/A	N/A	N/A
	K-Factor c = ___	N/A	N/A	N/A	N/A

LEG #2

FREQ. GHz	K-Factor a = ___	BLOCKED	BLOCKED	BLOCKED	BLOCKED
	K-Factor b = ___	BLOCKED	BLOCKED	BLOCKED	BLOCKED
	K-Factor c = ___	BLOCKED	BLOCKED	BLOCKED	BLOCKED
FREQ. GHz	K-Factor a = ___	N/A	N/A	N/A	N/A
	K-Factor b = ___	N/A	N/A	N/A	N/A
	K-Factor c = ___	N/A	N/A	N/A	N/A

LEG #3

FREQ. GHz	K-Factor a = ___	N/A	N/A	N/A	N/A
	K-Factor b = ___	N/A	N/A	N/A	N/A
	K-Factor c = ___	N/A	N/A	N/A	N/A
FREQ. GHz	K-Factor a = ___	N/A	N/A	N/A	N/A
	K-Factor b = ___	N/A	N/A	N/A	N/A
	K-Factor c = ___	N/A	N/A	N/A	N/A

LEG #4

FREQ. GHz	K-Factor a = ___	N/A	N/A	N/A	N/A
	K-Factor b = ___	N/A	N/A	N/A	N/A
	K-Factor c = ___	N/A	N/A	N/A	N/A
FREQ. GHz	K-Factor a = ___	N/A	N/A	N/A	N/A
	K-Factor b = ___	N/A	N/A	N/A	N/A
	K-Factor c = ___	N/A	N/A	N/A	N/A

10. ACCEPTABILITY OF CONSIDERED PATH:

    FREQ. (GHz) _____ [YES] ACCEPTABLE ☐ NOT ACCEPTABLE

    FREQ. (GHz) __N/A__ ☐ ACCEPTABLE ☐ NOT ACCEPTABLE

    (IF NOT ACCEPTABLE, INDICATE REASONS THEREFORE UNDER REMARKS)

11. REMARKS AND CLARIFYING NOTATION:

ENGINEER'S SIGNATURE _____

Fig. 5-8 continued from page 174.

Gunnysack/Kokomo leg is point P, (i.e. $P_1 P_2 P_3$ for the considered k-factors). Drawing a straight line (not shown here) from B_1 to the base of tower C_1, from B_2 to the base of tower C_2, and also from B_3 to the base of tower C_3, we find that the raw clearance is around 30.0 feet at ½ K-factor which is less clearance than 0.6 first Fresnel zone radius. We record this information as shown in this item. Drawing similar straight lines from the base of the tower at point A (Podunk) to B_1, B_2, and B_3, we find in all three k-factor cases that the direct radio ray would be blocked and we indicate this information accordingly as shown. Note that there is room on the form for additional path legs if needed. In our case, these were not needed so we marked them N/A (Not Applicable). There is also space reserved for another frequency in case more than one frequency is considered (recall that Fresnel Zone clearance is a function of operating frequency). This we have also marked as not applicable (N/A).

Item 6: Recommended Clearance. Providing that your field investigation turns up nothing new, this remains the same as for the office phase. Let's assume that this is actually the case, which it often is. First we record the operating frequency then the requisite clearance at some k-factor. In our case, this is 0.6 First Fresnel Zone Radius at a k-factor of ½. Next we measure the clearances in feet (in this case since we are working with the British system for the K-factors of 4/3 and ½. This comes out to be some 220.0 feet at 4/3 k-factor and some 340.0 feet at infinity k-factor. Substituting in the following formula and solving for n

$$F_n = 2280.0 \times \left(\frac{n\, d_1\, d_2}{fD} \right)^{½}$$

Where: f = frequency in Megahertz.
d_1 = distance from a given point on the path to one path (or path leg) end (Statute miles)
d_2 = distance from this same point on the path to the other path (or path leg) end (Statute Miles)
F_n = n-th Fresnel Zone radius in feet
n = Fresnel Zone number.
$D = (d_1 + d_2)$

we get approximately 9th Fresnel Zone radius for a k-factor of 4/3 and approximately 20th Fresnel Zone radius for a k-factor of infinity (or flat earth). It can thus be seen that in going from the ½ k-factor condition to that of k = infinity that the clearance has passed through several even-zone clearance conditions. It is these even Fresnel Zone clearances which cause difficulty if the ray is reflected from earth (or other surface) to the receiving antenna and might cause serious signal cancellation should a ray be reflected from a high reflection coefficient area. Here again we consider only our single frequency of 6.725 GHz. If another frequency were used in addition to this one, we would work out the problem for it in precisely the same manner. It might be of interest to point out here that a given k-factor, per se, is considered to be essentially the same (well within 0.5 %) from about 0.0 (D.C.) GHz to about 30.0 GHz. Also, when the radio-factor is 4/3, conditions are such that the distance to the radio horizon is about 7.0% greater than the distance to the optical (line-of sight) horizon and approximately 15.0% greater than the geometrical (true straight-line) horizon. This fact is the bane of the simple "line-of-sight" concept. Please see Appendix 8 entitled, "Line-of-Sight".

It is interesting to consider that when the k-factor radio-meteorological conditions are around 4/3, the k-factor for light rays is close to 7/6 to some 6/5. Thus, for a 4/3 radio k-factor on a 30.0 mile hop (path), the mid-path earth bulge is 112.5 feet. So at the "standard" radio-meteorological condition, one might "get away" with equating the two (4/3 = 1.33, 7/6 = 1.17 and 6/5 = 1.2). In a not unusual case, however, the radio k-factor changed to + 0.87 while the visual k-factor remained essentially the same (actually near 1.3). As already shown above, the earth bulge for a 30.0 mile path at 4/3 k-factor is 112.5 feet. The corresponding earth bulge for the + 0.87 k-factor at mid-path becomes 172.0 feet. It can be appreciated that the difference (172.0 − 112.5 = 59.5 feet) may be quite significant when we consider that 0.6 first Fresnel radius, in the 6.0 GHz microwave band, is of the order of 44.0 feet at midpath. In another random case, based upon actual meteorological measurement the k-factor at microwave frequencies computed out at +3.5 while that for vision was calculated to be +1.39. The +3.5 k-factor produces an earth bulge of 43 feet while the optical k-factor produces a bulge of 108.0 feet at midpath for a 30.0 mile path (or hop). In this case the radio clearance was more than the vision (or line-of-sight) clearance. The light-ray and the radio-ray simply do not always see "eye-to-eye" and the term Line-Of-Sight has frequently been misapplied and misunderstood. The reader might compare a radio k-factor with a corresponding visual k-factor. It will be recalled that formulas were given for computing radio refractive index and k-factor. The fundamental difference between radio k-factor and visual k-factor is that the latter is independent of tropospheric water vapor content, being reliant upon only the meteorological caprices of temperature and barometric pressure. Following is the formula for optical (Line-Of-Sight) refractivity:

$$N = (n - 1.0) \times 10^6 = \frac{79.0P}{T}$$

Where: N = Line-Of-Sight Refractivity
n = Line-Of-Sight Refraction Index
P = Barometric pressure in millibars
= Temperature in Kelvins

If the above mentioned horizon lines (viz: radio horizon, optical horizon, and geometrical horizon) were plotted to their true scales upon a spherical earth diagram, only the geometric horizon line would be straight. The radio horizon and optical horizon lines would be arcs. It is convenient and desirable then to rectify the curvatures of these arcs so that they may be represented as straight lines. This was touched upon previously, as will be recalled. The rectification is accomplished, for any given value of dn/dh (tropospheric refractive gradient), by subtracting the corresponding curvatures, thus representing the surface of the earth as a line having a different curvature from normal, while the desired rays become straight lines. The rectified earth curvature may be considered as parabolic with negligible error. It might also be of interest to mention here that since on the usual microwave path profile it is desired to represent distances of some 30.0 to 60.0 miles versus topographical heights or altitudes of only a mile or so, it becomes convenient to expand the height scaling with respect to the path distance. One limitation arising from this horizontal-versus-vertical scale distortion is the obvious non-reality of vertical angle representations, and it can thus be seen that it would be technically unsound to measure reflection angles from a k-factor terrain profile using a regular protractor or similar regular angle measuring device. Please see Appendix 26.

Plotting the earth's surface on a partial parabola rather than a circular arc produces an error of less than 0.3%. Another slight error, of course, arises from the fact that elevations on the profile are plotted along vertical lines rather than along radial lines. This error, however, is small since the angular difference between a vertical and a radial antenna tower, is normally less than 10.0 minutes. It can be appreciated, then, that the inaccuracies involved in employing the parabolic concept for k-factor terrain profiles are negligible, making these principles eminently applicable to microwave path problems.

ITEM 7:   Antenna heights off ground corresponding to steps 6A and 6B above under "Recommended Clearance." The antenna heights, whether achieved by towers, roof-top, or otherwise, must satisfy the most demanding Fresnel Clearance requirements at the worst k-factor expected for your judged amount of time. Please see Appendix 19. As already shown, the required antenna heights are also functions of frequency of operation. Since, for example, 0.3 first Fresnel zone radius clearance would be different for a given k-factor for two different frequencies (say 2.0 and 11.0 GHz), two sets of tower heights are provided for. In our simple case, for the sake of clarity, we do not employ this other frequency and thus item 6B is marked as N/A. At any rate, in accordance with well known trigonometric rules, it should be kept in mind that the antenna height at one end of a microwave hop (path) may be increased while that of the other end of the path is decreased, according to the various physical (and/or other) circumstances, maintaining all the while proper Fresnel clearance in relation to a given path obstruction at a particular k-factor under consideration. For example, tower heights at new installations (and/or upgradings) are not infrequently dictated by the availability of existing towers are existing older installations, and in completing this item these trade-off possibilities must be kept in mind. It's all a matter of meeting the path propagational requirements most economically.

ITEM 8:   Reflection points (actually areas) for k-factors under consideration. As will be noted, provision is made in this item for logging reflection point (actually area) locations as functions of k-factor and frequency and also for noting whether or not reflections are blocked, this for all (if more than one) legs of a path. Contrary to pseudo-technical folk-lore, a terrestrial reflection location is a function of k-factor. This is easily seen on Fig. 5-7 by noting the shift of the reflection location for the same profile for different k-factor plots. The point where in the incident and reflected rays of a reflected wave are equal in angle for given antenna heights and path distances will shift with changing k-factor weather/climatic conditions. As will be seen, the point (actually area) of terrestrial reflection will creep away from the lowest path end (in this case path end C) as the k-factor decreases. The degreee of such shift on a "nominal" microwave path might amount to some 5.0 miles or so between a k-factor value of ∞ (infinity or flat earth) and one producing 0.0 (zero) First Fresnel zone radius (grazing) clearance. The reflection location, as such, is considered frequency-independent, being purely dependent upon ray geometry, except, of course, to the extent that two differing sets of tower (or antenna) heights might have been recommended for two different frequencies under consideration in Item 7. In this item (Item 8) please note that the reflection point locations are given for convenience of reflection point location as distances from the A terminal end in statute miles. It is of interest to note also that the three locations A, B, and C are not "in-line". At any rate, if reflections are blocked, as they are here, they are normally dismissed as being non-harmful regardless of the reflection coefficient at the "would-be" reflection point. For the grazing angles (ray angle of incidence and reflection with the terrain) normally involved in most microwave paths (usually considerably less than 1.0°), we may consider for all practical purposes that a 180.0° phase reversal occurs upon terrestrial reflection for both horizontal and vertical wave polarizations.

All this information now puts us in a position of tying things together and presenting a more integrated picture of this rather capricious subject. Recalling that we defined the first Fresnel zone as an intangible ellipsoidal surface from which a reflected ray arrives at the receiving antenna one-half wavelength later than the direct ray, we might falsely conclude that such a terrestrial reflection (considering first Fresnel zone boundary contiguous at some path point with earth) might outphase the direct ray energy. This, however, is not so, owing to the above mentioned phase-reversal upon terrestrial reflection. The first and all other odd-numbered Fresnel Zones reflections are in close phase relationship with the direct ray and the direct ray and reflected ray components are thus considered additive. The second Fresnel Zone is defined by an ellipsoid of revolution, reflections from which could arrive at the receiving point a full wavelength later. This, of course, would be 360.0° later and might be erroneously considered in phase if we overlook that the reflection incurred 180.0° phase shift took place. Thus, it is easily seen that the "apple-cart" is upset again and we're actually out of phase. All even Fresnel zone terrestrial reflections cause outphasing, and, thus, it may be appreciated why these even zones are considered the bane of microwave propagation, especially with the higher values of terrestrial reflection coefficients. Theoretically, with odd Fresnal Zone reflections, a gain of 6.0 dB may be realized over and above the direct ray alone, whereas with even numbered Fresnel Zone reflections, the signal may be cancelled completely. Actually, however, the vector phase shift upon reflection from terra-firma (or terra-infirma) is not as neat as generally supposed in the classical "text-book" case. The following table gives an idea (not absolute in all cases) of what might occur to a microwave upon reflection from sea water, as a function of grazing angle and wave polarization. As will be seen from this figure (please see Fig. 5-9) the phase shift (reflection coefficient phase) remains practically the same (phase reversed) for horizontal polarization, but rotates from 180.0° through approximately 1.0° for vertical wave polarization, from a grazing angle of 0.0° to 90.0° (parallel to reflecting surface throught perpendicular to reflecting surface).

ITEM 9: For reflections which are not blocked, indicate reflection point (actually area) location and estimated reflection coefficient regarding all path legs, frequencies of operation, and k-factors shown. Also indicate additional necessary fade margin, in decibels, should unblocked terrestrial reflections emanate from even Fresnel zones within the range of k-factors considered. This may sound confusing at first blush, but really is quite uncomplicated taken a "chip"at a time. First of all, if terrestrial reflections are not blocked (in our sample problem they *are* blocked), we immediately know that some of the microwave energy impinging upon the refection point (actually area) will reach the receiving antenna. We also know that as the k-factor changes with weather conditions, even numbered Fresnel zone RF energy may be reflected to the receiving antenna. The remaining question is "How much energy will be reflected to the receiving antenna?" It is here that scrutiny of the terrain and logging of results during the field investigation phase becomes extremely important. It is very important that the location and nature of the terrain features along the microwave path be evaluated (e.g. rough rocky, salt flats, swamps, tree groves, etc. an whether or not the trees are deciduous) among a host of other possible terrain classifications. From these inputs, the microwave propagation engineer is able to assign a reasonably accurate reflection coefficient to any given reflection point (area). Whether trees or other plants are evergreeen or deciduous, to point out only one of the many possible pitfalls, has much effect upon the possible seasonal variation of terrestrial reflection coefficient.

Intimately related to terrestrial reflection coefficient, of course, is the scattering coefficient, and although many radio engineers tend to lump these concepts (since in varying degrees both these parameters tend to abet or inhibit the arrival of reflected energy at the receiving antenna), it should be pointed out that there is a basic difference in their mechanisms. A rough metallic ore area, for example, may be considered, material-wise, an excellent reflector, but being rough and craggy it may also lend itself to scattering. To this extent, the misdemeanor of equating scattering with absorption might not really be punishable by inflicting everlasting high reflection coefficients upon the guilty engineer. There are other cases where, for example, the reflection area could consist of smooth reflective terrain and still be usable due to its tilt causing the reflected ray to strike at a point away from the receiving antenna. Many factors must be taken into account when considering ray reflections. A guide which might prove helpful in deciding whether or not a given terrain reflector is rough or smooth is known as the "Rayleigh Criterion Of Roughness." The "crest-to-trough" terrain height undulations, which allow a classification of roughness, may be calculated from the following formula:

Where: $U = 11.8 \, (\lambda/A)$ Classification of rough in the crest-to-trough terrain undulations
α = Operating wavelength in meters
A = Grazing Angle in degrees

From the above, it might be supposed that microwave reflections for a given height, U, might be somewhat less severe at the higher microwave frequencies. Actually, however, in real life this is not true, again calling for judgement on the part of the propagationist. The reason is that a given reflection area has an effectively greater reflection surface (in terms of square wavelengths) due to the higher frequency and this "mechanism" tends to oppose and cancel the increased terrain roughness. Thus, at this writing, it is generally considered that a given reflection coefficient can be relatively the same between approximately 2.0 and 12.0 GHz.

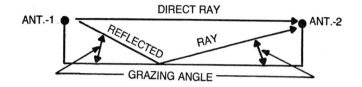

GRAZING ANGLE (Degrees)	APPROXIMATE PHASE LAG.	
	VERTICAL POLARIZATION	HORIZONTAL POLARIZATION
0°	180°	180°
1°	180°	180°
2°	175°	180°
3°	170°	180°
4°	166°	180°
5°	155°	180°
6°	130°	180°
7°	62°	180°
8°	28°	180°
9°	16°	180°
10°	13°	180°
20°	3°	180°
30°	2°	180.5°
40°	1.5°	181°
50°	1°	181.5°
90°	1°	182°

Fig. 5-9. Reflection coefficient phase vs. grazing angle and polarization.

Let's go on a little bit with this reflection coefficient subject. Cancellation of direct rays be reflected rays is at its worst, generally, when the reflecting surface is a clam body of water (especially salt-water) or other areas such as smooth moist earth. Not as widely comprehended, on the other hand, is that the thin tayer of hot air just over the surface of desert sand can be highly destructive during the daytime by returning the incident ray to the receiving antenna. This fact might be distressing when we realize that this hot air stratum may have the tendency of smoothing out (filling-in) roughnesses which may exist in the desert terrain in the form of sand ruts and dunes and to diminish the absorptivity of desert flora (albeit already sparce).

It might also be mentioned here that in addition to ground level reflections, a microwave propagation engineer must worry about possible reflective contributions from such sources as mountain sides along the microwave path, and as necessary, he assesses their slopes, roughnesses, and reflection coefficients. All this may be quite important as the Fresnel clearance ellipsoid is quite impartial as to the direction of arrival of its defilement which could even come from above (as aircraft traffic). Thus, at any rate, it can be seen that the statement "Reflected-Ray Fading (also known as Fresnel Fading) can be extremely important in path engineering" is, if anything, an understatement.

In addition to terrain characteristics already mentioned, the reflection coefficient magnitude may be, in some cases, dependent upon the grazing angle and wave polarization. Taking again the simple (over sea-water) situation as before we present the table in Fig. 5-10. From this figure and Fig. 5-9, it can be seen that at about 6.5° grazing angle, with vertical polarization, there occur two changes. One is a rather dramatic change in the "phase of reflectivity" while the other is a reduction (null) of the reflection coefficient magnitude.

In accordance with the principles of reflection of a plane electromagnetic wave from the boundary between air and an imperfect dielectric of a given inductivity and conductivity, the reflection coefficient for vertically polarized waves may be derived from the Fresnel vitreous reflection equation. This equation (not herein reproduced) shows that the complex reflection coefficient is a function of the conductivity and inductivity of the reflecting surface, the operating frequency, and the grazing angle. This indicates that there is a decrease in the magnitude and a change in phase in the reflected ray with respect to the ray of incidence. When this equation is solved for the reflection coefficient for various grazing angles from 0° to 90°, it is found that at some particular grazing angle, the reflection coefficient magnitude is at a minimum and that the vector phase shift is 90.0°. This point is called the "Pseudo Brewster Angle" which corresponds to the Brewster Angle of optics. If the ground (reflecting agent), however, were a perfectly isotropic dielectric of 0.0 (zero) conductivity, the reflection coefficient magnitude would be (zero) conductivity, the reflection coefficient magnitude would be 0.0 (zero) at the Brewster Angle. As may be inferred from the foregoing, the shape of the magnitude of reflection curve, as a function of grazing angle, would be 0.0 (zero) is not always as neat as the ideal case angle, indicates, however, with the exception of the smooth-sea-water case in which instance the reflection may be considered specular (mirror-like). Experiment has shown, however, to take one randomly chosen rough-sea example, that a deep reflection coefficient magnitude minimum (null) of -0.07 occurs at about 13.0° grazing angle as opposed to theory's -0.3 reflection coefficient magnitude at close to 4.0° grazing angle.

Situations such as these tend to stress the necessity for careful attention to the reflection qualities (and quantities) of the reflecting source. Each case is individual, as may be surmised, there being no neat "Ohm's Law Formula" to plug into many instances and frequently not even a rule-of-thumb for all propagational needs just as castor oil does not provide a panacea for everything from "bellyache" to falling dandruff. Accordingly herein we intersperse basics with the step-by-step "how to". In concluding this treatment on reflections, it may be of interest to note that the reflection coefficient may, for practical purposes, be computed if the difference in received signal strength in decibels (resulting from alternate direct ray combination with reflected adjacent odd and even numbered Fresnel zones) is known. This type of information may be obtained from a height/gain run

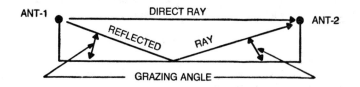

GRAZING ANGLE (Degrees)	APPROXIMATE REFLECTION COEFFICIENT MAGNITUDE	
	VERTICAL POLARIZATION	HORIZONTAL POLARIZATION
0°	−1.0	−1.0
0.1°	−0.98	−1.0
0.2°	−0.95	−1.0
0.3°	−0.92	−1.0
0.4°	−0.89	−1.0
0.5°	−0.86	−1.0
0.6°	−0.82	−1.0
0.7°	−0.80	−1.0
0.8°	−0.78	−1.0
0.9°	−0.74	−1.0
1.0°	−0.72	−1.0
2.0°	−0.53	−1.0
2.0°	−0.39	−0.99
4.0°	−0.28	−0.985
5.0°	−0.16	−0.98
6.0°	−0.07	−0.975
7.0°	−0.07	−0.97
8.0°	−0.13	−0.965
9.0°	−0.18	−0.96
10.0°	−0.21	−0.955
20.0°	−0.5	−9.94
30.0°	−0.7	−0.91
90.0°	−0.8	−0.8

Fig. 5-10. Reflection coefficient magnitude vs. grazing angle and polarization.

or test. In such a run or test the antenna is moved up and down from its normal height while the signal strength is recorded at a receiver in the far (or Fraunhofer) field. The formula for computing the "real live" reflection coefficient is as follows:

$$R = \frac{10^{x_{20}} - 10}{10^{x_{20}} + 1.0}$$

Where:
R = Reflection efficient
x = Received Signal Strengh Decibel Difference between
adjacent odd and even Fresnel zone reflections in combination with the direct ray.

Conversely, of course, the known (or estimated) reflection coefficient may be used in computing the expected decibel Fresnel fading depth.

$$x = 20.0 \, \text{Log}_{10} \left(\frac{1.0 + R}{1.0 - R} \right)$$

Where: R and x have the same meanings as above.

It should be readily apparent that this Fresnel fading mechanism is quite different from tropospheric interference fading or another type called attenuation or power fading. Attenuation or power fading can be caused by trapping of a wave within a duct which does not contain the receiving antenna and also by, excessive path blocking by an obstacle or k-factor earth-bulge. Generally, attenuation or power fading is not frequency sensitive or, as we say, it is "flat". Interference fading is fading resulting from an outphasing combination of the direct and reflected rays and can occur in a complex fashion of fading mechanisms by reflections from the ground or tropospheric layers and also from instantaneous deviations, along the microwave path, of the various rays from the direct ray. This causes a gamut of time lag among the various rays in a moving dynamic and changing manner in accordance with changes in the refractive index, air density, water vapor, temperature, etc. along the microwave path. This latter type of fading is usually of the "fast" variety.

If a microwave path is over water (e.g. lake, bay, etc.), and the geometry of the path is such that a specular (mirror-like) reflection point falls upon the water, we know that extreme fading may occur with minima not infrequently deeper than indicated by a simple two-ray model. This can be demonstrated by a model in which the secondary and/or tertiary (in addition to the two primary) rays are taken into account. Such suspected paths are normally avoided, if possible, other factors being equal, unless the reflections can be eliminated effectively by terrain feature blockage of the reflected wave, at all k-factors occurring in the locality of interest for a given time percentage. Please see Appendix No. 19. Even so, unfortunately, propagation is not infrequently untidy and simple ray analysis calls for much experience to augument it. As an example, experience on one over-water path as reported to CCIR indicated that it was extremely difficult to achieve transmission of complex signals (e.g. color television) under the conditions that all of the water was not completely "invisible" from at least one terminal. It would have to be evaluated how such a condition might affect a less (or more) complex signal. It might be injected here that path-testing is not the simplistic panacea it appears to be either since it is considerably more than a "one-shot" deal. To be meaningful, paths are frequently tested for more than a year. Another complication (there are millions of them, making the art/science of propagation most absorbing, albeit at times exasperating) not-infrequently encountered is the previously mentioned negative k-factor (exceeding infinity). In effect this k-factor passes through positive inifinity, through negative infinity and then to decreasing values of negative numbers. The negative k-factor is known as a concave-earth and represents a radio ray of ray curvature exceeding that of k-factor infinity. It can be seen that such a meteorological condition would produce multiple terrestrial reflection points (areas).

In some areas of the world, such as the Persian Gulf, the k-factor might, during certain months, reach values of some 1/5 (or less) for up to

1.0% of the time. This means that if a path in this radio-meteorological ambient of 30.0 miles is designed for 5/12 k-factor (already considered in some quarters conservative) for a full first Fresnel clearance, the path would be blocked for 1.0% of the total time during these months. For two months of this order of k-factor, this would amount to some 15.0 hours outage, and as might easily be appreciated, an amount upon which the CCIR might look askance. Stick around, it gets worse! Just to attain a grazing clearance on a path of such type might require 750.0 foot towers at each end of the hop (path). Clearly, it would be folly to "fix such hops later" if this can be avoided. We might at this juncture "kill" anoth misconception. It should be kept in mind that no profile, regardless as to how it is obtained (map, aircraft radar, etc.) can be considered valid unless taken in contest with prevailing k-factors. Obtaining a profile and then applying "corrections" at the desk for k-factors is out of the question lest it be discovered that at some k-factor (or k factor range) the required clearance is not attainable with reasonable tower heights. In this vein, we might "put to bed" another propagational old wives' tale. While it has in various quarters in the past been averred that a terrain profile must be centered about the apex of the k-paper parabola, this is not at all true. Any length terrain profile within range of the profile paper used may be plotted anywhere along the path distance axis and have equal validity. Please see Fig. 5-7 in which, for example, the 5.0-mile segment or path leg could have been plotted anywhere along the 40.0-mile path distance scale.

It might also be mentioned in this microwave field-phase portion that the importance of k-factor varies with path distance, affecting the clearance less as the path length becomes smaller. Please see Fig. 5-11. As will be seen, this figure shows the relationship between mid-path earth bulge in feet and path distance in statute miles, with K-factor as a parameter.

Regarding the subject of terrestrial reflection coefficient, see Fig. 5-12 for the graphical presentation of a guide. This graph shows the approximate reflection coefficients of some "real-life" types of terrain surface in microwave work, ranging from theoretical smooth-earth through theoretical knife-edge with corresponding terrestrial reflection coefficients. Your particular type of terrain might be compared to this graph data by process of best fit if it does not coincide with one of the classifications therein. You might even, as your experience grows, contribute to the curve. To use this data, suppose that your reflections could not be blocked by terrain features and that the reflection point (area) is on a smooth-water (for much of the time) salt-water estuary. This, if not a worst-case, would represent at least a very bad case for the microwave propagationist as for all practical purposes this type of reflection surface could be considered specular (mirror-like). From the graph, it can be seen that this might "cost" one some 50.0 decibels of fade margin. On the other end of the scale, very heavily wooded forested terrain might exact only a decibel or two of additional margin. If highly reflective terrain is encountered, space diversity is often employed to overcome such difficulty. Thus, while the signal is outphased at one receiving antenna, hopefully it is at a maximum of phase addition at the other.

That radio propagation and meteorology are so intimately related might militate toward the logic of touching, at this point, upon the later.

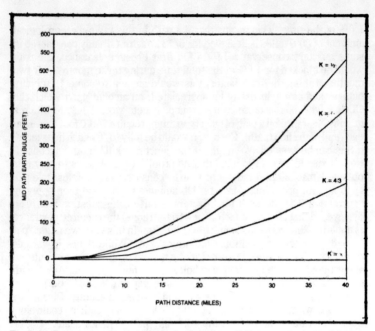

Fig. 5-11. Curves showing the relative importance of k-factor vs the path distance.

The task of weather bureaus, as might or might not be known, is staggering to contemplate. In the U.S.A. the weather bureau's task is to observe and analyze in excess of two billion (two thousand million) cubic miles of atmosphere that envelops the northern hemisphere in an evasively shifting labyrinth, a meteorological mosaic so complex that the bureau once estimated that the area of the United States of America may be simultaneously immersed in some ten thousand varities of weather, each sufficiently significant to be of local concern. In microwave and similar work, one must delve into the local weather-pocket micro-meteorology and not merely in the too often considered gross climatology. As may be seen, propagation is a continuing and on-going study and not merely something "once-learned-totally-mastered." The national, international, public, and private sources utilized by the bureau to accomplish its task is awesome. Throughout the northern hemisphere, some two thousand weather stations of many nations regularly transmit their weather data to national and regional collecting centers four time per day, seven days per week. Approximately five hundred of them are on the continent of North America and the surrounding seas. Some one thousand five hundred are land-based stations around the globe, including stations in China, Siberia, Sahara Desert, Scottish Highlands, Africa, South America, Alaska, and Hawaii. The U.S.N.M.C. (National Meteorological Center) located in Suitland, Maryland receives this vast body of weather information, codes it and feeds it into its array of computers. On an average day, it is said that the N.M.C. also receives approximately 3,200 reports originating from ships throughout the seven seas, some 1,000 reports from commerical aircraft in flight and approxi-

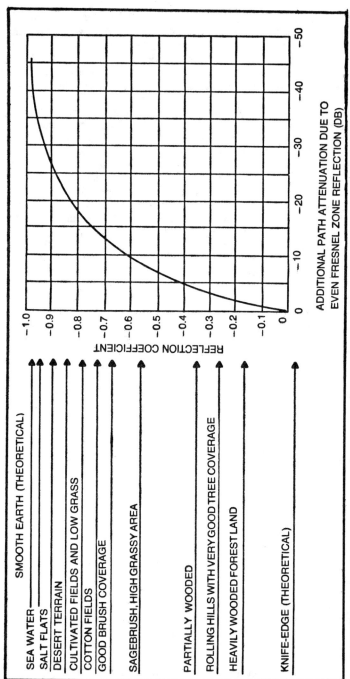

Fig. 5-12. Reflection coefficients for various real-life conditions.

mately 500 stations in the northern hemisphere report on the direction and speed of the wind. At least two times per day radio-sonde balloons (RAOBs) are sent aloft (this was already touched upon briefly in an earlier part of the chapter) over some 100 North American stations alone. These balloons ascend about 2,000.0 feet per minute, transmitting temperature, barometric pressure and humidity data earthward, until they burst at some 65,000.0 feet altitude. The services of the weather bureau are most probably used every day by more people than those of any other single agency, except the U.S. Post Office, of course. The people who earn their living in radio-meteorology are in this user group. Although the modus-operandi of the weather bureau is not exactly suitable to the science/art of radio propagation, it is the only such data available on such a great scale. In addition to the above sources of meteorological data, the weather bureau utilizes orbiting weather satellites, automatic meteorological observation stations some 10,000 climatological stations, about 300 or so government agency aviation stations at airports, some 200 upper air sounding stations, a multitude of radar stattions, military stations the world over, lighthouses and Coast Guard vessels in off-shore waters, approximately 2,000 foreign stations reporting (United Nations World Meteorological Organization), approximately 4,000 part-time stations reporting river stages and weather conditions, some twenty five ocean upper-air stations used fixed and mobile ships, three thousand U.S. and foreign ships reporting by radio and mail, and finally about 400 part-time airport and aviation reporting stations. The world-wide effort is vast indeed!

Yet, as has been touched upon with all the above, the degree of observation and reporting is woefully inadequate, especially as pertains to radio-meteorology-propagation. Even where observation and reporting stations are most dense, it has been said that there are too few to give more than an obscure outline of the radio-meteorological situation. Short of new theories and/or discoveries (not as yet available), current philosophy, which places emphasis upon sheer volume of weather data, seems the only available tool. Debunking, then, another prevailing technical old wives' tale, the desirable quantity and quality of weather data are simply not presently available. More observations and reports performed much more frequently and more unbiquitously are clearly indicated. It should, by now, have been evident to the student propagationist that the radio engineer dealing in radio path design is not particularly interested in averages or medians as he is in k-factor extremes occurring for periods of time commensurate with the outage figure percentages quoted in the various radio propagation reliability requirements, as laid down, for example, by the CCIR. For example, k-factors occurring for 0.001% of the time (or less) would be valuable information indeed to the microwave path design engineer. It would be dangerous to "engineer" a microwave hop/path) for say, 99.9999% propagational reliability without knowing whether that hop could be blocked by earth-bulge for up to 1.0% of the time. Such a 1.0% path blockage could render even a one-hop microwave system only 99.0% propagationally reliable, even under the assumption that the above-mentioned path blockage constituted the only fading mechanism (a rara avis indeed!). Two such hops then could render a propagational reliability of some 98.0%, again assuming that everything else in the system went perfectly. As the above indicates values such as 90.0% and 10.0% occurrence levels, while interesting, are not particularly useful in the design of paths of today's demanded high reliability systems. Extrapolation of refractivity and k-factor information into the wee numbers commensurate with path outages of some 0.0001% while at times the only tack available, can be a somewhat shaky venture since the wings of the curves tend to become somewhat erratic and unpredictable. To be sure it would take an inordinate amount of measurement and data reduction, a time and money consuming undertaking indeed. Extracting data for the curve wings (data extremities) much more costly as much more data must be measured and processed for this purpose to be meaningful. This is one of the reasons that this kind of data is not at this point in time available. While there is a basic lag between the needs and requirements of radio-meteorology and weather study, the gap is being closed and indeed one day it may not be at all uncommon for the microwave radio propagationist designing a microwave (or other) system in any part of the globe, to consult his "k-factor atlas" and from isopleths therein, simply read off the k-factors occurring for, say, 0.1 %, 0.01 %, 0.001 % and 0.0001 % of the time in his geographical area of interest. It is probable that computerization and general access to the computerized data might go a long way toward the realization of this goal.

ITEM-10: In this item, path acceptability, as a function of frequency, is indicated.

ITEM-11: This space is reserved for any clarifying notation and/or remarks. (Don't trust memory, write it down). There may be 50 hops or so in your system, and if you don't keep track of all the information in some convenient form, woe will surely follow you.

Chapter 6
Diffraction Propagation

The word diffraction comes from the Latin diffractus, meaning to break into parts. In propagation work it means the breaking up of an electromagnetic wave into an interference pattern, as when the wave is deflected at the edge of an electromagnetically opaque object which might be anything from a knife-edge type of an obstruction to the earth curvature. Diffraction propagation is not-infrequently employed when direct ray (WRH-Within The Horizon) propagation is blocked or obstructed.

The rules and operations described in this chapter have been found by the writer to produce results which are generally in quite good agreement with measured data. Later on in the chapter specific propagational examples will show how these are applied.

GENERAL RULES

We begin with two general rules for the application of solution procedures in diffraction propagation problems.

General Rule No. 1

When first Fresnel radius clearance obtains in the near vicinity of both receive and transmit antennas but does not obtain due only to the obstruction(s), the general rule is to use Free Space Loss for the entire path plus diffraction loss for each individual obstruction. Sum all these dB losses to obtain the total median loss for the path. Do NOT, in the case of multiple obstructions along the path, construct an "equivalent single knife-edge" for use in your computations as this tack tends to produce erroneous path loss results (too little path loss). The meaning of this will become clear as we proceed with the examples herein.

General Rule No. 2

When first Fresnel radius clearance does not obtain in the near vicinity of either or both terminal antennas, and/or anywhere along the obstructed path (the obstruction excepted), the general rule is to use plane earth loss for the entire path and obstruction loss in addition to plane earth loss. In the case of multi-obstruction paths DO construct and utilize an "equivalent single obstruction." This will become clear as we proceed with the examples herein.

OPERATIONS FOR DIFFRACTION PROBLEMS

We shall present a series of operations from which certain combinations will be selected for specific types of diffraction propagation problems. The reader should become familiar with them. Please take care to select in all cases the correct applicable heights (h, h_1, h_2, H, H',) and distances (d, d', d_1, d_2, etc.) as indicated in the following operations and instructions.

It might be mentioned here that it would be worthwhile to do your nomogram work on a thin transparent plastic sheet laid over the nomogram. Mark the nomogram with grease pencil so that at the end of work it may be easily erased with a piece of cloth and reused. Writing directly on the nomograms will quickly ruin them.

Operation No.1

Operation number one is the determination of the validity of the diffraction solution in OTH radio paths. On a given OTH (Over The Horizon) radio path which is blocked, we first decide whether or not the diffraction solution is probably valid, since in the post diffractive region of a radio path the tropospheric forward scatter propagation mechanism might be the ruling one. This requires a completely different approach and solution. We shall treat this tropospheric scatter propagation mechanism in a later chapter, as well as under certain conditions the applicability of a combined value of both the diffraction and forward scatter solutions. In the above mentioned post diffractive radio path regions, the diffraction solution might render erroneous results (excessive path loss). To determine whether or not our OTH path is probably diffractive, we employ the most pessimistic result of the two following methods.

Method-1 the U.S. Government NBS (National Bureau of Standards) Report 6767, entitled "Ground Performance Standards part 5 of 6," points out that in a preponderance of OTH (Over The Horizon) paths having a path angular distance of at least 0.02 radians, or 20.0 milliradians (slightly in excess of 1.0°), the diffractive solution would most probably render an incorrect solution. So, according to this, OTH paths having an angular distance below this amount should be amenable to the diffraction solution.

Method-2 U.S. Government document ERL-ITS-67 indicates that when the product of the actual path length in kilometers and the path angular distance in radians exceeds the value of 0.5, the diffraction calculation would not likely produce correct results.

Here, then, we have two sources indicating the limits of the diffraction solutions.

To determine whether or not either of these quantities violates our aforementioned criteria (path angular distance exceeding 0.02 radians or the product of this angle in radians and the total path distance in kilometers exceeding 0.5), we first determine the path angular distance, Θ. This is accomplished as follows (follow along on Fig. 6-1). The illustration in this figure shows in clear "skeleton" form the antenna heights of both terminals of the radio path. These antenna heights are shown as h_1 and h_2. Also shown is the height of the path obstacle obstruction, H. All heights in *this* figure are above mean sea level. Also shown are total path distance, d, and path segment distances, d_1 and d_2 from the antennas to the path obstacle. Also shown is the path angular distance, Θ. To compute this path angular distance, Θ, we use the following formula:

$$\Theta \cong \frac{H - h_1}{d_1} + \frac{H - h_2}{d_2} + \frac{d}{1.7 \times 10^7}$$

Where: Θ = Path angular distance (Radians)
 H = Obstacle height above mean sea level (Meters)
 h_1 = Antenna 1 height above mean sea level (Meters)
 h_2 = Antenna 2 height above mean sea level (Meters)
 d_1 = Distance from obstacle to Antenna 1 (Meters)
 d_2 = Distance from obstacle to Antenna 2 (Meters)
 d = Total inter-antenna path distance (Meters)

Now, if you prefer a graphical method, use Figure 6-1. Please keep in mind, however, that in using this graph the distances are all in kilometers while the heights h_1, h_2, and H, as before remain in meters. The angular distance, Θ, is in radians. Use the graph as follows:
Enter the bottom distance scale with your distance d_1 and proceed vertically to your $(H - h_1)$ value in radians. All distance and height values are, of course, from the path terrain profile of your particular propagational problem. From the (H-h_1) point, proceed horizontally to the left and read the $(H - h_1/d_1)$ value in radians. Record this value. Next, enter the bottom distance scale with distance d_2 and proceed vertically to your calculated $(H - h_2)$ value. From here proceed horizontally to the left and read off the formula $(H - h_2/d_2)$ term value in radians. Record this value also. Lastly, enter the bottom distance scale of Fig. 6-1 with the total radio path distance, d, and proceed straight down to read the formula's third term value in radians. Also record this value. Now, simply add up all three of the recorded values. This sum, then, is the value of your path angular distance, Θ, in radians.

Let's take an example. From our radio path terrain profile, we have the following values:

Obstacle height above mean sea level, H = 200.0 meters
Antenna 1 height above mean sea level, h_1 = 10.0 meters
Antenna 2 height above mean sea level, h_2 = 30.0 meters
Path segment distance, d_1 = 10.0 kilometers
Path segment distance, d_2 = 50.0 kilometers
Total path distance, d = 60.0 kilometers

Substituting in the formula we obtain $\Theta = 0.0259$ radians. Working out the problem graphically we obtain $\Theta = 0.026$ radians. This is, of course, 26.0 milliradians—good agreement.

So we see that by method 1 for deciding a blocked radio path's suitability for the diffractive solution, our path at best would be a borderline case, possibly becoming a tropospheric forward scatter problem or lending itself to a combination diffraction/tropospheric scatter solution.

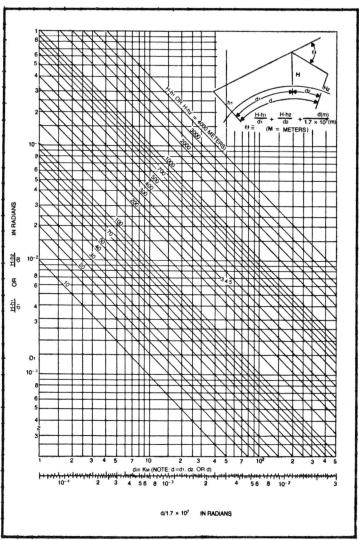

Fig. 6-1. Graphical means for solving for the path angular distance in a diffraction propagation problem (courtesy Radio Research Laboratories, Japan).

Now, let's try it with method 2. Please recall that in this method we multiply the path angular distance in radians by the total path distance in kilometers, thus:

$$\Theta - d = 0.026 \times 60.0 = 1.56$$

So again we see that the chances for a valid diffractive solution for our path dwindles to practically zero, since our Θd product exceeds 0.5. Since this

193

chapter is on diffraction, however, let's assume that our path passed the test and continue on. The methodology is the important item.

(Note: The above methods of deciding whether or not an OTH (Over The Horizon) path is diffractive or one of tropospheric scatter (or a combination of both) is only approximate and in case of doubt, both solutions, the diffraction one and the to-be-described tropospheric scatter one should be performed. If the difference between the two solutions is 18.0 dB or more, then the solution rendering the least path attenuation is to be chosen. Should the difference be less than 18.0 dB, then both solutions are combined according to Appendix 30, entitled "Nature's Combiner.")

Operation No. 2

Operation number 2 is the determination of the applicability of the diffraction knife-edge solution in obstacle blocked paths. If it is determined by operation number 1 that the diffraction solution is applicable, we must decide whether or not a prominent path obstacle (such as a mountain ridge)—or more than one such obstacle—lends itself (themselves) to the ideal knife-edge (touched upon later in this chapter) solution. The sharpness of an obstruction, and therefore its qualification as a radio knife-edge, is a complex function of the operating wavelength and the path angular distance, Θ. This operation is performed simply by first determining the radius of curvature of the peak of the obstruction by plotting the obstruction's peak on scratch paper, using the same scales (feet, meters, etc.) for distance and height (as opposed to using different distance and height scales as in normal k-factor paper). Do not be overly concerned if you do not come up with a perfect circle. Simply make your radius determining circle a best fit to the actual topography of the obstruction peak. The effect of moderate surface roughness of the obstruction peak is generally to reduce the loss below the amount accruing from a smooth, rounded obstacle peak of the same radius of curvature. Next, simply calculate the quantity:

$$A = 0.002 \times \frac{L}{\Theta^3}$$

Where: A = Limiting radius of curvature (Meters)
L = Operating Wavelength (Meters)
Θ = Path Angular Distance, Θ, (Radians) as determined in operation No. 1.

If you prefer a graphical method, consult Fig. 6-2. Simply enter Fig. 6-2 with your diffraction angle, Θ, in radians, slide to the right to the operating frequency in Megahertz. At this intersection, drop straight down and read off your limiting radius of curvature, A, in meters. Now compare your limiting radius of curvature, A, with the actual radius of curvature of the obstruction(s) in your path. If the actual radius of curvature is smaller than the limiting value, your path obstacle peak is a knife-edge and the knife-edge solution may be employed.

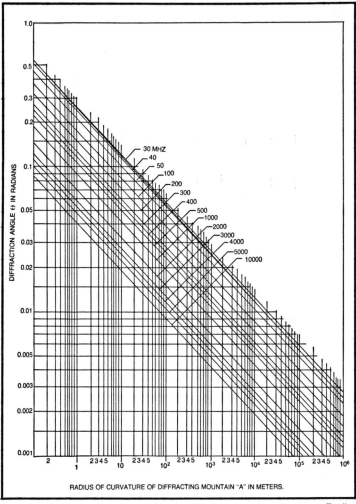

Fig. 6-2. Graph to solve for the limiting radius of curvature (courtesy Radio Research Laboratories, Japan).

Let's take an example:

Operating Wavelength, L,	=	2.0 Meters (150.0 MHz)
Path Angular Distance, Θ,	=	0.01 Radians
Obstacle-Peak Radius of Curvature	=	3,200.0 Meters

Determine the obstacle peak's suitability for the knife-edge diffractive solution. Substituting in the formula, we obtain:

$$A = 0.002 \times L/\Theta^3 \ = 0.002 \times 2.0/0.000001 \ = \ 4,000.0 \text{ Meters}$$

Working the example graphically as per above instructions, we also obtain 4,000.0 Meters.

Now, since the radius of curvature of our actual obstruction peak is 3,200.0 meters and our limiting value is 4,000.0 meters, our actual value is smaller than our limiting value and thus qualifies for the knife-edge solution. If it did not so qualify, there would be an additional propagational path-loss penalty called "Roundness Loss" as described in operation No. 3. This "Roundness Loss" would be *in addition* to knife-edge loss.

Operation No. 3

Operation number 3 is the determination Of Obstacle-Peak Roundness Loss. As indicated in Operation No. 2, should the path obstacle peak radius of curvature be too large to qualify as a radio knife-edge, we must suffer an extra path loss which we call, "Roundness Loss". This extra loss is determined as follows:

a. Calculate the quantity $(6.28 \times r/L)^{1/3} \times \Theta$

Where: r = Actual obstacle peak radius of curvature (Meters)
L = Operating Wavelength (Meters)
Θ = Path Angular Distance (Radians)

b. Enter the graph of Fig. 6-3 with the value determined above and read the corresponding additional loss (in excess of knife-edge loss) produced by the obstacle's roundness (departure from knife-edge).

The roundness loss is determined in step b above, we repeat, is in addition to the knife-edge loss. This roundness loss (dB) is then summed with the knife-edge loss to obtain the total loss for the obstacle.

Let's take an example:

Path Angular Distance, Θ, = 0.01 Radians
Operating Wavelength, L, = 2.0 Meters
Obstacle Radius Of Curvature, r, = 6,000.0 Meters

Substituting in the formula, we have:

$$(6.28 \times r/L)^{1/3} \times \Omega = (6.28 \times 6,000.0/2.0)^{1/3} \times 0.01 = 0.27$$

Now, entering Fig. 6-3 with the above 0.27 value (horizontal scale), we obtain an obstacle-peak roundness loss of close to 3.0 dB. This is to be added to the knife-edge loss for the total loss caused by the obstacle.

Operation No. 4

Operation number 4 is the determination of Obstacle Transverse Profile Loss. In addition to the regular longitudinal terrain path profile (from transmitting antenna to receiving antenna), it is advisable to also consider the transverse profile of an obstacle ridge, since it often can affect the total amount of obstacle diffraction loss. The transverse terrain profile of the

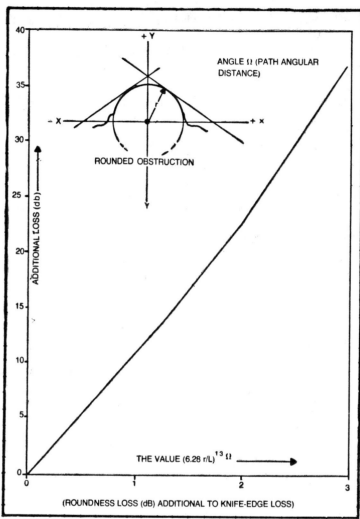

Fig. 6-3. Graph to solve for the roundness loss which is in addition to knife-edge loss.

obstacle is the profile perpendicular to, or crossing, the longitudinal one. If the obstacle transverse profile does not run exactly perpendicular to the longitudinal one, use its projection to the perpendicular. To make this transverse profile loss calculation, perform the following.

    a. Visualize the direct radio ray as passing perpendicularly through the page of Fig. 6-4, at origin, 0. From the best means available (good maps, aerial radar profile, etc.) plot the obstacle's transverse terrain profile.

    b. Calculate the ordinate, y, values from the formula

$$y = \pm \frac{Ld}{2h}$$

197

and sketch in the corresponding horizontal dotted lines intersecting the transverse profile as shown.

Calculate the abscissa, x, values from the formula

$$x = \frac{\pm (Ld)^{1/2}}{2}$$

and sketch in the corresponding vertical dotted lines intersecting the transverse profile as shown.

Where: L = Operating Wavelength (Meters)
 d = Total actual path distance (Great circle path) (Meters)
 h = Obstacle height (vertical distance in meters to origin, 0 from a straight line drawn between antennas)

d. The intersection of the horizontal dotted lines (as determined in step b above) with the transverse profile generates a length, E, on the x axis. Similarly, the intersection of the vertical dotted lines (determined in step c above) with the transverse profile generates a length, F on the x-axis. Now, should length E be smaller than the length F, then an additional obstacle loss would accrue. This additional obstacle loss is computed by:

$$B = 10 \, Log_{10} \, (E^2/dL) \quad \text{(dB)}$$

Where: B = The additional obstacle loss (dB)
 E = Length determined in step, d, above (Meters)
 d = Total actual radio path distance (Meters)
 L = Operating Wavelength (Meters).

Let's take an example:

Operating Wavelength, = 0.5 Meters (600.0 MHz)
Total Path Distance, d, = 50.0 kilometers (50,000 Meters)
Obstacle height to origin, 0, = 100.0 Meters
From a straight line drawn between antennas.

The solution is as follows: Solving for y and x, we have:

$$y = \pm (Ld/2h) = \pm 0.5 \times 50,000.0/2.0 \times 100.0$$
$$y = \pm 125.0$$

$$x = \pm (Ld)^{1/2}/2.0 = \pm (0.5 \times 50,000.0)^{1/2}/2.0$$
$$x = \pm 79.0$$

From Fig. 6-4, we can deduce that since y = ± 125.0, E = 250.0 Meters. Also, since x = ± 79.0, then F = 2.0 × 79.0 or 158.0 meters. So since E is 250.0 meters and F is 158.0 meters, E is not smaller than F and therefore no additional loss (in this case) accruing from transverse profile inclination occurs.

Let's take the methodology to its final phase now. Let's calculate the dB loss from the hypothetical (not above example) case shown in Fig. 6-4.

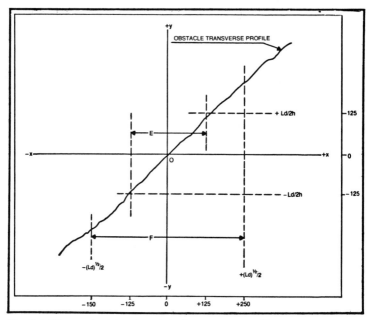

Fig. 6-4. Diagram to determine obstacle transverse profile loss.

Here we see that F is actually greater than E (1.0/1.0 scales). Thus, E is approximately 250.0 meters and F is approximately 500.0 meters. In this case, E is smaller than F and therefore an additional loss due to obstacle transverse profile inclination would occur. This would, as already pointed out, be calculated as follows:

$$B = 10 \text{ Log}_{10} (E^2/dL) \text{ (dB)}$$
$$B = 10 \text{ Log}_{10} (250.0^2/50,000.0 \times 0.5)\text{(dB)}$$
$$B = 10 \text{ Log}_{10} 2.5 \text{ (dB)}$$
$$B = 4.0 \text{ (dB)}$$

Operation No. 5

Operation number 5 is Free Space Propagation. Free-space propagation, according to one radio electronics dictionary, is radio energy propagation via a straight line path in a vacuum, removed from objects or material possibly affecting it. While this "ideal" concept may never be attained in real life (please see Appendix No. 7 entitled, "Free Space"), it does, not unlike the isotropic antenna (another bit of technical fiction), have a basic use in the radio propagation science/art.

Accordingly, we attach hereto in operation no. 5, Fig. 6-5A and 6-5B. The use of these figures is simplicity itself. Note that Fig. 6-5B is simply a metric version of Fig. 6-5A. Taking Fig. 6-5A first, we draw a line between the path distance in miles, of scale 1 and the operating frequency of scale 3. Extending this straight line through scale 4 then intercepts the received signal power in decibels below one watt. Since 1.0 watt is equal to 0.0 dBw (zero dB referred one watt), this scale is then automatically in dBw. This

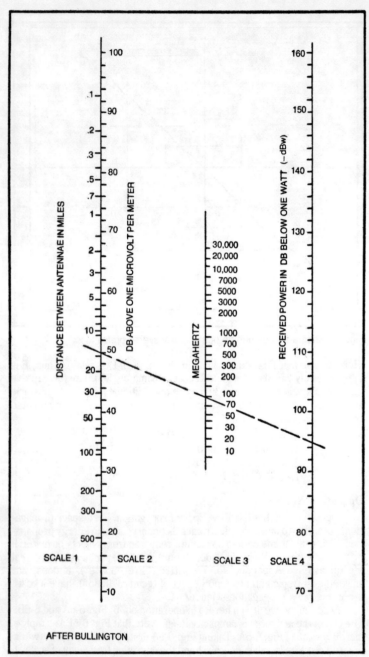

Fig. 6-5A. Nomograph for free space field intensity and received power between half-wave dipoles with 1 watt radiated (courtesy Bell Laboratories).

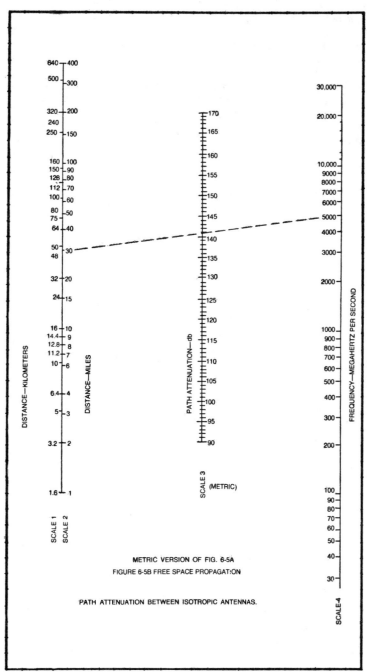

Fig. 6-5B. Metric version of Fig. 6-5A.

nomogram assumes one watt into the transmitting antenna, which in this case is a half-wave dipole. The signal is also assumed to be received by a similar half-wave dipole. Actually, then, this nomogram gives, under these conditions, free-space loss of the radio path in dB. Scale 2 of Fig. 6-5A as an "extra" shows the field strength in dB above 1.0 microvolt per meter at various distances from the transmitting half-wave dipole when one watt of radio frequency power is fed into it. Should you desire to convert this field strength to dBm delivered from an isotropic receiving antenna, consult Appendix No. 2 entitled,"DBuV/m to dBm (and vice versa) Conversions." As shown as an example in Fig. 6-5A, a path distance in free space of 14.0 miles (statute) renders, at a frequency of 100.0 MHz, a free-space path loss of approximately 95.0 dB. Now, as we have mentioned above, Fig. 6-5B is a metric version of Fig. 6-5A. The scales are somewhat differently arranged, but the same general rules hold.

Let's take the same example on both Fig. 6-5A and 6-5B and see what happens.
    a. Radio path length = 62 statute miles $\cong$ 100 kilometers
    b. Frequency of Operation = 300.0 MHz.
According to Fig. 6-5A, the answer is a free-space radio path loss of approximately 118.0 dB. But, according to Fig. 6-5B, the answer seems to be approximately 122.5 dB. Why the 4.5 dB disparity? Well, it's simply because Fig. 6-5A shows the free-space loss, but includes in the "system" the gain of the two half-wave dipoles, while Fig. 6-5B is based on the isotropic antenna and shows no such gains. Recall that an isotropic antenna has a gain of 0.0 dBi, while the gain of a half-wave dipole is 2.15 dBi. A half-wave dipole on each end of the path would thus show a total antenna gain of 4.30 dB. Our path loss, as derived from the two figures, showed a difference of approximately 4.5 dB which is in quite close agreement with our 4.30 dB.

Operation No. 6

Operation number 6 is Plane Earth Propagation. The first item we should clear up, herein, is the sometimes attendant confusion regarding the terms "Plane Earth" and "Smooth Earth". A smooth earth may be exemplified by a calm lake or salt flats. The smooth earth is spherical and can never, per se, be planar while a plane earth might be considered an artifice over which propagation suffers for a given path distance, with attenuation in excess of that for the case of free space. One source describes Plane-Earth Loss as the attenuation of a wave propagating over an imperfectly conducting plane, in excess of the corresponding attenuation over a fictitious perfectly conducting plane. The earth affects a propagating radio wave in a very complex manner by playing the combination role of a reflector and absorber according to its conductivity and permittivity (please see Appendix 1 entitled, "Earth Constants" and also Appendix 28 entitled, "Radiogeology").

The plane earth mode of propagation is employed if less than first Fresnel radius clearance does not obtain as follows:
    a. In the near vicinity of antenna 1.
    b. In the near vicinity of antenna 2.
    c. Anywhere along the path (the diffracting obstruction(s) excepted).
    d. Any one, two, or all three of the items listed in a, b, and c.

To calculate plane-earth loss, we utilize Fig. 6-6A (or the metric version Fig. 6-6B) and Fig. 6-6C. Please keep in mind when using Fig. 6-6A or 6-6B, that the results would not be valid should the indicated received power be greater (or equivalently the plane earth loss be less) than that for the free-space loss mode indicated by Fig. 6-5A or 6-5B. Regarding Fig. 6-6C, this shows the effective antenna height as a function of antenna polarization, earth constant characteristics, and frequency of operation. Above 1,000.0 MHz operating frequency, use actual antenna heights as explained in Fig. 6-6C. With Fig. 6-6A and 6-6B use acutal antenna heights or the effective height of Fig. 6-6C, whichever is greater.

Let's take an example: A ten watt (+ 10.0 dBw) radio frequency signal is fed to an antenna of 5.0 dBd (please see Appendix No. 6 entitled, "List of Decibel Terms") gain. Receiving antenna has a gain of 7.0 dBd. Frequency of operation is 200.0 MHz. Antenna polarization is vertical. Path distance = 30.0 statute miles. Actual height of transmitting antenna 1 is 50.0 feet as is that of antenna 2 (receiving). The earth beneath the antennas

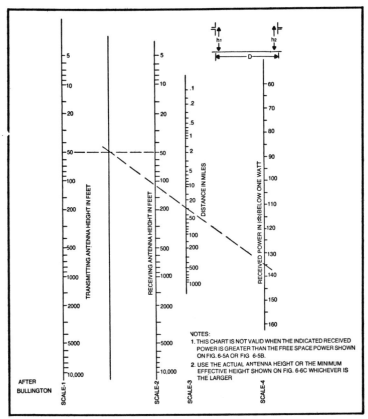

Fig. 6-6A. Received power over plane earth between half-wave dipoles with 1 watt radiated (courtesy Bell Laboratories).

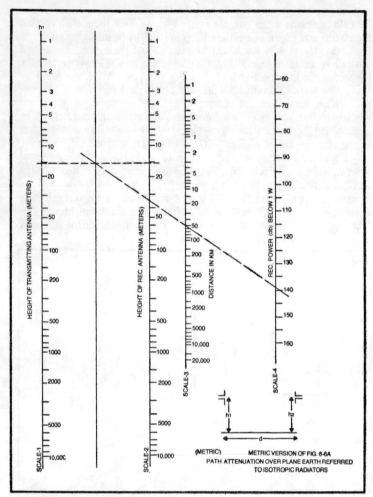

Fig. 6-6B. Metric version of Fig. 6-6A.

is salt-water swampland. Assume that our path terrain profile plot shows that first Fresnel radius clearance does not obtain in the near vicinity of antenna 1.

The solution is as follows. First compare the 50.0 foot actual antenna heights with their minimum effective heights, as shown in Fig. 6-6C. To do this, we enter Fig. 6-6C with our operating frequency of 200.0 MHz and proceed vertically upward along this frequency line to the curve marked "Sea Water, Permittivity = 80, Conductivity = 4.0 Mhos per meter, vertical polarization." At the intersection of this curve with the 200.0 MHz line, proceed straight left to read a minimum effective height of approximately 15.0 feet. Therefore, the actual antenna heights, being larger than the minimum effective heights, rule. So we have:

Antenna 1 height (h_1) = 50.0 feet.
Antenna 2 height (h_2) = 50.0 feet.

With the antenna heights, h_1 and h_2 (50.0 feet each) and the path distance, d (30.0 miles), enter the nomograph of Fig. 6-6A as shown. Draw a straight line from antenna 1 height to antenna 2 height. The intersection of this line with the index line (empty line) between scales 1 and 2 establishes a point. From this point to the path distance (scale 3), (path distance = 30.0 miles), draw another straight line and extend it to scale 4. At the intersection of our drawn straight line with scale 4, we read a received power of 135.0 dB below 1.0 watt (− 135.0 dBw). Now, if one watt, indeed, were fed to the transmitting dipole (antenna 1), this − 135.0 dBw would be the

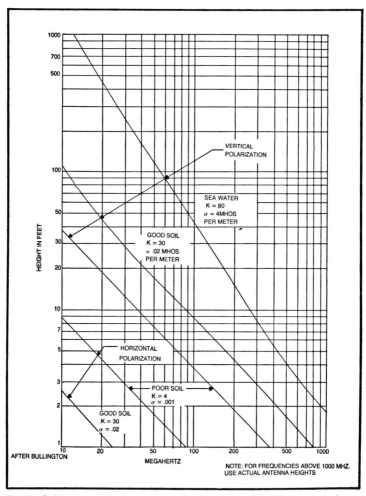

Fig. 6-6C. Minimum effective antenna height (courtesy Bell Laboratories).

received signal power available at the terminals of the receiving half-wave dipole. We now check the validity of this − 135.0 dBw received signal by comparing it with the result obtained in Fig. 6-5A, which is approximately −108.0 dBw. We see, thus, that our result of −135.0 dBw is valid since it is weaker than that of − 108.0 dBw, the free-space loss value. Or, another way of stating it would be that the free-space loss is less than the plane-earth loss. Now, recall that our transmitting antenna gain (antenna 1) is 5.0 dBd (five dB over a half-wave dipole). Similarly, the gain of our receiving antenna is 7.0 dBd while our actual power fed to the transmitting antenna is ten watts instead of one watt upon which our nomograms are based. This then is + 10.0 dBw (or 10.0 dB higher than 1.0 watt or 0.0 dBw). Thus, we have a total advantage of 5.0 dB + 7.0 dB + 10.0 dB = 22.0 dB. A 22.0 dB advantage over a received signal of − 135.0 dBw renders an actual received median signal of (− 135.0 dBw + 22.0 dB) − 113.0 dBw.

Operation No. 7

Operation number 7 is Obstruction Loss Additional To Plane Earth Loss. It is frequently expeditious to apply the following solution when first Fresnel radius clearance does not obtain in the near vicinity of one or both antennas and/or anywhere along the obstructed diffraction path (the obstruction(s) excepted). Please see general rule 2. To determine the path obstruction loss proceed as follows with Fig. 6-7.

With the path terrain profile drawn on infinite (flat) earth, draw triangle **Ant. 1, Ant. 2, 0.** The three sides of this triangle are as follows:

> The Line joining the phase center of antenna 1 through top(s) of highest peak(s) of the total obstruction, to the extension of a similar line from antenna 2. The intersection of these lines are at 0 (please see Fig. 6-7). The triangle is completed by drawing in a straight line between the phase centers of antennas 1 and 2.

From point 0 of the constructed triangle drop a vertical line. The intersection of this vertical line with the inter-antenna straight line establishes H, the obstacle "equivalent height," as well as path segment distances, d_1 and d_2 of the total path distance, d. In this case the smaller or equal path segment is designated as d_1 while the other is designated as d_2. Again, the total path distance is d.

Let's take an example. Distance, $d_1 = 25.0$ miles. Obstacle "equivalent height", H, = 500.0 feet. Frequency of operation = 300.0 MHz. Find the diffractive obstruction loss which should be added to the plane earth loss to obtain the total path loss.

The nomographic solution is as follows:

a. Draw, as shown in the nomogram of Fig. 6-7, a straight line from "25.0 statute miles" on scale 1 through "500.0 feet" on scale 2. Extend this straight line to the index line as shown.

b. From the point established by the intersection of the straight line constructed above and the index line, draw a straight line through our operating frequency, 300.0 MHz. Extend this line through scale 4. On scale 4 read the obstruction loss which is to be added to the plane earth loss for the total path loss used to obtain the median signal value for the entire radio path. In this case, the obstruction loss is 9.8 dB.

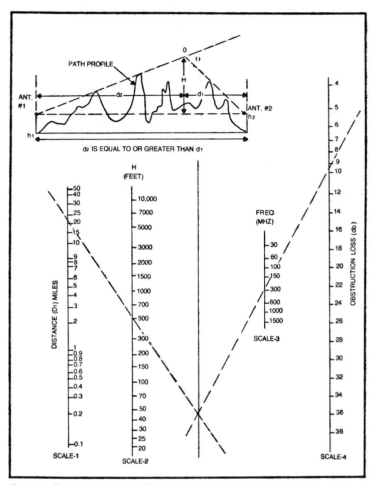

Fig. 6-7. Obstruction loss additional to plane earth loss.

Operation No. 8

Operation number 8 is Knife-edge Propagation Loss Additional To Free Space Loss. When the path obstruction is a knife-edge (zero reflection coefficient) at the frequency of operation (see operation 2 for determining whether or not a given obstruction is to be considered a knife-edge) and first Fresnel radius clearance obtains at all points (the knife-edge obstruction itself excepted) on the path, we may employ Fig. 6-8 (q.v.), or its metric version Fig. 6-8A. With your radio path terrain profile drawn on appropriate k-factor paper, draw in a straight line from antenna phase center 1 to antenna phase center 2. From the tip of the knife-edge obstruction drop a vertical line. The intersection of this dropped vertical line with the inter-antenna line establishes line H, the height line of the knife-edge ridge as well as path segment distances, d_1 and d_2 of the total path distance,

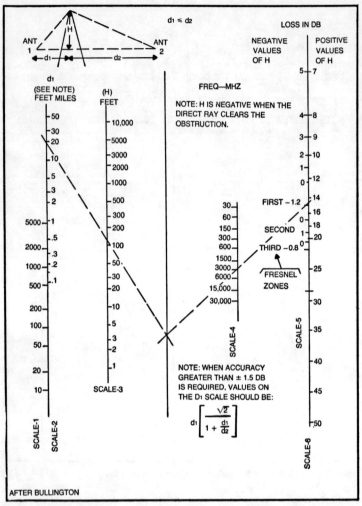

Fig. 6-8. Shadow loss relative to free space (courtesy Bell Laboratories).

d. The path segment distance d_1, in this case, is equal to or smaller than path segment distance d_2. All distances and heights, or course, may be read directly from the radio path terrain profile while the path angular distance, Θ, may be read directly from a profile only with the aid of a special protractor.

Let's take an example: Smaller path segment, d_1, = 20.0 Miles. Obstacle height, H, = 125.0 feet. Frequency Of Operation = 3.0 GHz (3,000.0 MHz). Find the obstruction's knife-edge loss which should be added to the total path free-space loss to obtain the total median path loss.

The nomographic solution is as follows (please follow along in Fig. 6-8):

a. Draw a straight line from "20.0 miles" on scale 1 through "125.0 feet" on scale 3. Extend this straight line to intersect with the index line (empty line).
b. From the intersection point established in step a above, draw a straight line through the operating frequency of 3.0 GHz on scale 4 and extend this line to intersect with scale 6. On scale number 6 read off our knife-edge obstruction loss in dB in excess (in this case approximately 14.5 dB) of that of free-space loss which is to be added to the knife-edge loss to obtain the total median path loss.

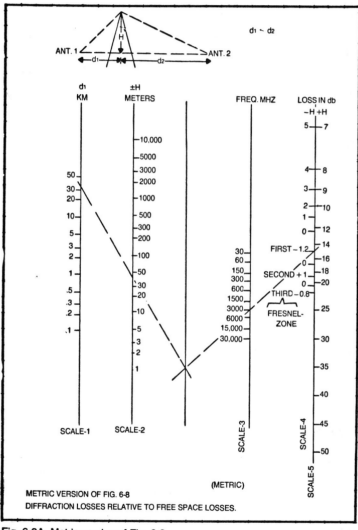

Fig. 6-8A. Metric version of Fig. 6-8.

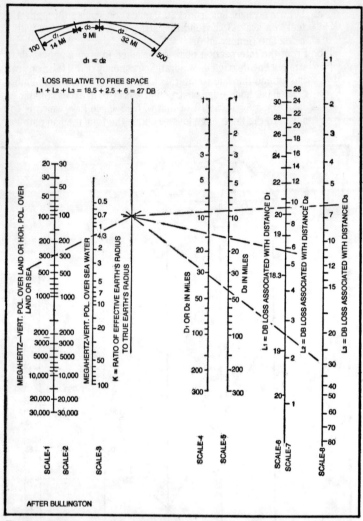

Fig. 6-9. dB loss relative to free space transmission at points beyond the horizon over a smooth earth (courtesy Bell Laboratories).

Operation No. 9

Operation number 9 is Diffraction Loss Over Smooth Earth Relative To Free Space Loss. Employing the nomograms of Fig. 6-9, 6-9A (metric version of Fig. 6-9), 6-9B, and 6-9C (metric version of Fig. 6-9B), we may determine the diffraction loss over smooth earth relative to free-space loss. (Please see general rule No. 1.) This diffraction phenomenon is sensitive to frequency of operation, antenna polarization, k-factor, and whether or not the antennas are over sea or land.

Let's take an example: We have an overland smooth earth path. The frequency of operation is 300.0 MHz. Polarization is vertical. The appropriate k-factor is considered to be 4/3 earth. Total path distance from antenna 1 to antenna 2 is 55.0 miles. Antenna 1 height = 100.0 feet and antenna 2 height = 500.0 feet. Find the smooth-earth diffraction attenuation which must be added to the inter-antenna free-space loss to obtain the total median path loss.

The nomographic solution is as follows (please follow along on Fig. 6-9, 6-9A, 6-9B, and 6-9C as appropriate): From Fig. 6-9B determine distances d_1, d_2, and d_3. These distances are found as follows:

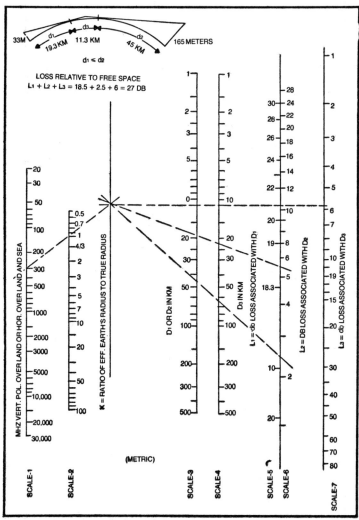

Fig. 6-9A. Metric version of Fig. 6-9.

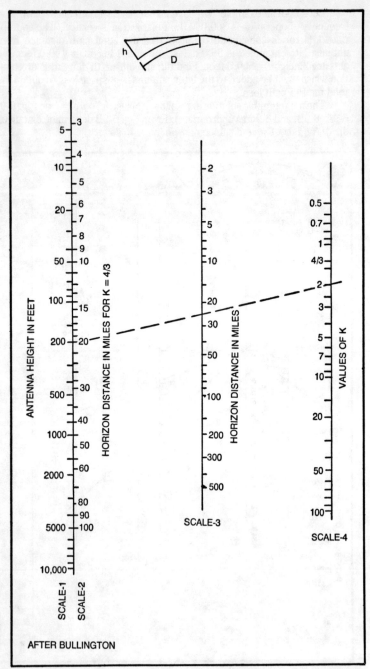

Fig. 6-9B. Nomograph for distance to the horizon (courtesy Bell Laboratories).

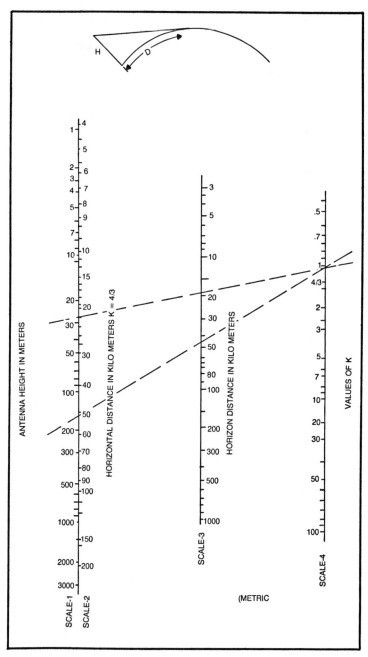

Fig. 6-9C. Metric version of Fig. 6-9B (courtesy Radio Research Laboratories, Japan).

a. To find distance, d_1, we enter Fig. 6-9B with our 100.0 foot antenna height. Draw a straight line from 100.0 feet on scale 1 to k = 4/3 on scale 4. This renders a horizon distance, d_1, of 14.0 statute miles, as read off on scale 2 or scale 3. If the k-factor were another value we would draw a straight line from scale 1 to the appropriate k-factor on scale 4. The point at which this line would intersect scale 3 would render the appropriate distance to the radio horizon.
b. To find distance d_2, we enter Fig. 6-9B with our 500.0 foot antenna height. Draw a straight line from 500.0 feet on scale 1 to k = 4/3 on scale 4. This renders the radio horizon distance, d_2, of 32.0 miles as read off scale 2 or scale 3. Again, if the k-factor were another value we would draw a straight line from scale 1 to the appropriate k-factor on scale 4. The point at which this line would intersect scale 3 would render the appropriate distance to the radio horizon.
c. To find distance d_3, simply subtract from d (total path length) the sum of d_1 and d_2 (path-end horizon distances), viz:

$$55.0 \text{ miles} - (14.0 \text{ miles} + 32.0 \text{ miles}) = 9.0 \text{ miles}.$$

d. On Fig 6-1, draw a straight line from 300.0 MHz on scale 1 (our antennas are vertically polarized and over land) to our k-factor of 4/3 earth. Extend this straight line to the index line (empty line). This intersection establishes a point.
e. From the point established on the index line, we next draw a straight line to our d_1 (14.0 miles) on scale 4. Extend this straight line to scale 6 and read off 18.5 dB as the loss component, L_1, of the smooth-earth diffraction loss.
f. From the point established on the index line (in step d above), we draw a straight line to our d_2 (32.0 miles) on scale 4. Extend this straight line to scale 7 and read off 2.5 dB as loss component, L_2, of the smooth-earth diffraction loss.
g. From the point established on the index line (in step d above) we draw a straight line to our d_3 (9.0 miles on scale 5. Extend this straight line to scale i and read off approximately 6.0 dB as loss component L_3 of the smooth earth diffraction loss.
h. Now simply add up L_1, L_2, and L_3 to obtain the smooth-earth diffraction loss. Viz:

$$18.5 + 2.5 \text{ dB} + 6.0 \text{ dB} = 27.0 \text{ dB}$$

Finally, of course, we add this 27.0 dB total diffraction loss to the inter-antenna free-space loss to obtain the total median path loss.

Operation No. 10

Operation number 10 is Diffraction Path Fading Allowance. As on any radio path, suitable fading allowances must be made in diffraction paths. Fading on diffraction paths is different from that of WRH (Within Radio Horizon) paths, generally conceded to be less. These fading allowances can

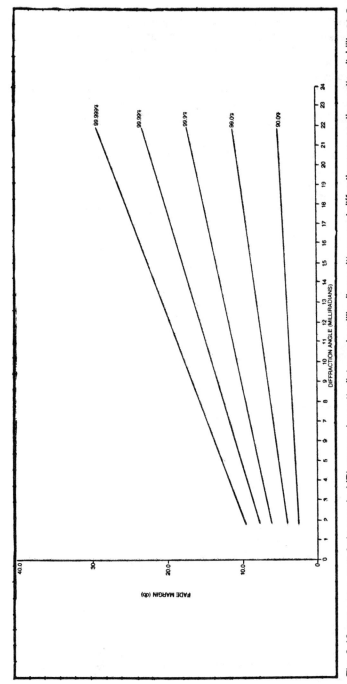

Fig. 6-10. Diffraction path fade margin (dB) vs. angular path distance in milliradians with percent diffraction propagation path reliability as a parameter.

be made, in your radio diffraction hop, by providing additional system gain (increased transmitter power, increased transmitting/receiving antenna gains, etc.) to absorb the fades. In the case of diffraction paths, the required fade margin for given percentages of propagational reliability may be obtained by simply consulting Fig. 6-10. To use this figure, enter it with Θ (path angular distance) in milliradians, as determined in operation No. 1. Note that this is a somewhat different quantity upon which to base fade margin than the usual great circle distance. Next, proceed vertically upward to the desired percentage of propagational reliability and proceed horizontally to the left to read the fade margin in dB.

Let's take an example. We have diffractive hop carrying digital information. The diffraction angle (another name for path angular distance in the case the hop is diffractive) is 16.0 milliradians. Our receiver threshold for the required BER (Bit Error Rate) of 10^{-6} (one bit per million bits) is an RF signal input of -85.0 dBm. Our receiver's actual unfaded signal input is -62.0 dBm. What is our diffraction path propagational reliability?

The solution is as follows: Firstly, our available fade margin under the above conditions is the difference between the required signal strength for the 10^{-6} BER (bit error rate) and the actual unfaded signal input to the receiver. This is the difference between, in our case, -85.0 dBm and -62.0 dBm which is equal to 23.0 dB. Our diffraction angle (path angular distance), Θ, as given, is 16.0 milliradians. Now, simply enter Fig. 6-10 with your diffraction angle of 16.0 milliradians and proceed vertically upward until you intersect a horizontal line drawn from 23.0 dB on the fade margin scale. The path propagation reliability would be approximately 99.999%. Conversely, we could state that a 23.0 dB fade margin in a diffraction path would be reasonably expected to render a 99.999% propagational reliability.

DIFFRACTION PROBLEMS

The following typical examples show which of the previously described operations and procedures should be employed in specific diffraction propagation problems.

Example 1

See Fig. 6-11. We have an obstructed path of total length, d. Triangle ABC establishes path segment distances, d_1 and d_2 as well as obstacle equivalent height, H. Obstacle height above mean sea level is h. Antenna A and C heights above mean sea level are respectively h_1 and h_2. First Fresnel radius clearance obtains everywhere along the path except at the obstruction itself. Find the total path median path loss.

Solution To Example 1

a. Plot your terrain path profile on appropriate k-factor paper
b. See General Rule No. 1
c. Perform operation No. 1
d. Perform operation No. 2
e. Perform operation No. 8, whether or not obstacle qualifies as a knife-edge.

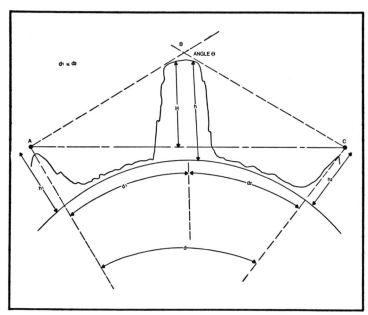

Fig. 6-11. Single obstacle path—first Fresnel radius clearance exists, except for the obstacle.

    f. If obstacle top radius of curvature is excessive for the knife-edge solution only (step d above) proceed to operation No. 3.
    g. Perform operation No. 4
    h. Perform operation No. 5 for the entire path distance, d.

Sum the decibel values of steps e, f, g, and h above for the median path attenuation. Then we must allow for any antenna gains or transmitter power differences. Also allow for fade margin, (Fig. 6-10) as desired by performing operation number 10.

Example No. 2

See Fig. 6-12. This example is fundamentally like the one of example No. 1. The exception is that two prominent path obstacles exist as shown (please keep in mind as we go along that any number of obstacles may be handled by the methods described herein). In this multi-obstacle procedure, we construct a triangle for each obstacle. The first triangle, as can be seen, is ABC (formed by straight lines drawn from Antenna A to obstacle 1 peak (B) to obstacle 2 peak (C). Completing this triangle is line AC. The second triangle uses obstacle peak No. 1 as a radio frequency energy source at B as the first triangle uses antenna A. This second triangle is BCD. The antenna heights above mean sea level are respectively h_1 and h_2 as shown in Fig. 6-12. The heights of the obstacles above mean sea level are h_1 and h_2 as shown while the equivalent heights from the obstruction peaks B and C to triangle bases respectively AC and BD are H and H'. Be certain to keep the differences between H and h in mind. In this case of two obstacles, d_1 is the equal or shorter of the path segments d_1 and d_2

217

Similarly, d_1' is the shorter of the path segments d_1' and d_2 of path portion d_1' and d_2. These values, d_1 and d_1' are to be employed in the nomographic solutions where d_1 is called for. The short path of the path or path segment is used with the obstruction loss nomogram. In multiple peaks of this nautre, where first Fresnel zone clearance obtains along the path (the obstacles themselves excepted), the total diffraction loss is the decibel sum of the individual obstruction losses. Find the total median path loss.

Solution To Example 2

    a. Plot your path terrain profile on appropriate k-factor paper.
    b. See general rule No. 1.
    c. Perform operation No. 1 for each obstruction.
    d. Perform operation No. 2 for each obstruction.
    e. Perform operation No. 8 for each obstruction (whether or not the obstacles(s) qualify as knife-edge)
    f. If the obstacles(s) radius (radii) of curvature is excessive for knife-edge solution only, as in step d above, proceed to operation No. 3 (in addition to operation No. 8) for each obstruction.
    g. Perform operation No. 4 for each obstruction.
    h. Perform operation No. 5 for the entire radio path, d.
    i. Perform operation No. 10 for the largest Θ (angular distance) angle involved in the multi-obstruction diffraction path.

Sum all the decibel values of steps e, f, g, and h above to obtain the median loss for the entire path. Do not forget to allow for the fade margin of step i above to bring the signal up to the desired propagational reliability percentage.

Example No. 3

See Fig. 6-13. Here we have a radio path obstructed by a multiple-obstacle irregular terrain. The path length is d with equivalent obstacle height H. The equivalent obstacle height location divides the radio path, d, into the two path segments, d_1 and d_2. The actual antenna heights are h_1 and h_2 while the path angular distance, Θ, is formed by triangle sides, ANT. 1 – 0 and ANT. 2 – 0. The inter-antenna line, ANT 1 – ANT. 2 completes the triangle.

Solution To Example No.3

    a. Plot the path terrain profile on $k =$ infinity paper.
    b. See general rule No. 2.
    c. Perform operation No. 1.
    d. Perform operation No. 6.
    e. Perform operation No. 7
    f. Perform operation No. 10.

Sum the dB values of steps d and e above to obtain the median path loss. Be sure to allow, as in all previous problems, for any system gain differences such as antenna gains and/or transmitter power output. Also allow for the diffraction path fade margin in step f above to increase path propagational reliability over median (50%) value, as desired.

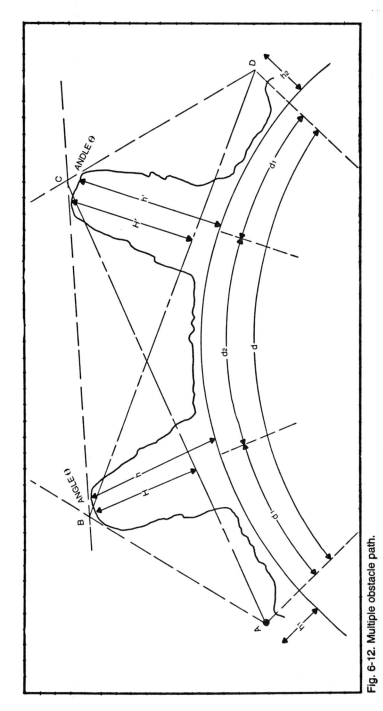

Fig. 6-12. Multiple obstacle path.

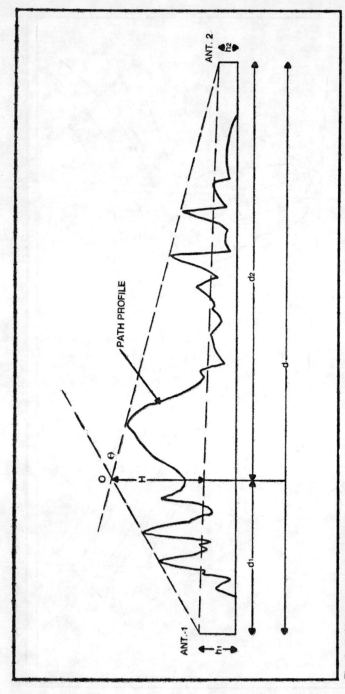

Fig. 6-13. Multiple obstacle path—first Fresnel radius clearance does not obtain at near vicinity of antenna 1 or near vicinity of antenna 2 and/or along the path/obstacles excepted.

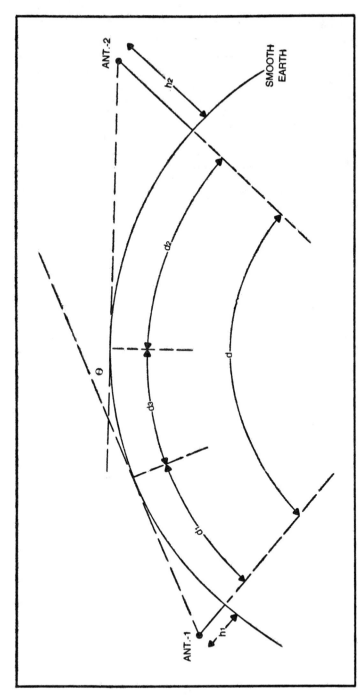

Fig. 6-14. Smooth earth path—first Fresnel radius clearance obtains in near vicinity of both antennas, but not at earth-curvature obstacle.

Example No. 4

See Fig. 6-14. Here we have a radio path obstructed by the curvature of the earth. First Fresnel clearance obtains in the vicinity of both antennas, 1 and 2. The total path length is d. For use with Fig. 6-9 the distance from the antenna 1 phase center to the radio horizon, at the particular k-factor considered, is d_1; the distance from the antenna 2 phase center to the radio horizon, at the particular k-factor considered, is d_2, and the path segment distance between radio horizons is d_3. Actual antenna heights in this example and Fig. 6-1 are h_1 and h_2. The straight line from each antenna to the radio horizons form the path angular distance angle, Θ. For use with Fig. 6-1, the ds have another definition, as explained previously, and H (not shown in Fig. 6-14 for clarity) must be measured from the profile or determined by other means. As already mentioned, these various "d" etc. definitions must be carefully observed.

Solution To Example 4

a. Plot terrain profile on appropriate k-factor paper.
b. See general rule No. 1.
c. Perform operation No. 1.
d. Perform operation No. 5.
e. Perform operation No. 9.

Sum up the dB values of steps d and e above to obtain the path median loss. Then allow for any differences of antenna gains or transmitter power output as well as for the desired fade margin, as shown in operation No. 10, for the desired % propagational path reliability above median (50%).

Chapter 7
Tropospheric Forward Scatter Propagation

Tropospheric forward scatter of radio energy is accomplished by beaming the radio energy into the troposphere (weather sphere) at low elevation angles (i.e. to the radio horizon) by a transmitting antenna. A small portion of this radio energy is "captured" via a common volume formed by the intersection of a similar beam originating at the receiving antenna see Fig. 7-1. Although not as yet fully understood, the propagation action within the common volume is generally supposed to consist of a complex varying mix of changing refractivity and reflectivity. While there is some radio energy in the transmitted beam which is scattered in all directions (isotropic scattering), we obviously are interested in that portion scattered in the forward direction (down the receiving antenna beam). The tropospheric forward scattering mechanisms are able to provide usable signal field at ranges exceeding those of diffraction. The tropospheric forward scatter field has this range (path distance) advantage over diffraction mainly because of the height of the scattering common volume. Some idea of the heights of these common volumes may be had from the following smooth earth, zero horizon angle examples:

Path Length	Common Volume Height
160.0 Km.	300.0 to 1,800.0 Mtrs.
320.0 Km.	600.0 to 3,000.0 Mtrs.
650.0 Km.	3,000.0 to 21,000.0 Mtrs.

Troposcatter (tropospheric forward scatter) is employed in point-to-point telecommunications and may be useful in transporting, in the FDM/FM mode, some 300.0 voice channels or a corresponding amount of digital information, at actual (great circle) path distances of some 80.0 to 500.0 kilometers. From a propagational point of view, the useful frequency range of tropospheric scatter is about 50.0 MHz to 10,000.0 MHz. Last minute information made available to the author by the CCIR director, Mr. Richard Kirby, indicates that a proposed optimum frequency order for troposcatter is now 2.0 GHz. Troposcat transmitter powers of 1.0 kilowatt to 50.0 kilowatts are generally employed, depending upon the specific path computations.

FORWARD SCATTER PATH CALCULATIONS

With this, then, let us launch into the specific procedure for calculating a tropospheric forward scatter path. This procedure for calculating a forward scatter path is based upon information from the CCIR (International Radio Consultative Committee) of the ITU (International Telecommunications Union)—the telecommunications arm of the United Nations and NTIA (National Telecommunications and Information Agency) of the USA Department Of Commerce. The author has modified these procedures to provide conservative solutions in a more direct and less abstruse manner.

THE PROCEDURE

The steps for solving the troposcat path problem are generally arranged as a series of graphs. The author has attempted to pre-calculate as much of the data as possible, condense, and to eliminate the usual abstruseness and drudgery involved in this type of computation. Let's begin then with the first step.

Step 1. Draw an accurate terrain profile of your path on appropriate parabolic k-paper. Generally, 4/3 k has been used. Please see the sample on Fig. 7-2. We shall touch upon all items indicated on the profile sheet as we go along.

Step 2. Draw an extended straight line from the phase center of the transmitting antenna through the tip of the highest path obstruction. This straight line is ABC in Fig. 7-2.

Step 3. Draw a similar extended straight line from the phase center of the receiving antenna, through the tip of the highest path obstruction. This is straight line EDC in Fig. 7-2.

Step 4. Determine from the path terrain profile (in our example Fig. 7-2) the heights AMSL (above mean sea level) of the transmitting antenna A and the receiving antenna E. These AMSL heights are shown in Fig. 7-2 respectively as h_t and h_r. Record same.

$$h_t = 1,000.0 \text{ Meters}$$
$$h_r = 500.0 \text{ Meters}$$

Step 5. Determine from the path terrain profile the heights AMSL (above mean sea level) of radio ray intersection point, C. Record same.

$$H = 1,500.0 \quad \text{Meters}$$

Step 6. Next we determine the *effective heights* of the transmitting and receiving antennas, respectively h_{te} and h_{re}. Antenna height is a tricky subject and its definition, to an extent, depends upon the application (please see Appendix 29 entitled, "Antenna Height"). For our purposes in this chapter, however, effective antenna height is determined as follows:

Fit an arc to the terrain (including topography, trees, buildings etc.) for 80.0% of the distance along the great circle path segment between the antenna and its horizon; of the excluded 20.0% of the path segment, 10.0% is at the antenna and 10.0% is at the antenna's horizon. This is illustrated on Fig. 7-2. This arc is fitted by simply dividing the 80.0% portion of the path segment into ten or more equal parts (the more parts, the better the

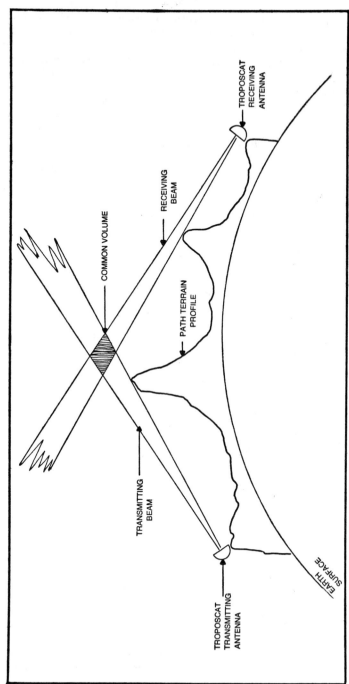

Fig. 7-1. Diagram showing the common volume for tropospheric forward scatter.

accuracy). The profile heights at the division lines are then simply averaged for the height of your arc. These are shown as dotted lines between antenna A and its horizon B and also between antenna E and its horizon D. Now then, the heights of the antennas' phase centers above their corresponding arcs constitute the antennas' effective heights. In our troposcat path example, in Fig. 7-2, the effective height of the transmitting antenna A is h_{te} and the effective height of the receiving antenna E is h_{re}. Record these values.

Effective transmitting antenna height, h_{te} = 650.0 Meters.
Effective receiving antenna height, h_{re} = 200.0 Meters.

Step 7. Determine actual path and path segment distances, d, d_1, and d_2 from the profile on the k-paper chart of Fig. 7-2. Record these values.

$$d = 300.0 \text{ Kilometers}$$
$$d_1 = 108.0 \text{ Kilometers}$$
$$d_2 = 192.0 \text{ Kilometers}$$

Step 8. The tropospheric scatter transmission loss not exceeded for a given percentage of a year may be estimated via the following formula:

$$L(\%) = 30.0 \text{ Log}_{10}f - 20.0 \text{ LOG}_{10}d + F(\theta d) - G_e + Z \text{ decibels.}$$

Where: f = Operating frequency in MHz
d = Actual (Great Circle) path distance in kilometers
θ = Path angular distance (scattering angle) in radians
G_e = Effective total antenna gain in decibels
Z = Factor for long and short term fading plus climate adjustment.

Now, remembering that you can eat an elephant only by going about it one bite at a time, let's touch upon each of the above formula's terms one at a time while working out the tropospheric scatter problem depicted in Fig. 7-2.

a. **L(%)**—This is simply the tropospheric scatter path attenuation in decibels, which is not exceeded for the part of the year as desired. Let's say that we want our troposcat circuit to be 99.999% propagationally reliable. Record this value.

Desired propagational reliability = 99.999%

b. **+ 30.0 Log$_{10}$f**: This term, as it says, is simply thirty times the common logarithm of the frequency in MHz and is part of the manner in which the propagational loss of a tropospheric scatter signal occurs. Let's assume that our operating frequency is to be 2,000.0 MHz (2.0 GHz). Substituting 2,000.0 MHz for f in the formula term, we obtain and record:

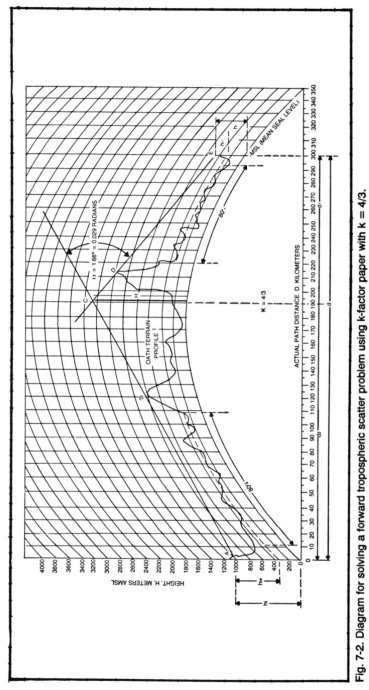

Fig. 7-2. Diagram for solving a forward tropospheric scatter problem using k-factor paper with k = 4/3.

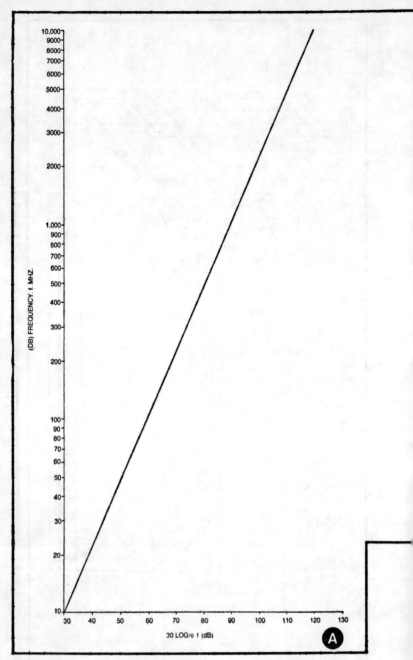

Fig. 7-3. Graphical means for solving parts of the forward scatter transmission loss problem.

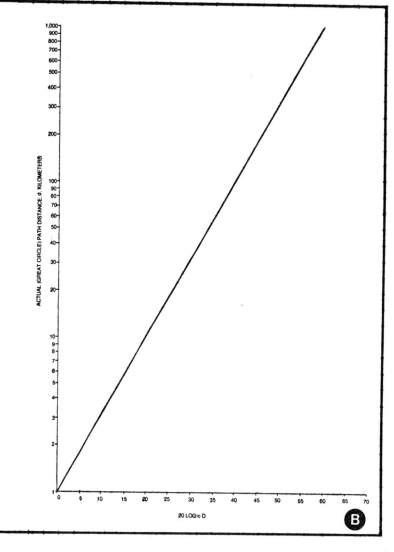

$+ 30.0 \, \text{Log}_{10} \, 2{,}000.0 = + 99.0 \, \text{dB}$
(Note: This can also be done graphically—see Fig. 7-3A.)

c. $-20.0 \, \text{Log}_{10} \, d$: This formula term is also just what it says—minus twenty times the common logarithm of the actual (great circle) path distance in kilometers. From step 7 we have, d 300.0 kilometers and so we solve and record:

$-20.0 \, \text{Log}_{10} \, 300.0 = -49.5 \, \text{dB}.$
(Note: This can also be done graphically—see Fig. 7-3B).

d. **+F(θd):** This is called the attenuation function and depends upon the scatter angle, θ, in radians and actual (great circle) path distance in kilometers as shown. Not shown is the fact that the term also depends upon the surface refractivity. This formula being a little "spookier" and less straight forward, we might do well it take it apart a "chip at a time," starting with the "visible".

1. We'll start with the scattering angle (path angular distance) θ. This scattering angle might be determined in several ways which are relatively simple. One way is to measure it directly—yes, measure it directly from the k-paper profile, but with a special protractor as described in Appendix 26. Another way to determine the scattering angle is to calculate it using the formula for this as given in chapter 6 on diffraction propagation, using the values H, h_1 (or h_t), h_2 (or h_r), d, d_1, and d_2 as we have already determined in steps 4, 5, and 7 of this procedure. Additionally, we might use the graph of Fig. 6-1 of the diffraction propagation chapter 6. The author prefers to use the special protractor but then, the reader's choice might be any of the other methods outlined above. At any rate, as indicated on the terrain profile of Fig. 7-2, the scatter angle is:

    Scatter Angle, θ, = 0.029 Radians

2. Now, since the value of actual (great circle) path distance is 300.0 kilometers, as already indicated in step 7 of this procedure, then our θd product is:

$$\theta d = 8.7$$

3. All we need now is our surface refractivity (N_s) value and the rule is as follows. Determine the N_s at each antenna horizon and take the average of these two, unless an antenna is more than 150.0 meters below its radio horizon, in which case the N_s value at the antenna location is used instead of that at the radio horizon. This, upon first reading might sound like a "can of worms" but it really is easy once you do an example or two. Let's go on. Now, to obtain our necessary N_s value, we first determine our corresponding N_0. This N_0 is simply the minimum monthly mean sea level refractivity from which we will determine our N_s (surface refractivity) value. The minimum monthly value is used since it represents a "worst case". Anyhow, continuing, we easily obtain our N_0 values from Fig. 7-4. Simply select your desired locations (antenna horizons or antenna locations as explained above) on Fig. 7-4 as accurately as possible and read off the corresponding N_0 values. Let's assume that the following values of N_0 were obtained from Fig. 7-4.

N_0 at transmitting antenna A = 340
N_0 at transmitting antenna horizon B = 350
N_0 at receiving antenna horizon D = 360
N_0 at receiving antenna E = 330

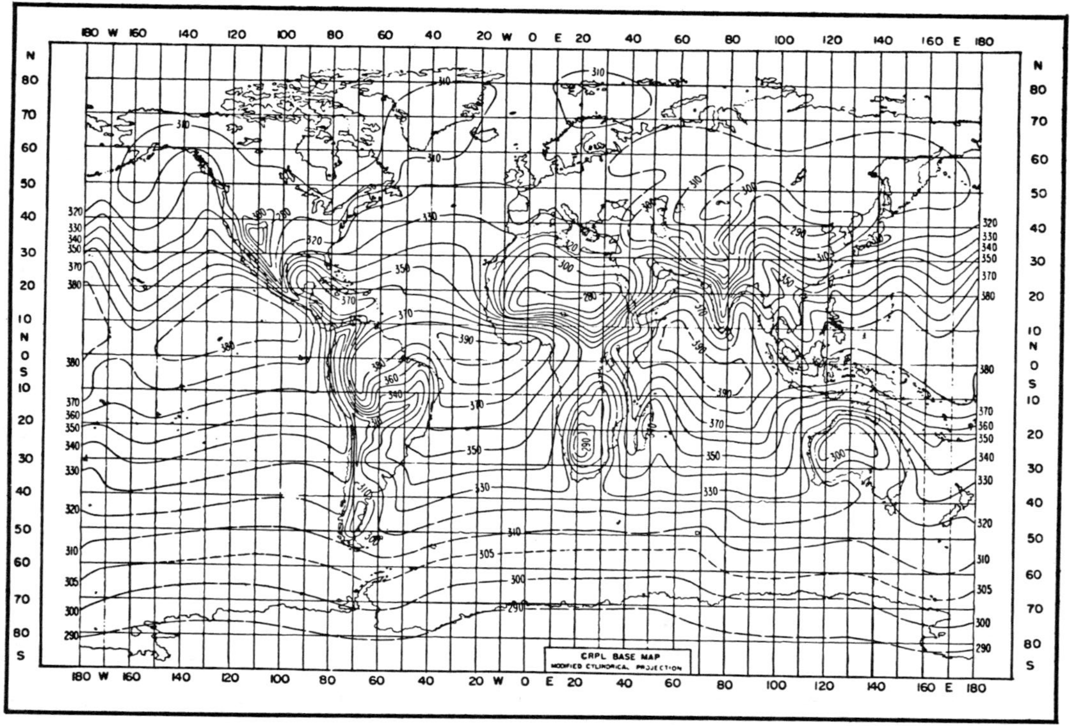

Fig. 7-4. Chart of the minimum monthly mean refractivity at sea

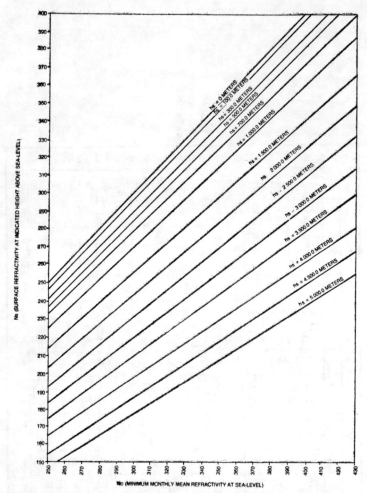

Fig. 7-5. Chart for converting mean refractivity at sea level to actual antenna height.

Now, since antenna A, as seen from the path terrain profile of Fig. 7-2 is not lower than its horizon at B, we select location B. Also, since antenna E is some 700.0 meters below its horizon D, we select the location of antenna E.

At this point we enter Fig. 7-5 to convert the above N_o values to those of N_s (or mean sea level refractivities to refractivities at actual surface heights, h_s). We obtain and record:

$$N_s \text{ at location B} = 322$$
$$N_s \text{ at location E} = 313$$

232

Taking the average of these N_s values renders

$$\frac{313 + 322}{2} = 317.5.$$ We record this path N_s value.

$$\text{Path } N_s = 317.5$$

With our θd value of 8.7 and N_s value of 317.5 we can enter Fig. 7-6 to obtain our $F(\theta d)$ term. We obtain and record:

$$+ F(\theta d) = + 165.0 \text{ dB}.$$

e. $-G_e$: This term represents the decibel summed free-space gains of the transmitting and receiving antennas minus their summed aperture-to-medium-coupling-loss (see Appendix 31). This $-G_e$ quantity is readily determined from Fig. 7-7 if the free-space gain of each antenna does not exceed 50.0 dB. Simply enter the graph with the decibel sum of the antennas' free-space gains and read

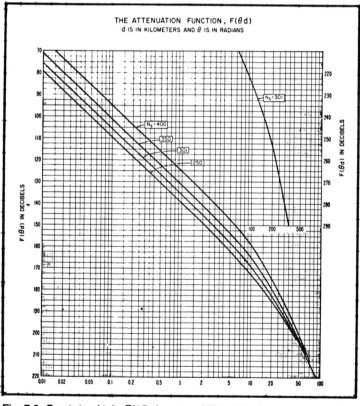

Fig. 7-6. Graph to obtain $F(\theta d)$ (courtesy NTIA, U.S. Department of Commerce).

the total effective path antenna gain, G_e, in dB. Let's assume that our transmitting and receiving antennas have free-space gains of 45.0 dB each. Entering the graph of Fig. 7-7 with this free-space antenna gain sum, or 90.0 dB, we obtain and record:

$$- G_e = - 79.0 \text{ dB}.$$

f. **+Z:** This interesting factor embodies the effects of short term and long term fading and a climatic correction, as well as adjustments for various time periods and different propagation mechanisms.

The two fading components, short term and long term, may be envisioned as the former "riding" on the latter. Short term may be thought of as rapid "within the hour" signal strength variation, whereas long term fading of the hourly median signal strength values throughout the year.

The above information is distilled into a series of graphs of tropospheric scatter path effective distance versus the factor Z (dB) with precentage propagation reliability as a parameter, these percentages running from 90.0% to 99.999%. They cover the various climates with dual and quadruple diversity reception configurations. The reader should leaf through these graphs and become familiarized with them. Please see Fig. 7-8A through 7-8N.

To use these graphs, we need to know the path effective distance. This topic of path effective distance can be perplexing indeed and in view of the various existing explanations of it, a concept not exactly easy to grasp. (Please see Appendix 32 entitled, "Effective Distance".) Suffice it is to point out here that to derive the path effective distance, we proceed as follows:

Calculate the quantity, $x = 65(100.0/f)^{1/3} + (3\sqrt{2h_{te}} + 3\sqrt{2h_{re}})$ kilometers.

Where f = Operating frequency in MHz.
h_{te} = Effective transmitting antenna height in meters.
h_{re} = Effective receiving antenna height in meters.

Substituting this example's f value of 2,000.0 MHz (see step 8b) and also h_{te} and h_{re} respectively of 650.0 meters and 200.0 meters from step 6, in the above formula, we obtain and record:

$$x = 192.1 \text{ kilometers}.$$

Now, if the above derived value of x (192.1 kilometers in our example) is equal to or greater than the actual (great circle) path distance of d (300.0 kilometers in our example), the effective troposcat path distance (d_e) is determined by the following formula:

$$d_e = (130.0 \text{ d/x}) \text{ Kilometers}$$

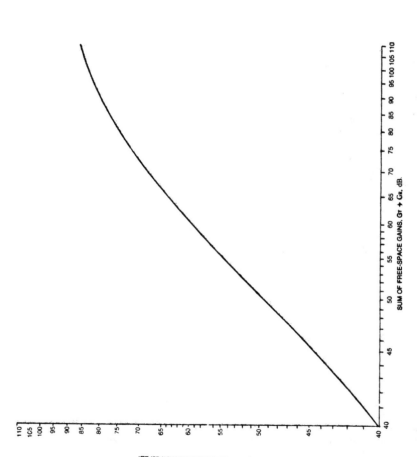

Fig. 7-7. Chart for determining effective antenna gain.

235

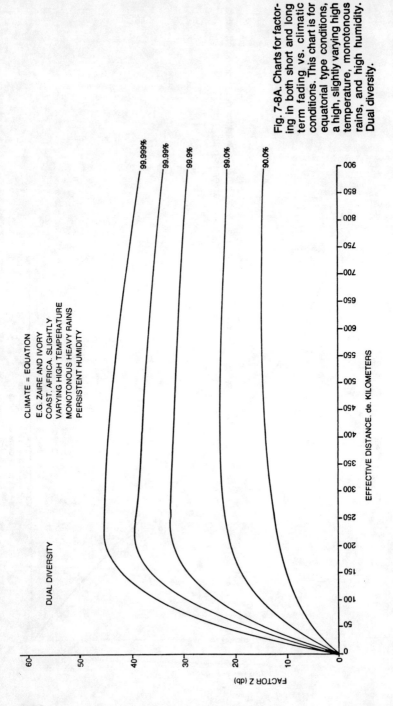

Fig. 7-8A. Charts for factoring in both short and long term fading vs. climatic conditions. This chart is for equatorial type conditions, a high, slightly varying high temperature, monotonous rains, and high humidity. Dual diversity.

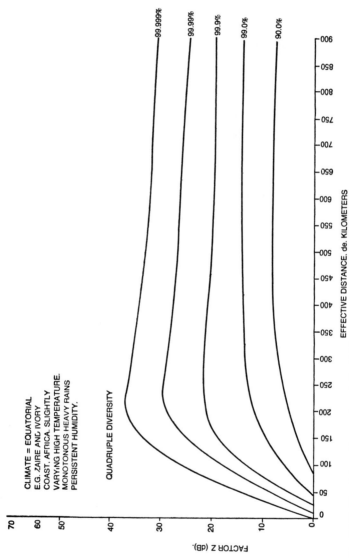

Fig. 7-8B. Quad diversity version of Fig. 7-8A.

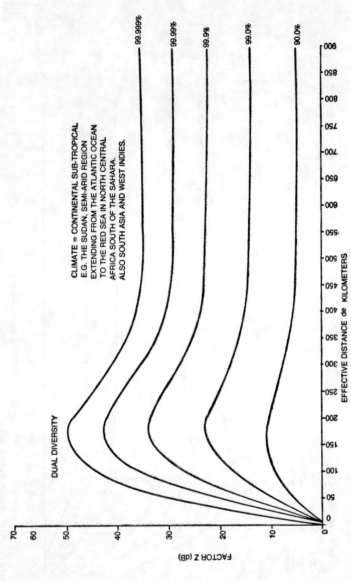

Fig. 7-8C. Continental sub-tropical, a semi-arid climate. Dual diversity.

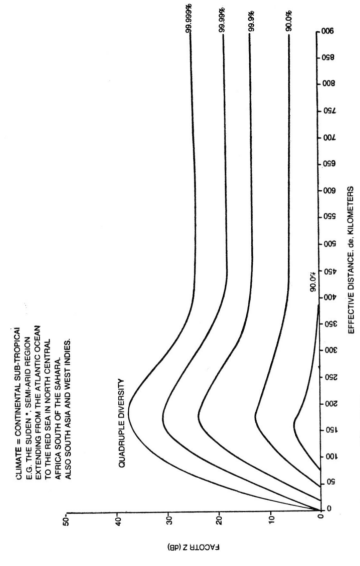

Fig. 7-8D. Quad diversity version of Fig. 7-8C.

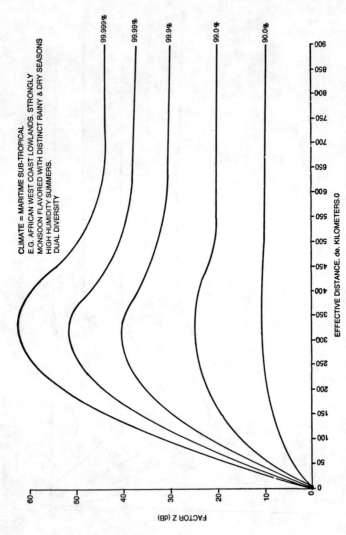

Fig. 7-8E. Maritime sub-tropical, strongly monsoon favored with distinct rainy and dry seasons. Dual diversity.

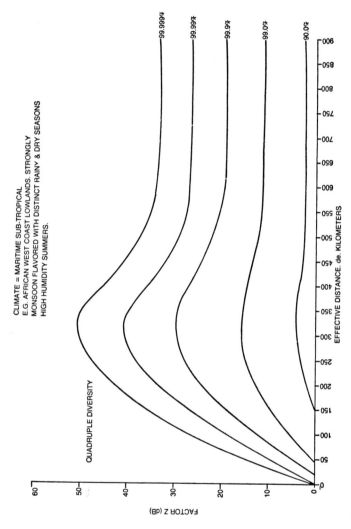

Fig. 7-8F. Quad diversity version of Fig. 7-8E.

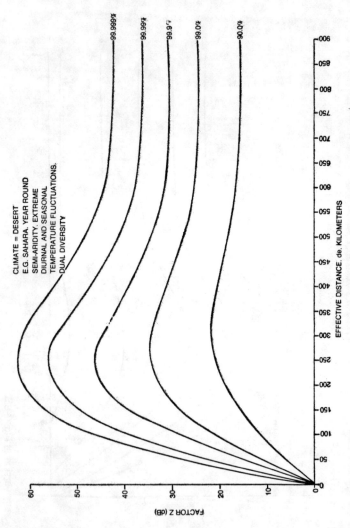

Fig. 7-8G. Desert, year round semi-arid, extreme diurnal and seasonal temperature fluctuations. Dual diversity.

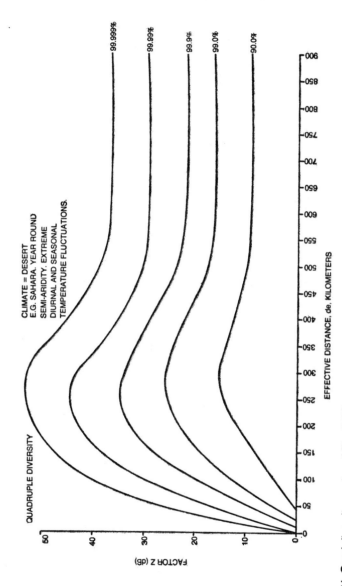

Fig. 7-8H. Quad diversity version of Fig. 7-8G.

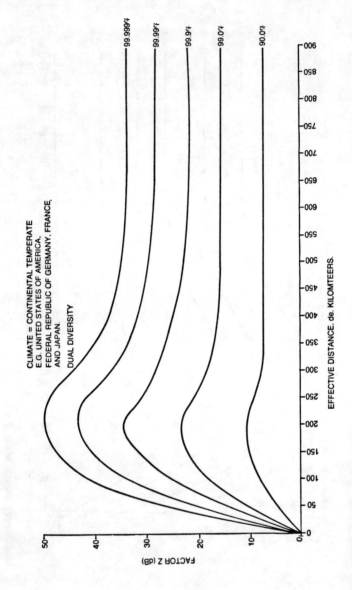

Fig. 7-81. Continential temperate. Dual diversity.

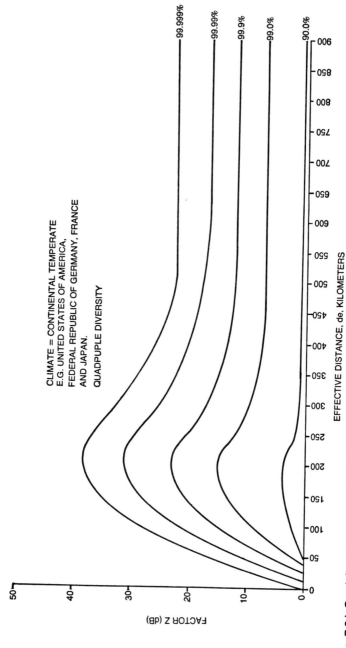

Fig. 7-8J. Quad diversity version of Fig. 7-8I.

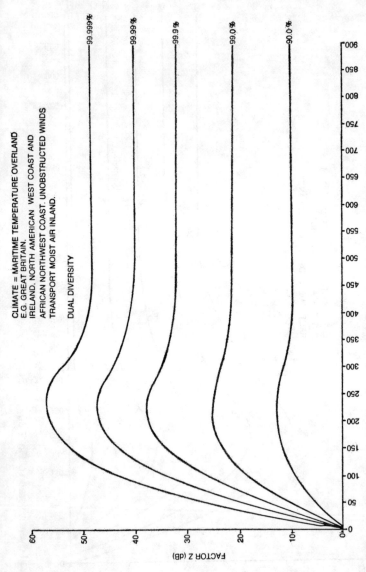

Fig. 7-8K. Maritime temperare overland, unobstructed winds transport moist air inland. Dual diversity.

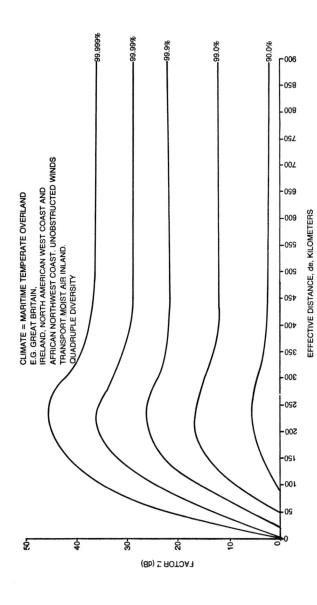

Fig. 7-8L. Quad diversity version of Fig. 7-8K.

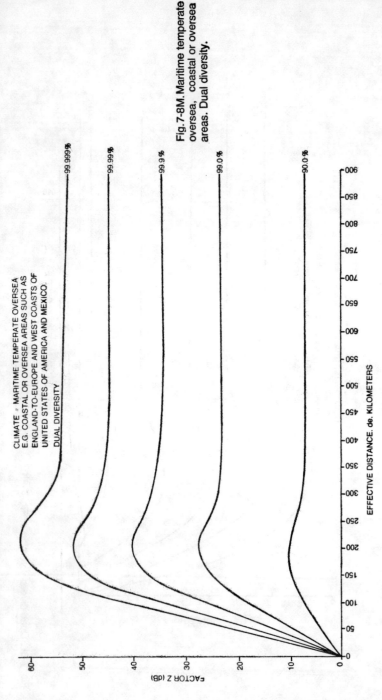

Fig. 7-8M. Maritime temperate oversea, coastal or oversea areas. Dual diversity.

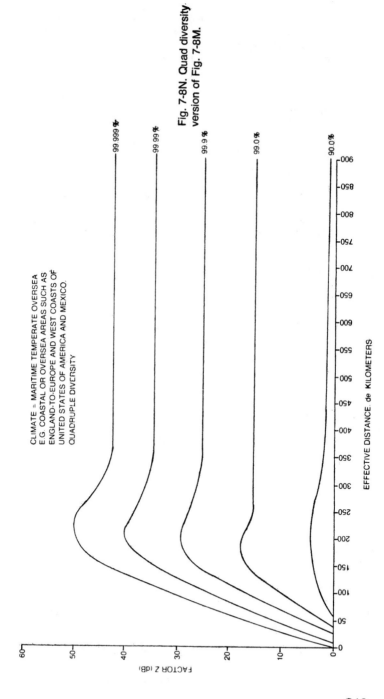

Fig. 7-8N. Quad diversity version of Fig. 7-8M.

If, however, the above derived x value (192.1 kilometers in our example), is less than the actual (great circle) path distance, d, (300.0 kilometers in our example), the effective distance (d_e) is determined by the following formula:

$$d_e = (130.0 + d - x) \text{ kilometers}$$

So, we see that the above recorded value of × (192.1 kilometers in our example) is less than the actual (great circle) path distance, d of 300.0 kilometers. Consequently, we use the formula $d_e = (130.0 + d - x)$ for our path effective distance. Substituting we obtain:

$$d_e = (130.0 + 300.0 - 192.1) = 237.9 \text{ Km.}$$

Now, we are ready to enter the appropriate graph of the set of graphs from Fig. 7-8A through 7-8N. Let's say that our climate is desert and we desire 99.999% propagational reliability using quadruple diversity. The graph fulfilling these requirements is, of course, Fig. 7-8H. Entering this figure with our effective distance, d_e, of 237.9 kilometers, we proceed vertically to intersect the 99.999% curve. At this point, proceed straight left to read (and record):

$$+Z = +52.5 \text{ dB.}$$

OK now, breathe a sigh of satisfaction of a job well done—you're just a few lines away from the end of the tropospheric scatter chapter. Ready? Let's wind it up and get it all together.

Our troposcat transmission attenuation formula is, again:

$$L(99.999\%) = 30.0 \text{ Log}_{10}f - 20.0 \text{ Log}_{10}d + F(\theta d) - G_e + Z(99.999\%) \text{dB.}$$

(Note—It might be clarified here that this troposcat loss is from the input to the transmitting antenna to the output of the receiving antenna).

Our formula term values as we already derived, are:

$$+ 30.0 \text{ Log}_{10}f = + 99.0 \text{ dB (From step 8b)}$$
$$- 20.0 \text{ Log}_{10}d = -49.5 \text{ dB (from step 8c)}$$
$$+ F(\theta) = +165.0 \text{ dB (From step 8d(3))}$$
$$- G_e = -79.0 \text{ dB (From step 8e)}$$
$$+ Z (99.999\%) = + 52.5 \text{ dB (From step 8f)}$$

Substituting and algebraically summing up our example's answer is:

Troposcat transmission loss = 188.0 dB. (Value not exceeded for 99.999% of the time)

In certain cases, it is advisable, as also pointed out in chapter 6, to combine the tropospheric scatter propagation solution with diffraction propagation solution. Please see Appendix item No. 30.

Chapter 8
Millimeter Wave Propagation

It is assumed in presenting the material in this millimeter-wave propagation chapter that the reader has familiarized himself with the contents of Chapter 5. The millimeter-wave band is defined by the CCIR (International Consultative Committee for Radio) as the frequency span of 30.0 to 300.0 GHz (10.0 to 1.0 millimeters wavelength) and is designed EHF (Extremely High Frequency). This band represents a frequency expanse of 270.0 GHz. This tremendous virgin frequency spectal territory lies practically dormat and ready to be awakened. The vast bandwidth available in this spectral area is NINE TIMES the total available spectrum below it. Please reread and contemplate this for a moment. The intelligence transmission capacity is fantastic. Coherent bandwidths of over 6.0 GHz have been achieved. This tremendous capacity makes the millimeter-wave band attractive, to say the least, in view of the saturation existing in our currently used spectrum below it. Today's technology makes millimeter-wave systems a practical possibility. Let's look at some of this band's characteristics.

Trophoshperic temperature, water content (both liquid and water vapor), oxygen, barometric pressure, and their caprices exert profound influences upon the propogation of millimeter-wave terrestrial paths X (this chapter like all other chapters in this book deals with terrestrial radio paths). While much more study and experiment are presently indicated to provide improved practical engineering data for path design in this band, we do have a respectable experience base and some fairly tractable tools with which to make some practical use of this vast EHF (millimeter-wave) communications frontier.

To begin with, "free-space" loss affects millimeter-waves just as it does other frequencies, while propagation through our trophosphere basically affects millimeter-wave in a manner not-unlike that of regular microwaves; there are some gross and some subtle differences. We shall break down the comparisons and contrasts into sub-titular classifications (although for convenience and clarity of instruction we present the various millimeter-wave characteristics separately, it is to be understood that to varying extents these classifications are interdependent).

REFRACTIVITY AND K-FACTOR

As detailed in Chapter 5, the troposphere's temperature, pressure, and relative humidity affect its dielectric properties and therefore its refractivity and k-factor. At millimeter-wave (EHF) frequencies, refractivity, at the present state of the art, appears to be quite similar, in many respects with some frequency effect. Ray bending, therefore, does not appear to be much different at the millimeter-wave frequencies than at the lower microwave band, and the values of k-factor and refractivity remain, for all practical purposes, the same. Accordingly, there appears to be no reason why k-factor plotting paper cannot be used for path profiles at the millimeter-waves frequencies, and the validity of its use should be essentially equal.

WINDOWS

As seen in Fig. 8-1, there are certain frequency spans in the millimeter-wave region where path attenuation is less that at others. These are called "windows" since it is easier for the millimeter-waves to penetrate ("see through") in these frequency regions. The gaseous water vapor or oxygen absorption frequency span may be looked upon as the area between the sill of one window and the sill of the next window. The entire accept/reject curve for the band has also been variously described as a tropospheric selective bandpass (or band elimination depending upon your point of view) filter. In actuality, the attenuation peaks shown may consist of multiple individual attenuation lines; generally, however, at the present state of the millimeter-wave path engineering art, these entire peak areas are avoided in hop design (except in cases such as "security" communications over short paths where high attenuations outside the intended communications area is desired). As is seen in Fig. 8-1, windows exist in the frequency spans of 30.0 to 50.0 GHz, 70.0 to 110.0 GHz, 130.0 to 160. GHz, and 220.0 to 280.0 GHz of the millimeter-wave (EHF) band.

This basic type of tropospheric attenuation is due to resonant and non-resonant trophospheric molecular absorption and the mechanism of scattering. Absorption may be of the true or pseudo type (please see Appendix 33.) Scattering may be considered a "squandering" of radio-wave energy from its intended target (the receiving antenna) by tropospheric particulate material. Obviously, then, tropospheric attenuation will be at least in a hot/dry/clear (no fog, clouds, pollutants, etc.) troposphere. Smoke and similar colloidal troposhperic particles appear to exert minimal attenuation, at least in the 30.0 to 50.0 GHz and 70.0 to 100.0 GHz windows. Millimeter-wave (EHF) attenuation rises with tropospheric water-vapor content and with water-vapor partial pressure. The propagational mechanism of scattering is unlike that of the beneficial forward troposcatter, considered a waste or a "squandering". Particulate matter in the propagation path extracts energy from the passing radio wave and the particles deflect this energy in various isotropic and anisotropic ways, into directions other than the one desired (that is, away from the receiving antenna). There is generally no net change in the particles' internal energy states. Scattering is generally considered a frequency-flat phenomenon.

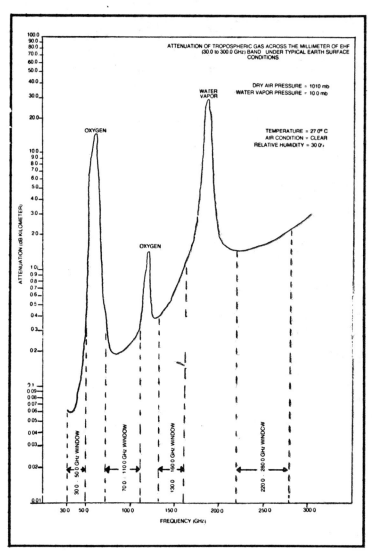

Fig. 8-1. Attenuation of tropospheric gas across the millimeter or EHF (30 to 300 GHz) band under typical earth surface conditions: dry air pressure-1010 mb, water vapor pressure-10 mb, temperature-27 degrees C, air condition relative humidity-30%.

FRESNEL CLEARANCE

Fresnel zone radius clearance does not appear to be the problem at millimeter-wave frequencies that is at the lower microwave frequencies (e.g. at 2.0 GHz). This is natural enough considering the formula for calculating of first Fresnel zone radius.

$$Z_1 = 17.3 \sqrt{\frac{d_1 d_2}{FD}}$$

Where: Z_1 = First Fresnel zone radius in meters.
d_1 = Distance from path point where Fresnel zone radius is desired, to terminal 1 of the millimeter wave path (kilometers).
d_2 = Distance from path point where Fresnel zone radius is desired, to terminal 2 of the millimeter wave path (kilometers).
D = Total actual millimeter-wave path distance in kilometers.
F = Operating frequency in GHz.

Working out the above formula under the conditions that $F = 30.0$ GHz, $D = 20.0$ kilometers, $d_1 = 10.0$ kilometers, and $d_2 = 10.0$ kilometers, we obtain:

$$Z_1 = 7.06 \text{ meters.}$$

For the same millimeter-wave path conditions but for the frequency at the other end of the band (300.0 GHz), we obtain:

$$Z_1 = 2.23 \text{ meters.}$$

Thus, it is seen that even in the worst case (middle of the path) the problem of Fresnel zone radius clearance requirement is not overly demanding. Such clearances are generally available from existing towers, structures, office building roofs, etc.

TERRESTRIAL REFLECTIONS

All in all, it appears at the time of writing, that the problem of terrestrial reflections for the millimeter-wave band are essentially the same as for the lower microwave and VHF/UHF frequencies. However, reflections from the ground (or water) may be exacerbated by the greater reflection efficiency of smaller areas in terms of square wavelengths. This accrues due to the very small wavelengths with which we are dealing (0.1 to 1.0 centimeters). To counter balance this to some extent is the terrain characteristic of roughness. As can be appreciated, a given reflection area would appear rougher (worse reflector) and less specular to a shorter wavelength than to a longer wavelength. As in regular microwave and VHF/UHF, terrestrial reflections should be avoided whenever possible.

FADING

In a clear troposphere millimeter-waves are "annoyed" by what is called scintillation fading. This type of fading has also variously been know as quiet or quiescent fading. This fading type is basic in millimeter-wave propagation and is practically ever-present. The effect is not-unlike that of twinkling of stars as observed through the "roiling" troposphere. Scintillation fading may be described as a random and "relatively" small fluctuation in received signal strength about its mean value. The more intense of this ubiquitous scintillation fading is of the order of several decibels at the lower frequency end of the "millimeter" band, (please see Fig. 8-3). In this figure, only the downward (weaker signal) fades are shown (upward fading does exist). Amplitude/phase scintillation is generally attributed to mini-scale

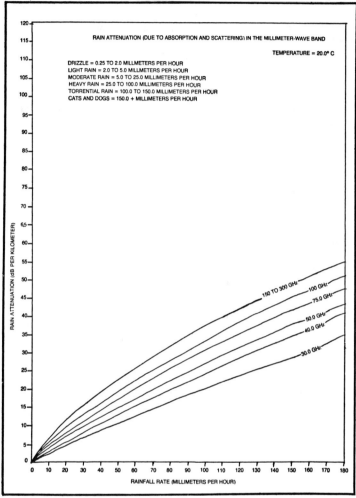

Fig. 8-2. Rain attenuation due to absorption and scattering in the millimeter band.

variations of trophospheric refractive index exacerbated by absorption/scattering. Scintillation type fading is generally expected to become worse as we proceed across the 30.0 GHz to 300.0 GHz "millimeter-wave" (EHF) band. As for XX "regular fading" in the millimeter-wave band, it appears that it may be reasonably described by the Rayleigh-type distribution (please see Appendix 25 and Appendix 13. This Rayleigh-type fading is sometime super-imposed upon the above described scintillation fading.

RAIN

Rain in millimeter-wave path work is perhaps the greatest single bugaboo and most destructive item. This "focuses in" when we realize that raindrops are, in a general fashion, comparable in size to the wavelength. Typical raindrop sizes vary from approximately 1.0 millimeter to 6.0 millimeters while the 30.0 GHz to 300.0 GHz band wavelength span is from 1.0 to 10.0 millimeters. Thus, the ensuing interaction between millimeter electromagnetic energy and resonant raindrops is devastating. One of the important considerations in the subject of rain attenuation is the thunderhead (alias convective cell, convective cloud, thunder cloud, thunder cell, or cumulonimbus cloud). An "average" thunderhead might have according to some estimates an effective rainproducing diameter of some 1.0 to 4.0 kilometers or so, but it is interesting to note that thunderhead dimensions are relatively smaller for the higher rain rates of fall. On short millimeter-wave paths, of the order of 1.0, 2.0 or so kilometers, path attenuation due to rain-fading is more directly proportional to hop lengths than at microwave and VHF/UHF since the path length in this case is comparable to the rain-producing diameter of the thunderhead. In longer paths, uncertainties increase due to lack of precise knowledge of the exact percentage of the path immersed in the rain. This is best treated statistically with the help of your *local* weather people. For proper evaluation of thunderhead distributions and rain rates of fall, it is practically necessary for the millimeter-wave path designer to consult with his local weather authorities and then take it from there. It is not as difficult as it sounds and will be covered in greater detail in our practical millimeter-wave path design example at the the end of this chapter. To give this rain subject a little perspective, the liquid water density for the heavier rains may be approximately 4.0 grams per cubic meter and this corresponds to an average distance between raindrops of some 10.0 centimeters. The larger raindrops tend to take on the geometrical shape of oblate or prolate spheroids and thus tend to lose their polarizational circularity or isotropy.

In general, rain attenuation in the millimeter-wave band is less for vertical polarization than for horizontal polarization. The cross-polarization discrimination, at a given attenuation, for transmitted vertical polarization is better than for transmitted horizontal polarization. In addition to depolarization, rain inflicts interference, system noise and deterioration of antenna performance. Hogg points out increased system noise can accrue due to raindrop black-body radiation as well as from water layers formed on antenna radomes. Interference may occur through the action of interfering signal scattering into a receiving antenna reception pattern by rain. Rain attenuation in the millimeter-wave band is shown in Fig. 8-2.

POLARIZATION

This is actually closely related to precipitation in the path. As pointed out above signal depolarization (change of polarization) may result from ambient raindrops. If the raindrops are so canted (slanted or tilted in their downward progression) that an incident millimetric wave, at a given polarization, undergoes greater attenuation and phase shift than at another polarization, the initial polarization will be changed. This phenomenon accrues because of differential absorption and/or scattering by the rain droplets due to their departure from sphericity. Circularly polarized incident millimeter-waves have been demonstrated to undergo some 16.0 decibels less scatter degradation from rain than linearly polarized ones. In a

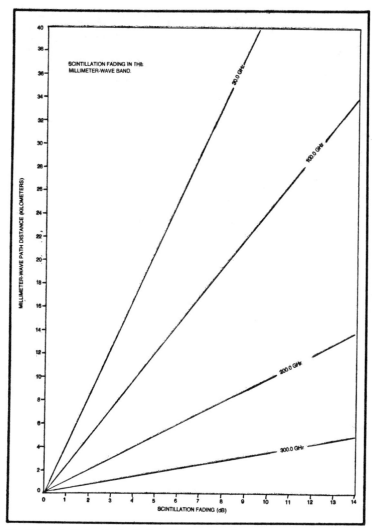

Fig. 8-3. Scintillation fading in the millimeter band.

clear trophosphere, random molecular action produces minimal depolarization and may generally be neglected under these conditions. Water and snow on antenna surfaces foment some cross-polarization.

FOG

Although more work is needed in this realm, as well as others in the millimeter-wave field, we herein produce two generic fog-attenuation curves which may be used for water-clouds also when required. See Fig.

257

8-4. Ice clouds exhibit much less attenuation than water clouds. It might also be stated here that fog and clouds manifest approximately two orders of magnitude less scattering than rain.

SNOW AND HAIL

Attenuation from dry snow in the millimeter-wave band is relatively low compared with rain of equal liquid water content. Wet snow or melting hailstones, however, have been shown to manifest attenuations in excess of those exhibited by rain drops of equal volumes.

DUCTING (TRAPPING)

Ducting is the trapping of an electromagnetic wave in a wave-guide like action between two tropospheric layers or between one such layer and the earth's surface.

Ducting or trapping may take place in the millimeter-wave band, just as in the lower microwaves, when the refractivity gradient becomes -157/km or less (greater negative number). In the 30.0 GHz to 50.0 GHz and 70.0 to 110.0 GHz millimeter-wave windows, for example, ducts of the order of 4.0 to 6.0 meters and 1.0 to 3.0 meters respectively could entrap millimeter-wave energy and possibly cause signal attenuation (or enhancement) at the receiving antenna. If the transmitting antenna and the receiving antenna are both in the duct, the signal could be enhanced; if not, signal loss might be large. Ducts of greater dimensions than those mentioned above might also entrap significant millimeter-wave electromagnetic signal energy.

DIVERSITY

Diversity reception may be of aid in the millimeter-wave band just as it is in the lower microwave bands. However, due to the fact that rain fading is considered "flat" (i.e. it is power or attenuation fading), the usual frequency and/or space diversity employed at the lower microwave frequencies would not work in this case. Route or path diversity has generally been indicated with each of two routes (or paths) separated several kilometers from the other. Thus, while one of the millimeter-wave routes or paths was "under attack" by a thunder cloud(s) rainfall, the chances are good that the other route or path would not be and carry the traffic "home-free".

MILLIMETER-WAVE PATH DESIGN

In millimeter-wave work, at the present state of the science/art, the path losses are basically comprised of free-space loss, tropospheric gas loss (mainly oxygen and gaseous water-vapor, although many other gases and man-made miasmata exact their attenuations), rain or precipitation loss, and scintillation loss. These are the millimeter-wave "basic" losses, but additionally there is "regular" fading, fog loss, etc. It should be kept in mind that much more work is needed in the millimeter-wave band for the provision of "solid" engineering-type path design models since the above may vary with barometric pressure, partial water-vapor pressure, temperature, relative humidity, dew point, location, etc. At any rate, in spite of all this, it *is* feasible now to make quite respectable calculations and path performance predictions. Let's try an example.

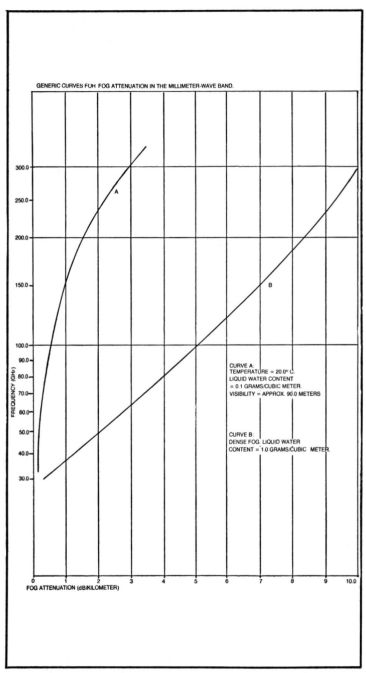

Fig. 8-4. Generic curves for fog attenuation in the millimeter band.

Example:

Path Actual Distance	= 20.0 Kilometers
Frequency	= 100.0 GHz
Location	= Urbana, Illinois (USA)
Required Propagational Reliability	= 99.0%

Find the path attenuation.

Solution:

Step 1. With the path distance of 20.0 kilometers and operating frequency of 100.0 GHz, we employ the free-space attenuation formula:

$$FSL(dB) = 92.4 + 20 \log_{10} F_{GHz} + 20 \log_{10} D_{Km}$$

Where: $FSL(dB)$ = Free Space Loss in decibels.
F_{GHz} = Operating Frequency In GHz.
D_{Km} = Millimeter-wave path actual distance (Kilometers).

By substitution in the formula, we then have:

$$FSL(dB) = 92.4 + 20 \log_{10} 100 + 20 \log_{10} 20$$
$$= 92.4 + 40.0 + 26.02$$
$$= 158.42 \text{ dB}$$

Record this value.   Free Space Attenuation = 158.42 dB

Step 2. With the geographical location of our millimeter-wave path and the required propagational reliability, we enter Appendix No. 21. Under our geographical location, Urbana, Illinois, (USA) we see that the rain rate of fall exceeded for 1.0% of the time is 2.1 millimeters per hour. This means that for 100% − 1.0% = 99.0% of the time this rate of rainfall is *not* exceeded. Entering Figure 8-2 with our 2.1 millimeters per hour rate of rainfall and our operating frequency of 100.0 GHz, we obtain a rain attenuation of 1.5 dB per path kilometer. Now, let's assume that after alot of "praying" and "pulling teeth" we determine via the best weather services available in a given location that it would be reasonably probable that a maximum of three thunderheads of rain producing diameters of 1.5, 2.0, and 3.0 kilometers would be crossing our millimeter-wave path at any one time during a rainstorm. This makes an effective rain attenuated path length of 1.5 + 2.0 + 3.0 = 6.5 kilometers. For our 20.0 kilometer path the rain then, the rain attenuation for 99.0% of the time is equal to or less than 1.5 × 6.5 = 9.75 dB. Record this value.

Rain Attenuation For The Path = 9.75 dB

Step 3. Tropospheric gas loss is determined as follows. Enter Figure 8-1 with our frequency of 100.0 GHz and read the tropospheric gas attenuation of 0.24 dB per kilometer of path. Now, with our path distance of 20.0 kilometers, our total path gas attenuation becomes 20.0 × 0.24 = 4.8 dB. Record this value.

Tropospheric Gas Attenuation = 4.8 dB

Step 4. To determine scintillation fading loss allowance required for this path, enter Fig. 8-3 with our path distance and operating frequency (respectively 20.0 Kilometers and 100.0 GHz) and read on the "SCINTILLATION FADING dB" scale the value of 8.2 dB. Record this value.

Path Scintillation Fading Loss = 8.2 dB

Step 5. From Appendix 25 we enter graph 25-A at a path distance of 12.0 miles (equivalent of our 20.0 kilometers). From the 99.0% propagation reliability curve, we read a fade margin of 5.25 dB for non-diversity reception. (We assume non-diversity operation herein, if, however, the reader wishes to calculate for diversity receiving conditions, simply use graph 25-B.) Record our 5.0 dB value.

Rayleigh-type fading allowance = 5.0 dB

(Note—For simplicity, we have herein assumed that the rain attenuation and Rayleigh-type fading take place simultaneously. A "worst-case" would dictate an assumption of their taking place at totally different times in which case it might seem as though the total outage would be 2.0% of the time, making a path propagational reliability of 100.0% - 2.0% = 98.0%. In such a case, however, of non-simultaneity, the Rain path attenuation pad could "double in brass" and more than cover for Rayleigh type fading. There are many possible trade-offs in such cases and the final decision is up to the design engineer and his particular situation. The "extra" fade margin available in the above case (9.75 dB instead of 5.25 dB) would then increase the percentage reliability during Rayleigh-type fading.) Remember that your ideas of outages should be well jelled. Decibels cost money. Please see Appendix 23. One's usage of a system and acceptable percentage of outage must be realistically faced. If you need alot of reliability, and you have the money to spend, design it in. Usually though, this is not the case and there is little to be gained realistically, by "using golden wheelbarrows for hauling sand."

At any rate, returning to our millimeter-wave problem, let's add up our path losses (between isotropic antennas).

Free space attenuation	= 158.42 dB
Path Rain Attenuation	= 9.75 dB
Tropospheric gas loss	= 4.80 dB
Path scintillation loss	= 8.20 dB
Rayleigh-type fading	= 5.00 dB

Total path loss for 99.0% propagational reliability = 186.17 dB.

If there existed fog loss, this would have been appropriately added in from estimates obtained from Fig. 8-4.

Appendices

Appendix-1
Earth Constants

Earth conductivity (δ) and earth permittivity (ϵ) exert profound influences upon radio telecommunications (e.g. ground wave propagation, terrestrial reflections, etc). Earth conductivity is the conductance between opposite faces of a unit cable (usually 1.0 cubic meter) of a given earth material (e.g. rock, sand, clay, loam, water, etc.).

Earth permittivity is the ratio of a capacitor's capacitance using a given earth sample as a dielectric, to that using air as a dielectric. Earth permittivity is also known as earth dielectric constant, earth inductive capacity and earth capacitivity.

The geologic factors determining our earth constants (conductivity and permittivity) are complex and copious indeed! Earth characteristics vary in accordance with, for example, physio-chemical composition and moisture absorbing/retaining capacity. The thermal ambient additionally complicates the problem. For example, measurements have revealed earth conductivity and permittivity coefficients of relatively small percentages at "normal" temperature ranges while at the freezing point both these constants manifest dramatic variations. There are additional earth-constant dependencies. These are operating frequency and antenna polarization. While measured values of ground constants are doubtless the best for a given application, this procedure is too often excessively time consuming and expensive. We accordingly present the following earth-constant table reflecting the conductivity and permittivity values employed in the telecommunications industry.

Courtesy TAI Inc. (subsidiary of E-Systems Inc.) from Consuletter International Vol. 6 No. 5 entitled "EARTH CONSTANTS" and Vol. 5 No. 4 entitled "RADIOGEOLOGY."

Table A-1. Earth-Constant Table.

Earth Type	Conductivity(δ) in Mhos (or Siemens) per meter.	Permittivity (ϵ)
Poor	0.001	4.0 to 5.0
Moderate	0.003	4.0
Fair	0.01	15.0
Average	0.005 to 0.03	10.0 to 15.0
Good	0.01 to 0.02	4.0 to 30.0
Dry	0.00001 to 0.001	2.0 to 5.0
land	0.002	15.0
Desert	0.01	3.0
Dry, sandy, flat—typical of coastal land	0.002	10.0
Flat, wet, coastal region	0.01 to 0.02	4.0 to 30.0
Dry sandy soils	0.001	10.0
Pastoral Land	0.005	15.0
Pastoral hills, rich soil	0.003 to 0.01	14.0 to 20.0
Pastoral medium hills and forestation	0.004 to 0.006	13.0
Fertile soil	0.002	10.0
Rich agricultural land (low hills)	0.01	15.0
Rocky land (steep hills)	0.002	10.0 to 15.0
Highly moist soil	0.005 to 0.02	30.0
Wet	0.001 to 0.1	5.0 to 30.0
Marshy	0.1	30.0
Marshy, flat (densly wooded)	0.0075	12.0
Marshy, forested, flat	0.008	12.0
Low hills with unforested rich soil	0.01 to 0.02	4.0 to 30.0
Mountainous, hilly (to approx. 1,000.0 meters	0.001	5.0
Urban industrial (average attenuation)	0.001	5.0
Urban industrial (maximal attenuation)	0.0004	3.0
Urban industrial	0.0001	3.0
Fresh water	0.001 to 0.01	80.0 to 81.0
Fresh water (10° Celsius at 100.0 MHz)	0.001 to 0.01	84.0
Fresh water (20° Celsius at 100.0 MHz)	0.001 to 0.01	80.0
Sea water	3.0 to 5.0	80.0 to 81.0
Sea water (10° Celsius at up to 1.0 GHz)	4.0 to 5.0	80.0
Sea water (20° Celsius at up to 1.0 GHz)	4.0 to 5.0	73.0
Sea ice	0.001	4.0
Polar ice	0.000025	3.0
Polar ice Cap	0.0001	1.0
Arctic land	0.0005	3.0 to 5.0

Appendix 2
dBµV/m to dBm Conversions

The following three graphical representations make it a very simple matter to convert from electromagnetic field strength in decibels relative to 1.0 microvolt-per-meter (dBuB/m) to decibels relative to 1.0 milliwatt (dBm) as delivered by an isotropic antenna immersed in the corresponding dBuV/m field, and vice versa. Some may feel more comfortable working in terms of one of these than the other.

The radio physics involved are interesting indeed. A total transmitted radio electromagnetic field is comprised of equal quantities of mutually orthogonal (perpendicular) E and H energy (respectively electric and magnetic radio frequency energy) fields. A radio wave's intensity, therefore, can be appraised in terms of a radio wave front's per-unit-length electric or magnetic field strength as these fields propagate through free-space impedance (Z) which amounts to 120π or 377 ohms. In other words, if one divides the spatial electric field by the spatial magnetic field (respectively roughly analogous to voltage and current in Ohm's Law) of a radio frequency wavefront, the result will be 120π ohms or 377 ohms. Now, should a measuring antenna's effective length be 1.0 meter, our reference area would be 1.0 meter squared (or 1.0 square meter)—1.0 meter side of the spatial capture square parallel to the electric (E) field and the orthogonal (perpendicular) 1.0 meter parallel to the magnetic field. But since the power may be deduced by proportionality to either E^2 or H^2 (by analogy in Ohm's Law to E^2/R and I^2/R) and we know E^2 and Z (377 ohms), then we need only measure the E component of the passing radio wave to know its power density. Now the radio power extracted from a passing wave is found by multiplying the power flow in one square meter by the effective area in square meters of the particular antenna in question. In our case, we are speaking of an isotropic antenna (a fictitious antenna receiving equally well—or transmitting—in all directions). The effective or capture area of such an isotrope (isotropic antenna) is $(\lambda^2/4\pi)$. From here it is easy to see that the spatial power or radio frequency wave power captured by an isotrope can be calculated by the simple algebraic formula $(E^2/120\pi \times \lambda^2/4\pi)$.

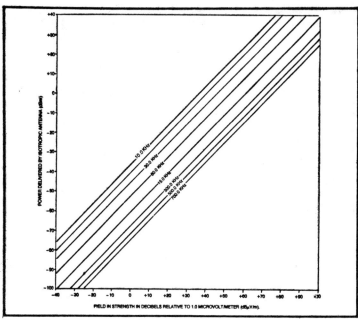

Fig. A2-1A. Chart to convert dBm to dBuV/m from 10 KHz to 700 kHz.

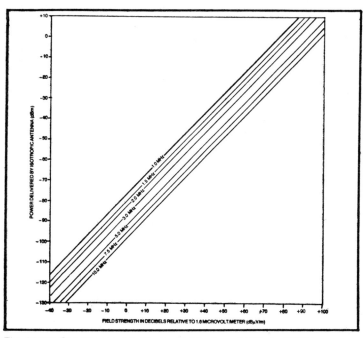

Fig. A2-1B. Chart to convert dBm to dBuV/m from 1 MHz to 10 MHz.

267

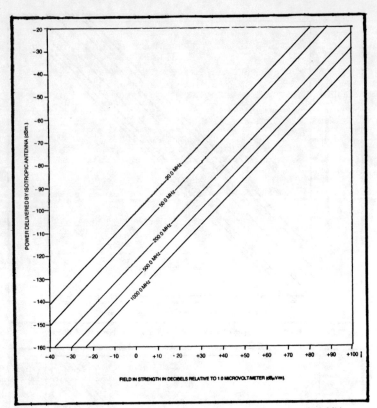

Fig. A2-1C. Chart to convert dBm to dBuV/m from 20 MHz to 1000 MHz.

Appendix 3
CCIR Noise Maps

In radio-telecommunications systems, noise is a constraint upon communications efficiency. If noise did not exist, it would require zero transmitter power to achieve a signal-to-noise ratio of 1:1 (0.0 dB). In propagation problems of all sorts, it becomes important to become cognizant of noise electrical characteristics, levels, and sources to determine the signal's competition. In the LF (30.0 to 300.0 KHz) band, atmospheric noise is usually dominant while at HF (3.0 to 30.0 MHz) atmospheric, galactic (cosmic), and man-made noise might conspire in various mixes to harass the desired signal information. At VHF (30.0 to 300.0 MHz) galactic and internal-receiver noise might conspire to plague our signal.

The data in this appendix item is made available through the courtesy of the CCIR (International Radio Consultative Committee) of the ITU (International Telecommunications Union-communications branch of the United Nations) in Geneva, Switzerland, from its Report-322, Geneva 1963, entitled, "World Distribution and Characteristics of Atmospheric Radio Noise." This data embodies, in addition to atmospheric noise, the man-made noise at a quiet receiving location and galactic noise, all having relative importance in various types of propagation problems.

The material in this appendix presents the noise values and types in a series of map/graph sets (three per set), each set identified alphanumerically by a number (1 through 24) and A,B, or C. As may be seen, the noise at any geographical location depends upon the time of day, the season of the year, frequency of operation, and receiving system bandwidth.
For convenience, the time of day is divided into six LOCAL STANDARD TIME four-hour blocks as follows:

0000 to 0400 Local	Standard	Time
0400 to 0800 Local	Standard	Time
0800 to 1200 Local	Standard	Time
1200 to 1600 Local	Standard	Time
1600 to 2000 Local	Standard	Time
2000 to 2400 Local	Standard	Time

The seasons of the year are as follows:
The noise power is identified as ATMOSPHERIC, MAN-MADE at a quiet receiving location, and GALACTIC (cosmic) and is presented as a function

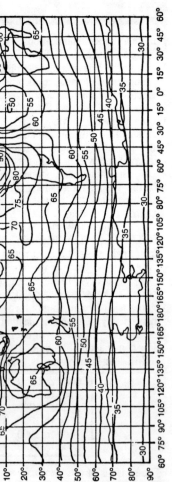

Fig. A3-1A. Expected values of atmospheric radio noise, F_{am} (dB above kT_ob at 1 MHz). Winter 00-04.

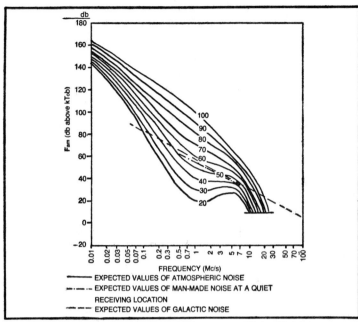

Fig. A3-1B. Variation of radio noise with frequency. Winter 00-04.

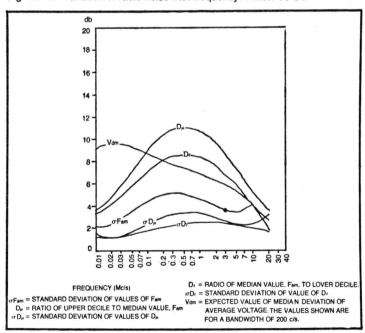

Fig. A3-1C. Data on noise variability and character. Winter 00-04.

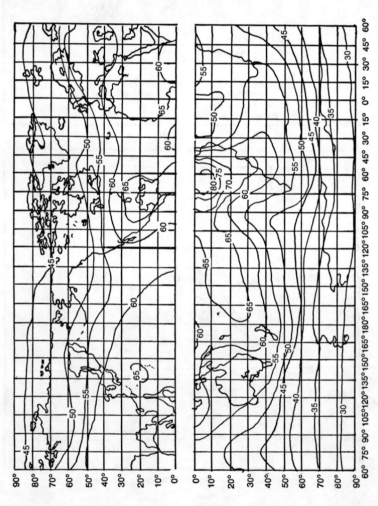

Fig. A3-2A. Expected values of atmospheric radio noise F_{am} (dB above kT_o at 1 MHz). Winter 04-08.

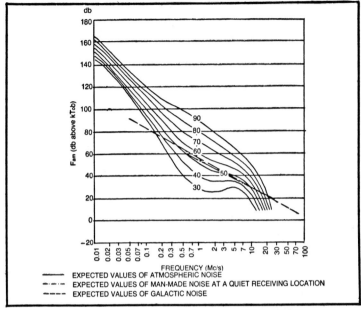

Fig. A3-2B. Variation of radio noise with frequency. Winter 04-08.

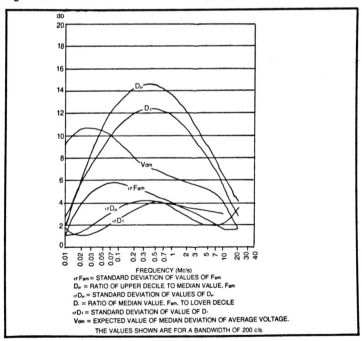

Fig A3-2C. Data on noise variability and character. Winter 04-08.

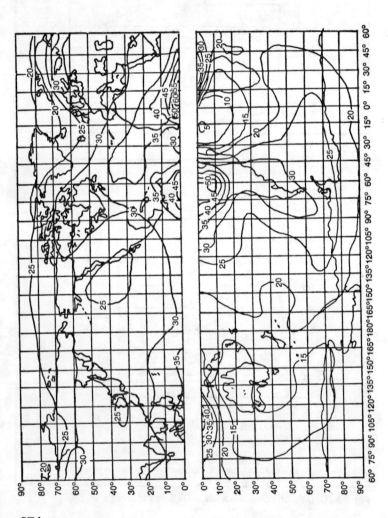

Fig. A3-3A. Expected values of atmospheric radio noise, Fam (dB above KTo at 1 MHz) Winter 08-12.

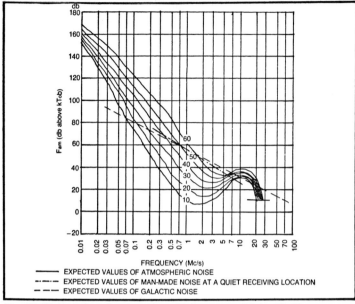

Fig. A3-3B. Variation of radar noise with frequency. Winter 08-12.

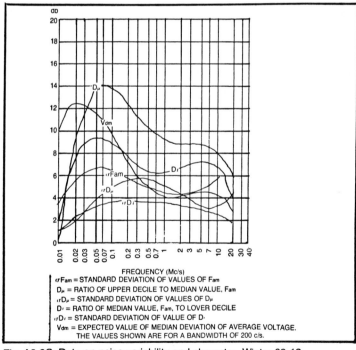

Fig. A3-3C. Data on noise variability and character. Winter 08-12.

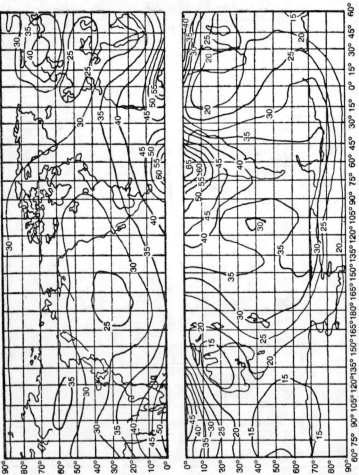

Fig. A3-4A. Expected values of atmospheric radio noise, F_{am} (dB above KT_o at 1 MHz) Winter 12-16.

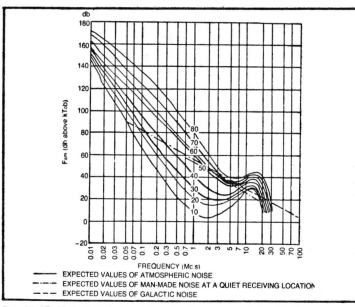

Fig. A3-4B. Variation of radio noise with frequency. Winter 12-16.

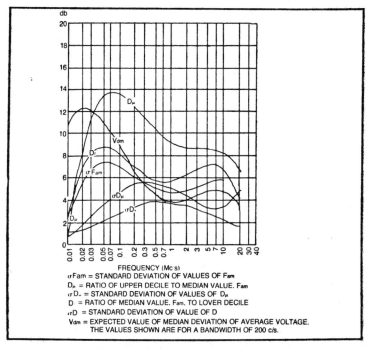

σFam = STANDARD DEVIATION OF VALUES OF Fam
D_μ = RATIO OF UPPER DECILE TO MEDIAN VALUE. Fam
σD_μ = STANDARD DEVIATION OF VALUES OF D_μ
D = RATIO OF MEDIAN VALUE, Fam, TO LOVER DECILE
σD = STANDARD DEVIATION OF VALUE OF D
Vdm = EXPECTED VALUE OF MEDIAN DEVIATION OF AVERAGE VOLTAGE.
THE VALUES SHOWN ARE FOR A BANDWIDTH OF 200 c/s.

Fig. A3-4C. Data on noise variability and character. Winter 12-16.

277

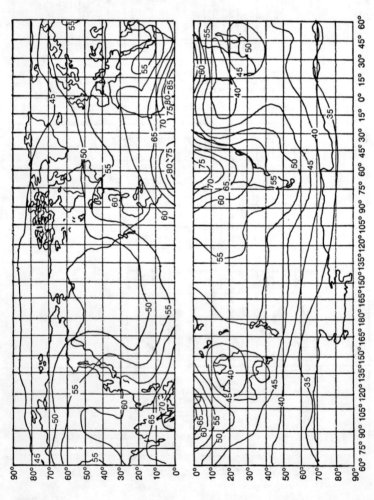

Fig. A3-5A. Expected values of atmospheric radio noise, F_{am} (dB above KTo at 1 MHz) Winter 16-20.

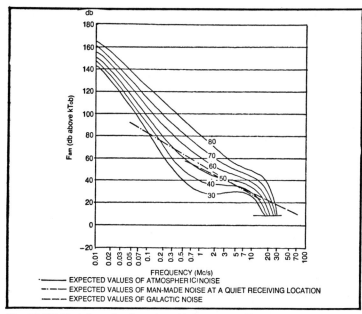

Fig. A3-5B. Variation of radio noise with frequency. Winter 16-20.

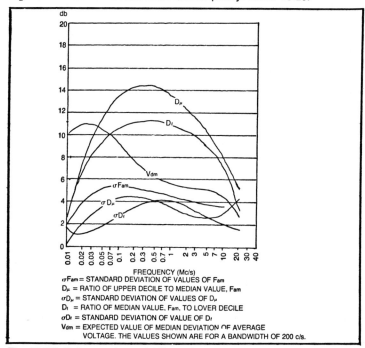

Fig. A3-5C. Data on noise variability and character. Winter 16-20.

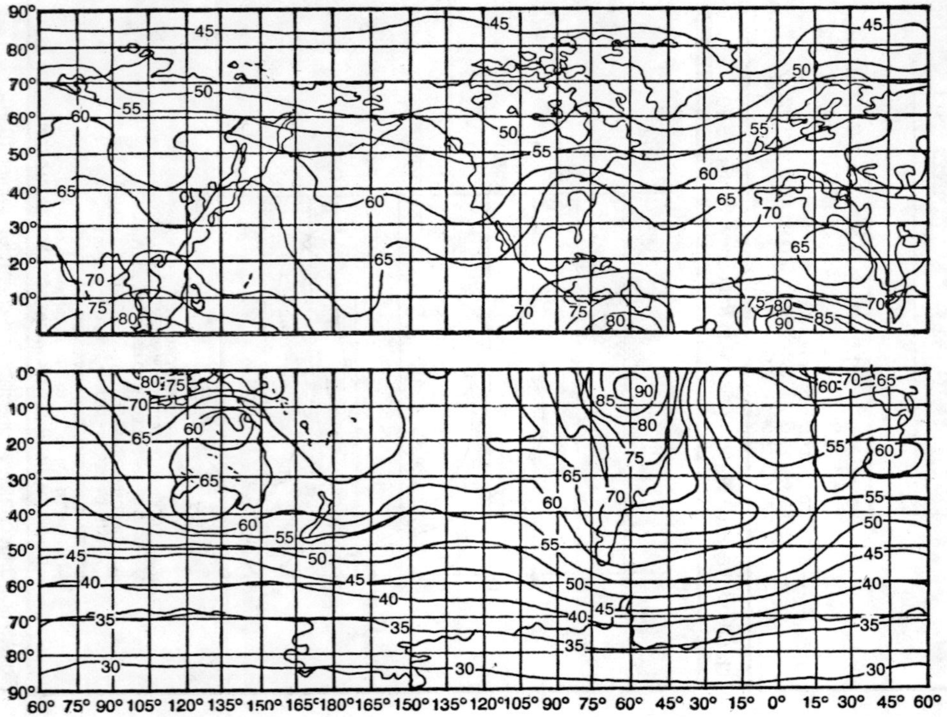

Fig. A3-6A. Expected values of atmospheric radio noise, F_{am} (dB above KT_ob at 1 MHz) Winter 20-24

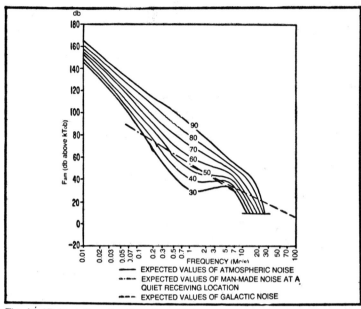

Fig. A3-6B. Variation of radio noise with frequency. Winter 20-24.

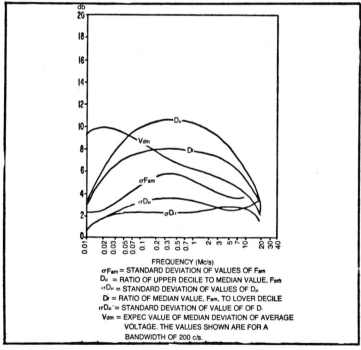

Fig. A3-6C. Data on noise variability and cnaracter. Winter 20-24.

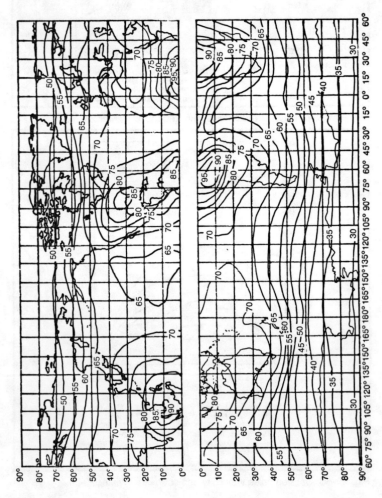

Fig. A3-7A. Expected values of atmospheric radio noise, F_{am} (dB above KT_ob at 1 MHz Spring 00-04.

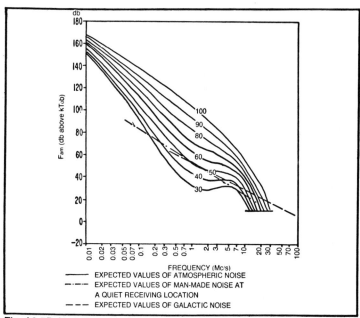

Fig. A3-7B. Variation of radio noise with frequency. Spring 00-04.

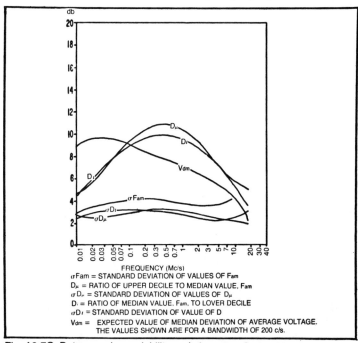

σF_{am} = STANDARD DEVIATION OF VALUES OF F_{am}
D_μ = RATIO OF UPPER DECILE TO MEDIAN VALUE, F_{am}
σD_μ = STANDARD DEVIATION OF VALUES OF D_μ
D_l = RATIO OF MEDIAN VALUE, F_{am}, TO LOWER DECILE
σD_l = STANDARD DEVIATION OF VALUE OF D
V_{dm} = EXPECTED VALUE OF MEDIAN DEVIATION OF AVERAGE VOLTAGE.
THE VALUES SHOWN ARE FOR A BANDWIDTH OF 200 c/s.

Fig. A3-7C. Data on noise variability and character. Spring 00-04.

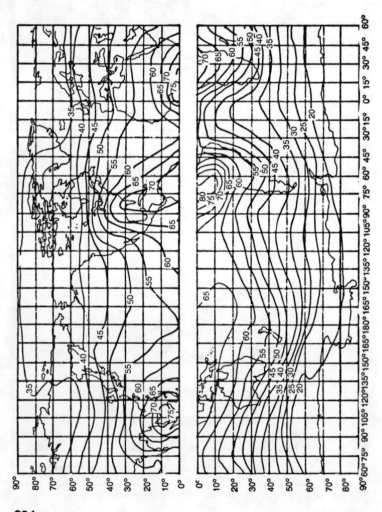

Fig. A3-8A. Expected values of atmospheric radio noise, Fam (dB above KTob at 1 MHz) Spring 04-08.

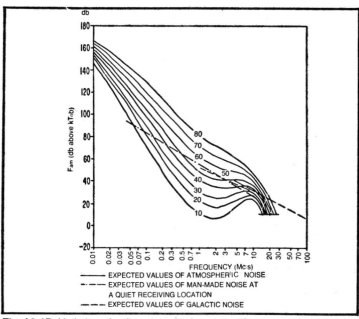

Fig. A3-1B. Variation of radio noise with frequency. Spring 04-08.

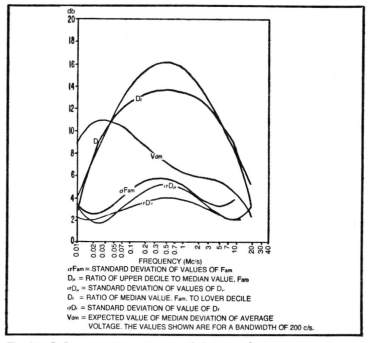

Fig. A3-8C. Data on noise variability and character. Spring 04-08.

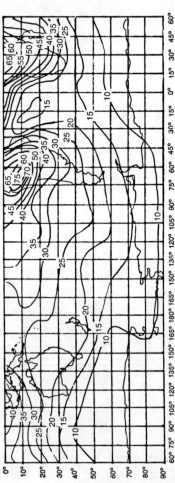

Fig. A3-9A. Expected valued of atmospheric radio noise, Fam (dB above KT_0 at 1 MHz) Spring 08-12.

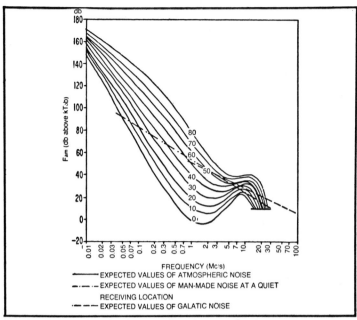

Fig. A3-9B. Variation of radio noise with frequency. Spring 08-12.

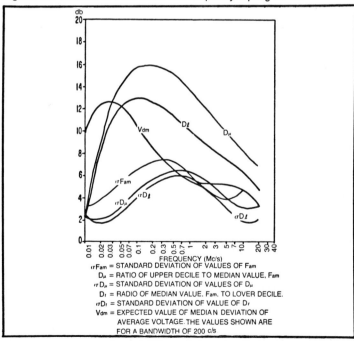

Fig. A3-9C. Data on noise variability and character. Spring 08-12.

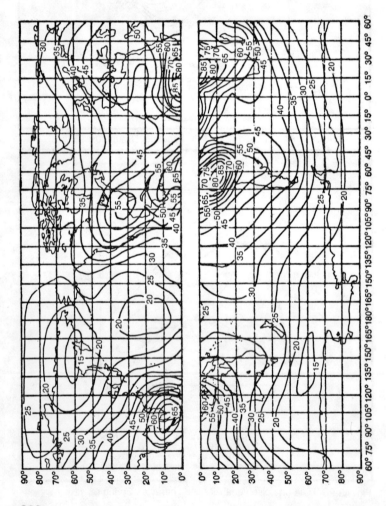

Fig. A3-10A. Expected values of atmospheric radio noise, F_{am} (dB above KT_o at 1 MHz) Spring 12-16.

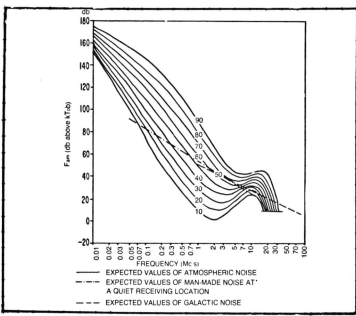

Fig. A3-10B. Variation of radio noise with frequency. Spring 12-16.

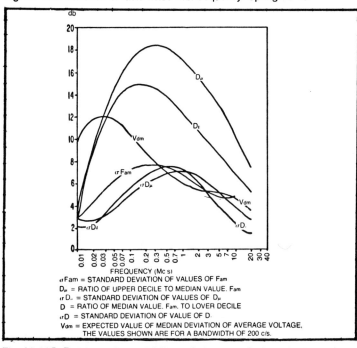

Fig. A3-10C. Data on noise variability and character. Spring 12-16.

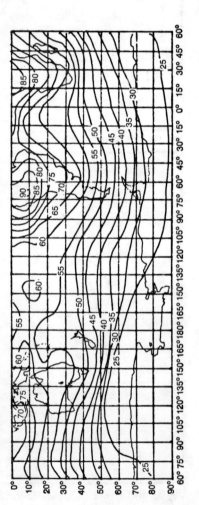

Fig. A3-11A. Expected values of atmospheric radio noise, F_{am} (dB above KT_ob at 1 MHz) Spring 16-20.

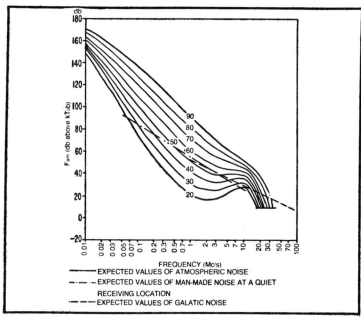

Fig. A3-11B. Variation of radio noise with frequency. Spring 16-20.

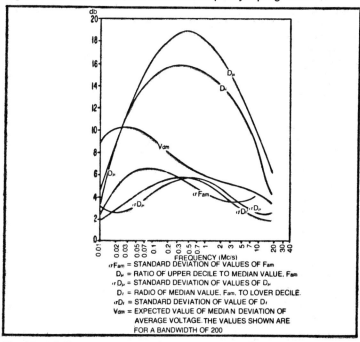

Fig. A3-11C. Data on noise variability and character. Spring 16-20.

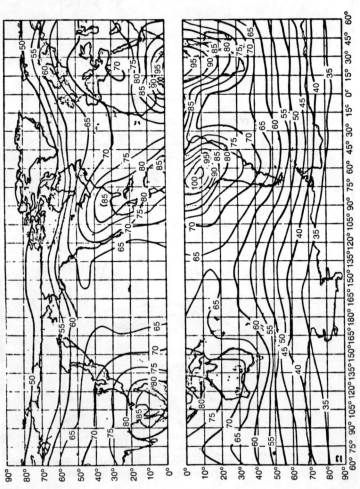

Fig. A3-12A. Expected values of atmospheric radio noise, Fam (dB above KTo at 1 MHz) Spring 20-24.

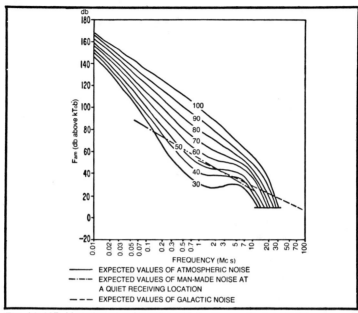

Fig. A3-12B. Variation of radio noise with frequency. Spring 20-24.

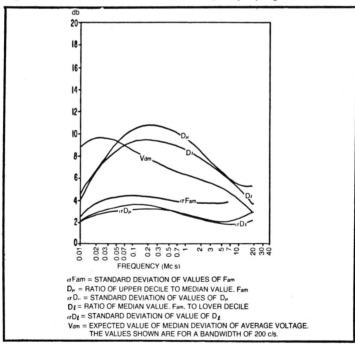

Fig. A3-12C. Data on noise variability and character. Spring 20-24.

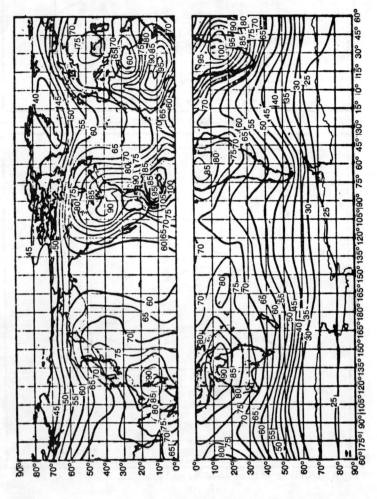

Fig. A3-13A. Expected values of atmospheric radio noise, F_{am} (dB above KT_ob at 1 MHz) Summer 00-04.

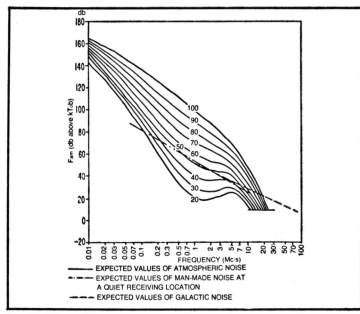

Fig. A3-13B. Variation of radio noise with frequency. Summer 00-04.

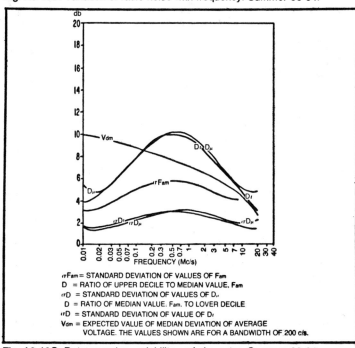

Fig. A3-13C. Data on noise variability and character. Summer 00-04.

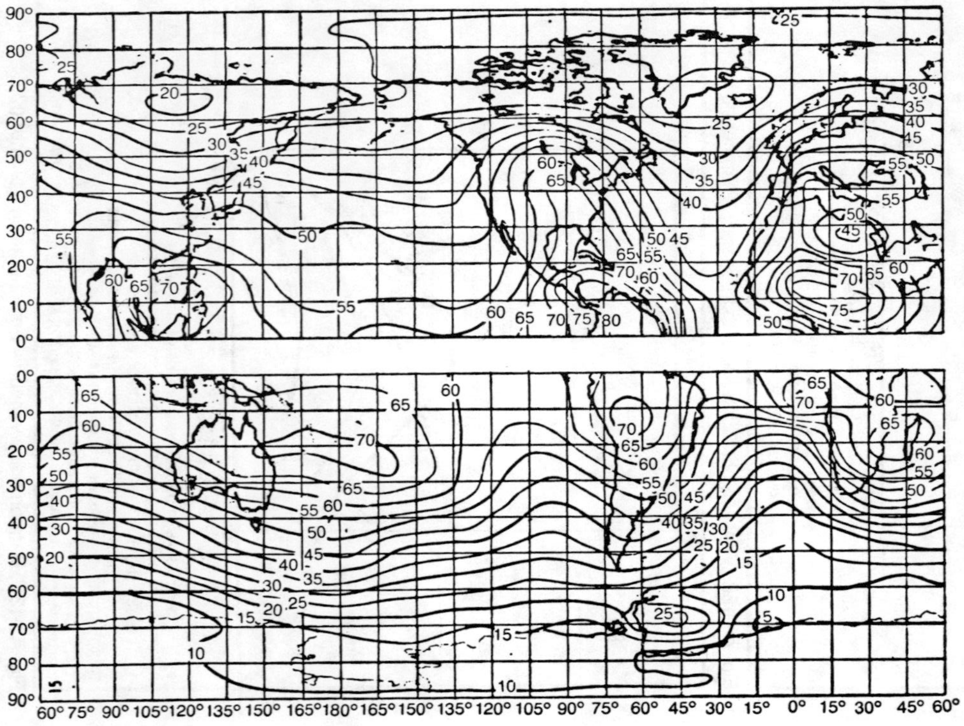

Fig. A3-14A. Expected values of atmospheric radio noise, F_{am} (dB above KT_0b at 1 MHz) Summer 04-08.

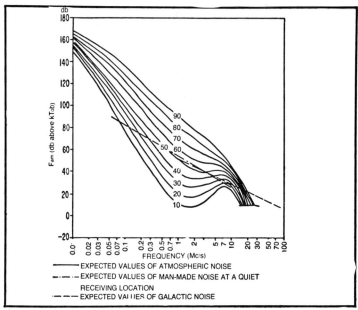

Fig. A3-14B. Variation of radio noise with frequency. Summer 04-08.

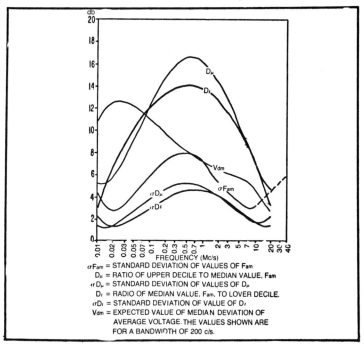

Fig. A3-14C. Data on noise variability and character. Summer 04-08.

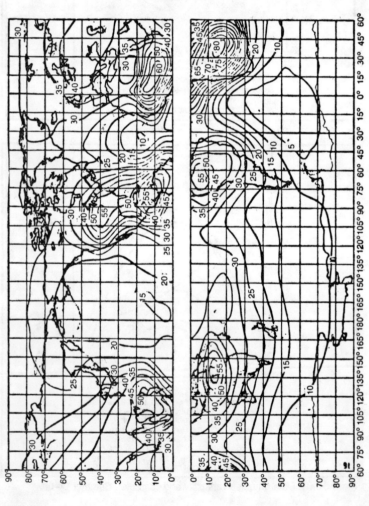

Fig. A3-15A. Expected values of atmospheric radio noise, Fam (dB above KT₀b at 1 MHz). Summer 08-12.

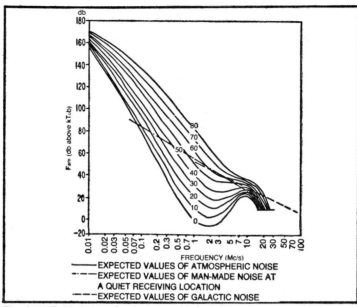

Fig. A3-15B. Variation of radio noise with frequency. Summer 08-12.

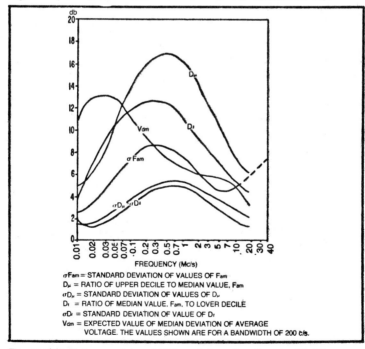

Fig. A3-15C. Data on noise variability and character. Summer 08-12.

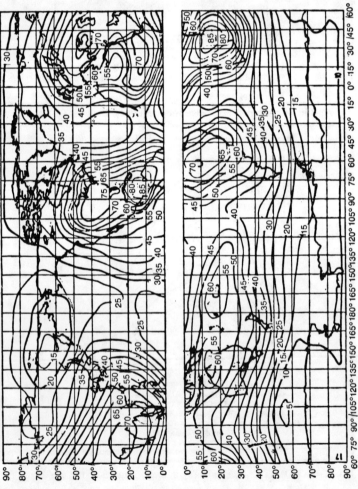

Fig. A3-16A. Expected values of atmospheric radio noise, F_{am} (dB above KT_0b at 1 MHz). Summer 12-16.

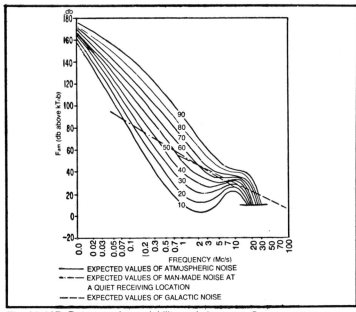

Fig. A3-16B. Data on noise variability and character. Summer 12-16.

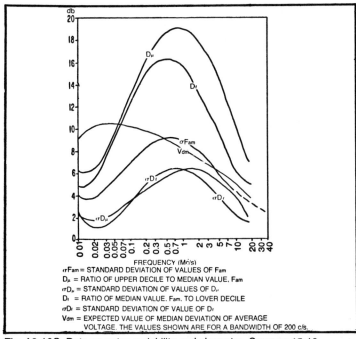

Fig. A3-16C. Data on noise variability and character. Summer 12-16.

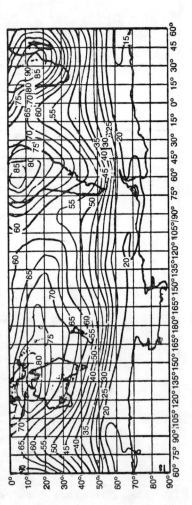

Fig. A3-17A. Expected values of atmospheric radio noise, F_{am} (dB above KT_ob at 1 MHz). Summer 16-20.

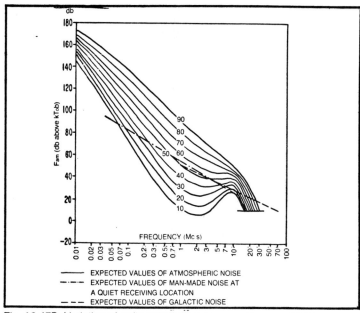

Fig. A3-17B. Variation of radio noise with frequency. Summer 16-20.

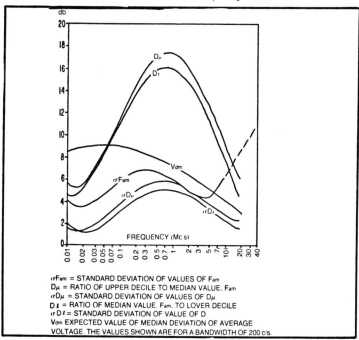

σF_{am} = STANDARD DEVIATION OF VALUES OF F_{am}
$D\mu$ = RATIO OF UPPER DECILE TO MEDIAN VALUE, F_{am}
$\sigma D\mu$ = STANDARD DEVIATION OF VALUES OF $D\mu$
Dl = RATIO OF MEDIAN VALUE, F_{am} TO LOWER DECILE
σDl = STANDARD DEVIATION OF VALUE OF Dl
V_{dm} EXPECTED VALUE OF MEDIAN DEVIATION OF AVERAGE VOLTAGE. THE VALUES SHOWN ARE FOR A BANDWIDTH OF 200 c/s.

Fig. A3-17C. Data on noise variability and character. Summer 16-20.

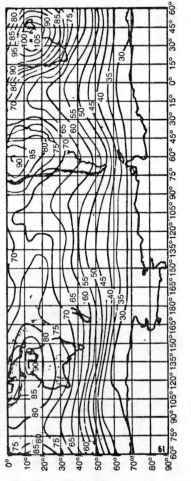

Fig. A3-18A. Expected values of atmospheric radio noise, F_{am} (dB above KT_0b at 1 MHz). Summer 20-24.

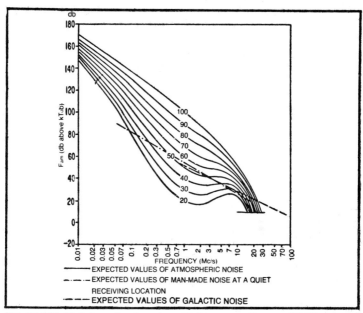

Fig. A3-18B. Variation of radio noise with frequency. Summer 20-24.

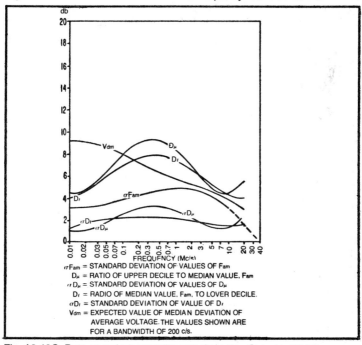

σF_{am} = STANDARD DEVIATION OF VALUES OF F_{am}
D_μ = RATIO OF UPPER DECILE TO MEDIAN VALUE, F_{am}
σD_μ = STANDARD DEVIATION OF VALUES OF D_μ
D_l = RADIO OF MEDIAN VALUE, F_{am}, TO LOVER DECILE.
σD_l = STANDARD DEVIATION OF VALUE OF D_l
V_{dm} = EXPECTED VALUE OF MEDIAN DEVIATION OF
AVERAGE VOLTAGE. THE VALUES SHOWN ARE
FOR A BANDWIDTH OF 200 c/s.

Fig. A3-18C. Data on noise variability and character. Summer 20-24.

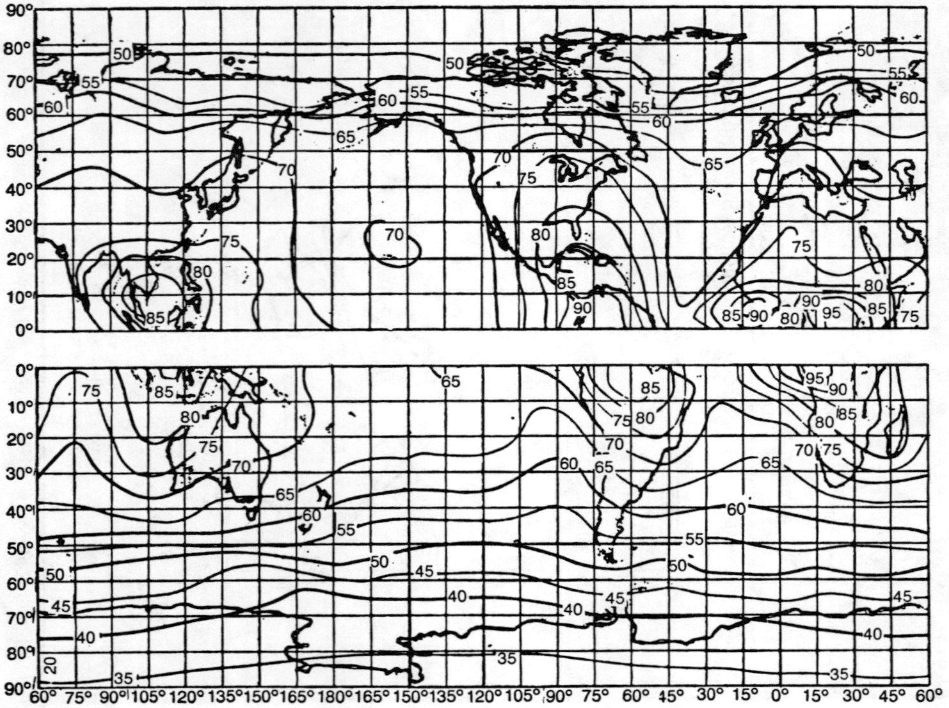

Fig. A3-19A. Expected values of atmospheric radio noise F_{am} (dB above KT_0 at 1 MHz) Autumn 00-04.

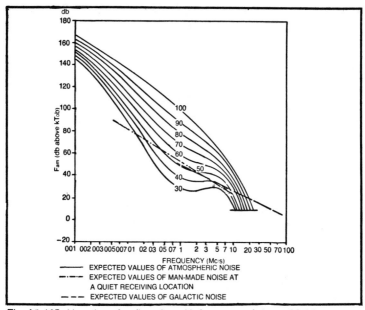

Fig. A3-19B. Variation of radio noise with frequency. Autumn 00-04.

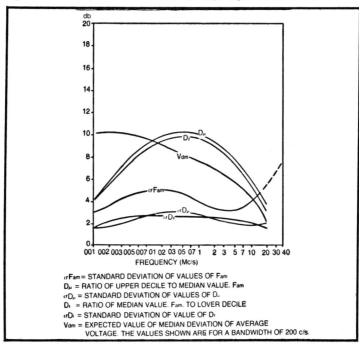

σF_{am} = STANDARD DEVIATION OF VALUES OF F_{am}
D_u = RATIO OF UPPER DECILE TO MEDIAN VALUE. F_{am}
σD_u = STANDARD DEVIATION OF VALUES OF D_u
D_l = RATIO OF MEDIAN VALUE. F_{am} TO LOVER DECILE
σD_l = STANDARD DEVIATION OF VALUE OF D_l
V_{dm} = EXPECTED VALUE OF MEDIAN DEVIATION OF AVERAGE VOLTAGE. THE VALUES SHOWN ARE FOR A BANDWIDTH OF 200 c/s.

Fig. A3-19C. Data on noise variability and character Autumn 00-04.

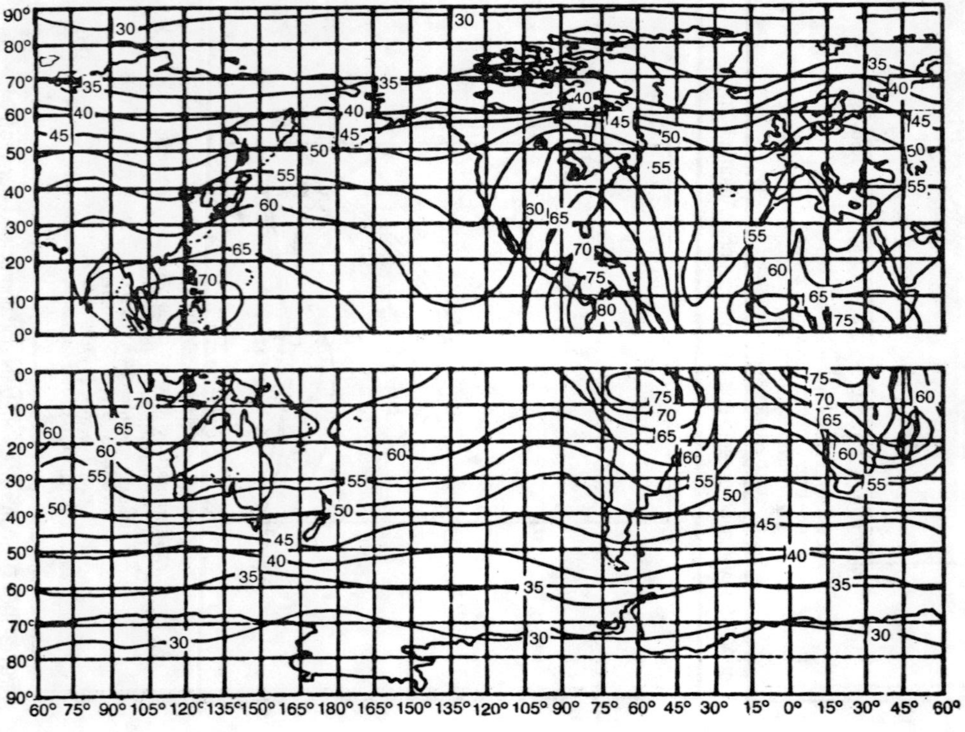

Fig. A3-20A. Expected values of atmospheric radio noise, Fam (dB above KTob at 1 MHz) Autumn 04-08.

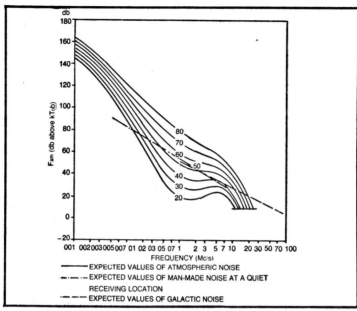

Fig. A3-20B. Variation of radio noise with frequency. Autumn 04-08.

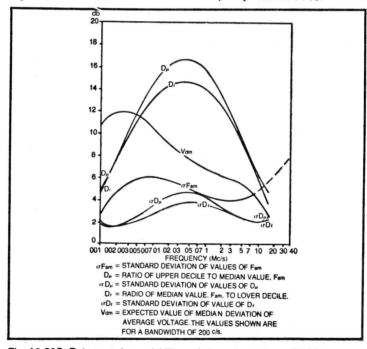

Fig. A3-20C. Data on noise variability and character. Autumn 04-08.

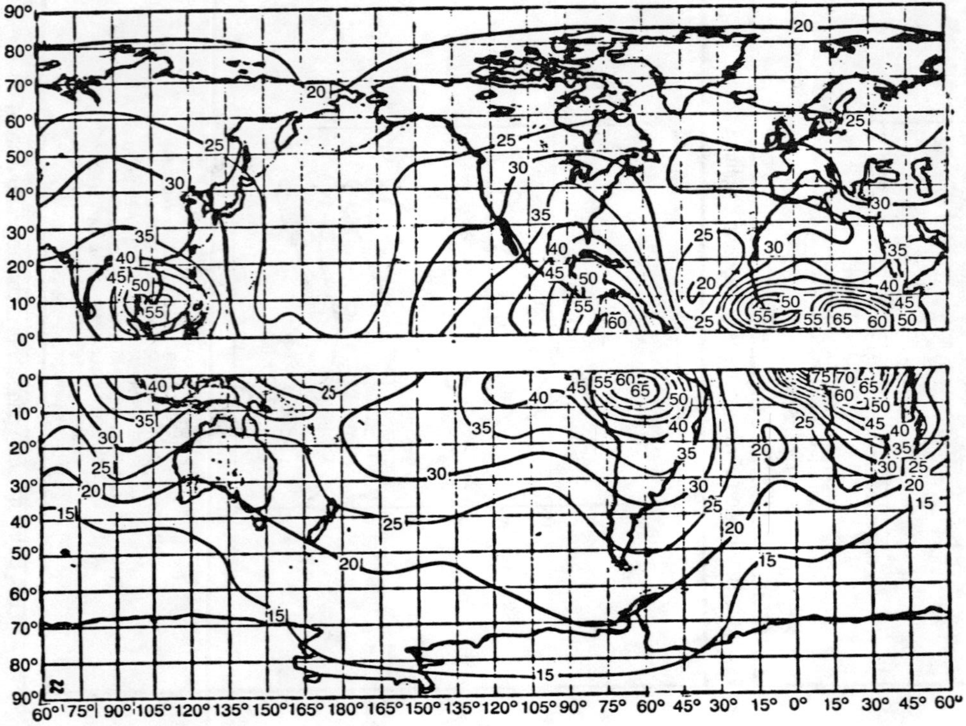

Fig. A3-21A: Expected values of atmospheric radio noise, Fam (dB above KTob at 1 MHz) Autumn 08-12.

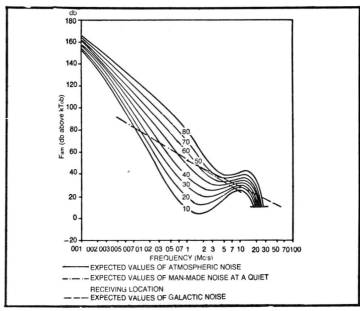

Fig. A3-21B. Variation of radio noise with frequency. Autumn 08-12.

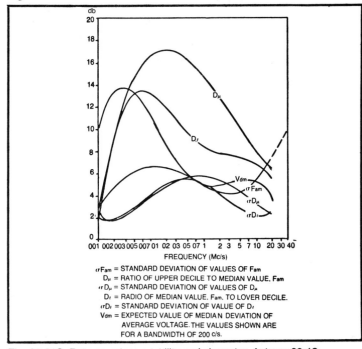

Fig. A3-21C. Data on noise variability and character. Autumn 08-12.

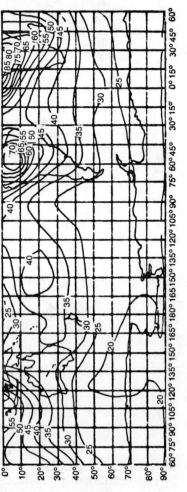

Fig. A3-22A. Expected values of atmospheric radio noise, F_{am} (dB above KT_0b at 1 MHz) Autumn 12-16.

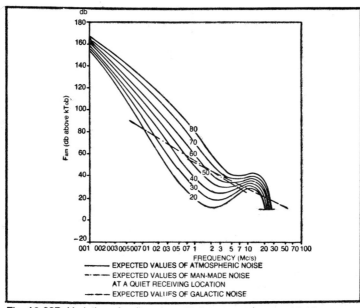

Fig. A3-22B. Variation of radio noise with frequency. Autumn 12-16.

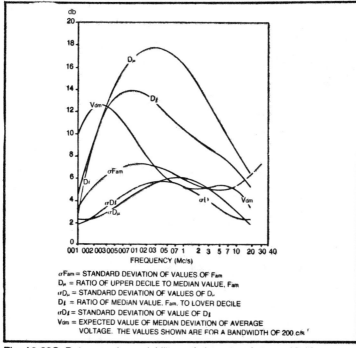

σF_{am} = STANDARD DEVIATION OF VALUES OF F_{am}
D_μ = RATIO OF UPPER DECILE TO MEDIAN VALUE, F_{am}
σD_μ = STANDARD DEVIATION OF VALUES OF D_μ
D_ℓ = RATIO OF MEDIAN VALUE, F_{am} TO LOWER DECILE
σD_ℓ = STANDARD DEVIATION OF VALUE OF D_ℓ
V_{dm} = EXPECTED VALUE OF MEDIAN DEVIATION OF AVERAGE
VOLTAGE. THE VALUES SHOWN ARE FOR A BANDWIDTH OF 200 c/s

Fig. A3-22C. Data on noise variability and character. Autumn 12-16.

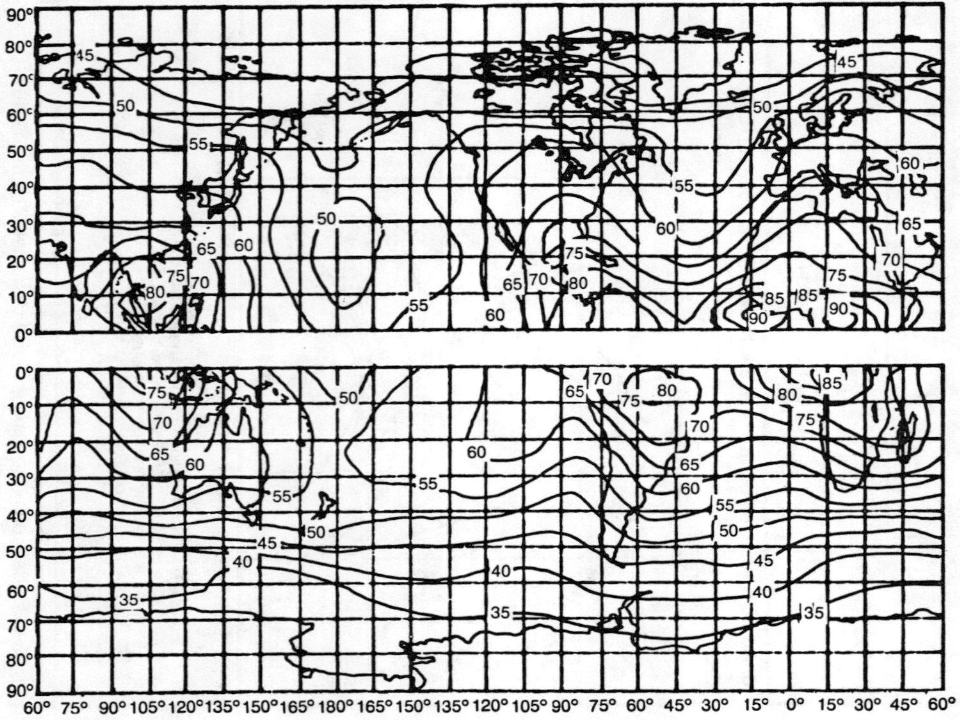

Fig. A3-23A. Expected values of atmospheric radio noise, F_{am} (dB above kT_0 at 1 MHz). Autumn 16-20.

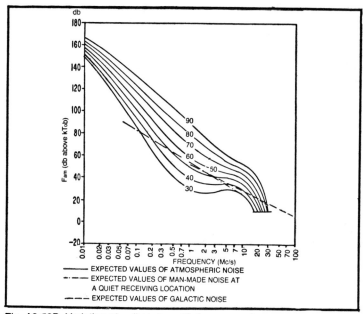

Fig. A3-23B. Variation of radio noise with frequency. Autumn 16-20.

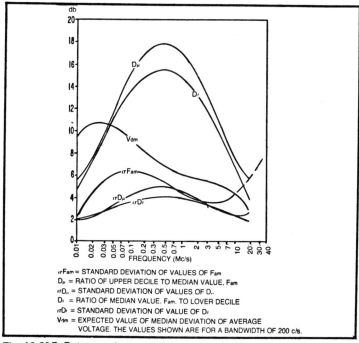

σF_{am} = STANDARD DEVIATION OF VALUES OF F_{am}
D_μ = RATIO OF UPPER DECILE TO MEDIAN VALUE, F_{am}
σD_μ = STANDARD DEVIATION OF VALUES OF D_μ
D_l = RATIO OF MEDIAN VALUE, F_{am} TO LOWER DECILE
σD_l = STANDARD DEVIATION OF VALUE OF D_l
V_{dm} = EXPECTED VALUE OF MEDIAN DEVIATION OF AVERAGE VOLTAGE. THE VALUES SHOWN ARE FOR A BANDWIDTH OF 200 c/s.

Fig. A3-23C. Data on noise variability and character. Autumn 16-20.

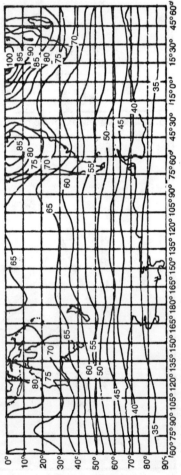

Fig. A3-24A. Expected values of atmospheric radio noise, Fam (dB above KT_ob at 1 MHz) Autumn 20-24.

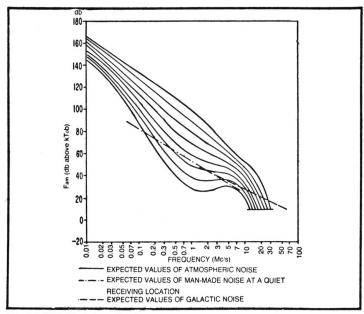

Fig. A3-24B. Variation of radio noise with frequency. Autumn 20-24.

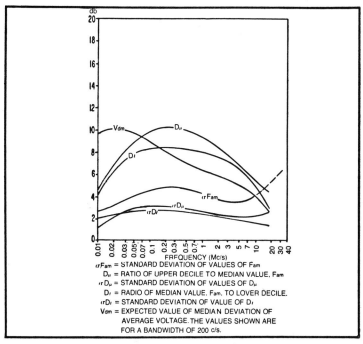

σF_{am} = STANDARD DEVIATION OF VALUES OF F_{am}
D_μ = RATIO OF UPPER DECILE TO MEDIAN VALUE, F_{am}
σD_μ = STANDARD DEVIATION OF VALUES OF D_μ
D_l = RADIO OF MEDIAN VALUE, F_{am}, TO LOWER DECILE.
σD_l = STANDARD DEVIATION OF VALUE OF D_l
V_{dm} = EXPECTED VALUE OF MEDIAN DEVIATION OF AVERAGE VOLTAGE. THE VALUES SHOWN ARE FOR A BANDWIDTH OF 200 c/s.

Fig. A3-24C. Data on noise variability and character. Autumn 20-24.

of frequency from 10.0 KHz through 100.0 MHz in terms of F_{am} (external median noise power) above kT_ob in a 1.0-Hz bandwidth, with representing Boltzmann's constant. T_o is the terrestrial reference temperature in Kelvins. Our kT_ob in a 1.0 Hz bandwidth (amounting to kT, or noise density) works out for our practical purposes to the number -204.0 dBw, or what is the same thing, -174.0 dBm. Consequently when we say that a certain noise is so much above kT_ob, we mean that it is so much in relation to -204.0 dBw (or -174.0 dBm).

Recapitulating then, world map A of each set gives the noise at a given geographical location for given seasons and time blocks (local time) in dB above kT_ob at a frequency of 1.0 MHz. Entering graph B with this 1.0-MHz noise value, we obtain a new noise value for our frequency of interest. With this new noise value, we enter graph C which gives corrections.

δ F_{am} = Standard deviation of F_{am} values. This is a measure of error introduced by smoothing procedures used in measuring and plotting.

D_u = Upper decile of the statistical distribution of the values, F_a (noise) about its median value, F_{am}.

D_l = Lower decile of the statistical distribution of the values, F_a (noise) about its median value, F_{am}.

Graph B also gives the man-made and Galactic (cosmic) noise values.

Appendix 4
Standard Man-Made Noise and Galactic Noise in 1.0-Hz Bandwidth

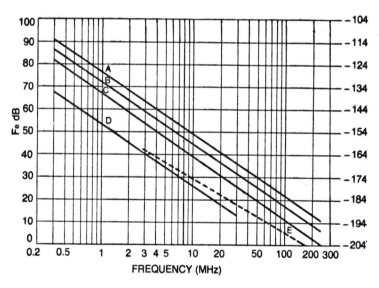

Fig. A4-1. Median values of man-made noise power for a short vertical lossless grounded monopole antenna: A: Business B: Residential C: Rural D: Quiet rural E: Galactic.

Courtesy CCIR, Geneva, Switzerland. Vol. VI *Ionospheric Propagation*, XIII, Plenary Assembly).

Appendix 5
dB Powers—Addition and Subtraction Curves

Obviating mathematical drudgery with its concomitant error potential, our graph depicts a simplified direct procedure for both **adding** and **subtracting** decibel power values (in lieu of conversion to "watts", performance of these addition/subtraction operations and reconversion to dB powers), as shown in the following detailed examples.

DB POWER ADDITION:

Example-1: Add 6.0 dBw and 2.0 dBw (answer **not** 8.0 dBw). Graphical solution: Take the difference between the larger and smaller quantities (6.0 − 2.0 = 4.0). With this difference value (4.0) enter graph's horizontal (Difference between dB powers) scale. Proceed vertically to **dB power addition** curve and upon intersection therewith, left to the vertical (dB to be added to or subtracted from larger value) scale to read 1.5 which is added to the larger value) (6.0 + 1.5 = 7.5). Answer is 7.5 dBw.

Example-2: Add 10.0 dBm and 10.0 dBm (answer **not** 20.0 dBm). Graphical solution: Take the difference between the larger and smaller quantities . . . in this case either from the other (10.0 − 10.0 = 0.0). With this difference value (0.0) enter graph's horizontal scale. Proceed vertically to **dB power addition** curve and upon intersection therewith, left to vertical scale to read 3.0 which is added to the larger value . . either, (10.0 + 3.0 = 13.0). Answer is 13.0 dBm.

Example-3: Add 0.0 dBw (recall that 0.0 dBw = 30.0 dBm) and 20.0 dBm (answer **not** 50.0 dBm). Graphical solution: Take the difference between the larger and smaller quantities (30.0 − 20.0 = 10.0). With this difference value (10.0), enter the graph's horizontal scale. Proceed vertically to the **dB power addition** curve and upon intersection therewith, left to the vertical scale to read 0.42 which is added to the larger value (30.0 + 0.42 = 30.42). Answer is 30.42 dBm.

dB POWER SUBTRACTION:

Let's now take the same examples but in the subtraction mode.

Example-1: Subtract 2.0 dBw from 6.0 dBw (answer **not** 4.0 dBw). Graphical solution: Take the difference between the larger and smaller quantities (6.0 − 2.0 = 4.0). With this difference (4.0) enter graph's horizontal scale. Proceed vertically to **dB power subtraction** curve and upon intersection therewith, left to the vertical scale to read 2.1 which is to be taken from the larger value (6.0 − 2.1 = 3.9). Answer is 3.9 dBw.

Example-2: Subtract 10.0 dBm from 10.0 dBm (answer **not** 0.0 dBm). Graphical solution: Take the difference between the larger and smaller quantities...in this case either from the other (10.0 − 10.0 = 0.0). With this difference value (0.0) enter graph's horizontal scale. Proceed vertically to **dB power subtraction** curve and upon intersection therewith, left to the vertical scale to read an asympottic ∞ (infinity) which is to be taken from the larger (either) value to yield -∞dBm or zero watts power. (Answer is -∞dBm).

Example-3: Subtract 20.0 dBm from 0.0 dBw (30.0 dBm) ...answer **not** 10.0 dBm. Graphical solution: Take the difference between the larger and smaller quantities (30.0 - 20.0 = 10.0). With this difference (10.0) enter graph's horizontal scale. Proceed vertically to dB **power subtraction** curve and upon intersection therewith, left to the vertical scale to read 0.5

Table A5-1. dB Power Addition and Subtraction.

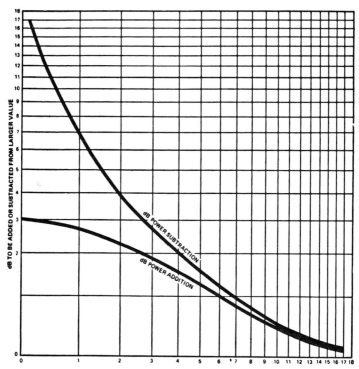

Courtesy of TAI Inc. From Consuletter-International Vol. 6 No. 9, entitled "dB Power Addition and Subtraction."

which is to be taken from the larger value (30.0−0.5 = 29.5). Answer is 29.5 dBm.

CONCLUDING NOTE:

It is interesting that as the difference between the dB powers to be added or subtracted approaches the order of some 18.0 dB, the answer approximates the larger dB value (e.g., 20.0 dBm + 2.0 dBm = 20.1 dBm and 40.0 dBk − 22.0 dBk = 39.9 dBk).

Appendix 6
List of Decibel Terms

dB: The well known power ratio expressed by, $dB = 10\, \text{Log}_{10}\, P_1/F_2$.

dB: Technically "archaic," this term was used by the broadcast/telephone industry and was based upon 0.006 watts across 500.0 ohms.

dB: An acoustic expression, it is based upon 20.0 micronewtons/square meter. The newton is the unit of force imparting an acceleration of one meter per second per second to a mass of one kilogram.

dB(A): An accoustic term based upon the 40.0 phon equal loudness contour. The phon is a unit reflecting apparent loudness and equals the sound pressure intensity of a 1,000.0-Hz tone relative to 0.0002 microbars (dynes per square centimeter) as judged by listeners to match the loudness at other frequencies.

dB(A)$_{90}$: The "A" weighted sound levels exceeded 90% of the time.

dB(A)$_{10}$: The "A" weighted sound levels exceeded 10% of the time.

dBa: An Fl-A weighted term employed by the USA telephone industry to indicate channel noise. It is based upon a −85.0 dBm annoyance threshold and is similar to the CCIR psophometric weighting. The dBa term is now all but obsolete.

dBa0: Related to the dBa, it indicates the noise intensity at the 0.0 dBm TLP.

dB(B): Same as dB(A) but based upon the 70.0 phon equal loudness contour.

dB(C): Same as dB(A) and dB(B) but based upon the 100.00 phon equal loudness contour.

dBc: Standing for "dB Collins" (USA), this is a technical sobriquet referenced to "0.0 dB" equalling 0.775 volts across any impedance.

dB/c: Denotes attenuation in decibels per 100 feet (e.g. as in waveguide).

dB(D): Refers to an acoustical equal noisiness contour.

dBd: Indicates antenna gain relative to that of a dipole (doublet).

dBe: A Siemens (West Germany) sobriquet based upon 0.775 volts across a low impedance.

dBE/N$_0$: Decibel ratio of energy per binary digit to spectral noise density—used in digital transmission.

dB$_{\text{eff}}$: A Rhode & Schwarz (West Germany) epithet referenced to 0.7 volts rms (TV).

dBer: Relates to 0.775 volts across a low impedance (Siemens-Germany).

dBF: Decibels relative to one femtowatt (10^{-15} watts).

dBHz: This term has the dimension of a frequency (e.g.: used to define bandwidth in decibels). Not to be confused with dB(W/Hz).

dB/Hz: Signal-to-noise power density ratio.

dBi: Antenna gain relative to an isotropic antenna.

dBj: A Jerrold Electronics (USA) epithet expressing a radio signal's intensity relative to 1,000.0 microvolts.

dBk: A decibel power level relative to 1.0 kilowatt.

dBK: Decibel unit expressing the quantity $10 \, Log_{10}$ (G/T), where G/T is a receiving station (e.g. in space communications) figure of merit (antenna gain divided by receiving system noise temperature in Kelvins).

dB/K: Decibel figure of merit (dB per Kelvin). dB referred to isotropic antenna and K referred to 1.0 Kelvin.

dB/°K: Decibel earth station figure of merit equal to the ratio of antenna gain in dB relative to isotrope, to system noise temperature in Kelvins.

dB/KM: Decibels attenuation per kilometer.

dB/Km/Mm/Hr: stands for decibels/kilometer per millimeters/hour. It is a rain attenuation coefficient in terms of decibels per kilometer of microwave path per millimeters per hour of rainfall in the path.

dBm: A decibel power level referred to 1.0 milliwatt (0.001 watts).

dB/m: Decibels attenuation per meter.

dBma: Power level based upon 1.0 milliwatt, Fl-A weighted.

dBmc: Power level based upon 1.0 milliwatt, c-message weighted.

dBm/Hz: Receiver noise power density in decibels referenced to 1.0 milliwatt per 1.0 Hz bandwidth.

dBMI: decibels attenuation per mile.

dBm0: Power level in dBm referred to the 0.0 dBm TLP. This term has been sometimes employed to indicate decibels relative to a given TLP (e.g. 12.8 dBm0 white noise loading at a -33.0 dBm TLP would denote a white noise loading power level of -20.2 dBm).

dBm0p: Based upon 1.0 milliwatt at the 0.0 dBm TLP, psophometrically weighted.

dBm0s: A CCITT reference to "zero" program level (s = sound for TV audio or sound radio broadcasting).

dBm0t: A CCITT reference to "zero" telephone level as opposed to sound TV audio or sound radio broadcasting.

dBmV: Decibels above 1.0 millivolt. In CATV work this term is defined in terms of 1.0 millivolt across 75.0 ohms.

dBmV/m: Decibels relative to 1.0 millivolt per meter.

dBuV/m: Decibels relative to 1.0 microvolt per meter.

dBu/m: Same as dBuV/m.

dBpp: A Rhode & Schwartz (Germany) epithet (TV) based upon 0.7 volts peak-to-peak.

dBr: Decibels relative (e.g. 12.8 dBr noise loading at a -33.0 dBm0 TLP would render a -20.2 dBm noise power).

dBrn: Decibels above reference noise. It is based on 0.0 dBrn = -90.0 dBm as employed in the "ancient" desk stand telephone (144 weighting).

dBrn0: A 144-weighting term based on a -90.0 dBm power level at the 0.0 dBm TLP.

dBrnc: Decibels reference noise c-message weighted (0.0 dBrnc = −90.0 dBm).

dBrnc0: A c-message weighted value based upon −90.0 dBm measured at the 0.0 dBm TLP.

dBrnp: Decibels relative to 1.0 picowatt per channel with CCITT (Psophometric) weighting.

dBrap: Decibels reference acoustic power. It is based upon a sound power of −160.0 dBw.

dBrms: A Rhode & Schwartz (Germany) "TV decibel" referenced to 0.7 volts rms.

dB SPL: Sound pressure level referred to hearing threshold (0.0002 dynes per square centimeter).

dBss: Same as dBpp, (q.v.)

dBu: Siemens (Germany)—based upon 0.775 volts.

dBu: dB micro. It is based upon 1.0 microvolt.

dBur: Siemens (Germany)—referenced to 0.775 volts.

dBv: Decibels relative to 1.0 volt.

dBvg: Decibels voltage gain.

dBw: Decibels gain or loss referred to 1.0 watt.

dB/W: Same as dBw above.

dB(W/Hz): Decibels relative to watts per Hertz. Absolute spectral power density level with watts per Hertz as the spectral reference power density.

dB(W/K): Absolute power density level in decibels referenced to watts per Kelvin.

dB(W/m^2): Decibels relative to watts per square meter. Delineates flux density.

dB(W/m^2 Hz): Absolute spectral density of flux. Decibels relative to watts per square meter per Hz of bandwidth.

dBx: Decibels reference cross coupling.

EPN dB: Effective perceived noise level, taking into account noise duration. This quantity is more exact than "A" weighting in relating man's perception of noise to its physical parameters and has been adopted by the USA FAA (Federal Aviation Administration) for aircraft noise specification.

PN dB: Decibels based upon acoustic perceived noise.

Courtesy TAI Inc. From Consuletter-International Vol. 5 No. 12.

Appendix 7
Free Space

The unembellished truth regarding "free-space" is that albeit by diligent design this state might be approached, it is something of a myth. Nevertheless, since it is apparently possessed of a certain Circean fascination, we shall dwell a "tittle" upon this rather chimerical theme prior to passing on to pragmatism.

"Free-space" attenuation (assuming "free-space's" existence) would not accrue from some mysterious absorptive property or enigmatic dissipation mechanism but simply that a spherically radiated wavefront (further assuming the existence of the fictitious, albeit conceptually utilitarian, electromagnetic isotrope . . . by way of an aside the acoustic isotropic source *does*, however, exist in the form of a pulsating sphere) would exhibit a continuous energy decrease with an expanding sphere radius (transmission range), thus making ever less signal available to a given receiving-antenna aperture.

An "amusing" way of envisioning "free-space" loss is via a radiating isotrope concentrically juxtaposed, in vacuo, within a metal sphere. Under this condition the sphere would intercept the total energy and regardless of the isotrope/sphere separation, there would be *no* "free-space" loss.

Most of the attenuation of "free-space" loss occurs close-in to the radiating antenna. The loss between two isotropic antennae separated by one wavelength is 22.0 dB, thereafter increasing by 6.0 dB for each doubling of the inter-antenna distance. At 4.0 GHz (Wavelength = 7.5 Centimeters), for example, this means that increasing the interval to 15.0 centimeters would involve a penalty of 28.0 dB (22.0 dB + 6.0 dB). At the other end of the range scale this rather fortunate circumstance aids in achieving modern tremendous-range extra-terrestrial communications—e.g., multiplying the Earth/Pluto distance by two would entail an additional attenuation of only 6.0 dB . . . a more favorable purchase indeed!

"Free-space" has been variously accorded an impedance of 120π ohms, a permeability of 1.275×10^{-6} Henry/Meter, a permittivity of $(36\pi \times 10^9)^{-1}$ Farad/Meter and an electromagnetic energy propagation constant of 2.998×10^8 Meters/Second.

But so much for technical chastity; by way of "cushioning" the transition to practicality, we might state that after all, there is nothing sacrosanct about "free-space" loss for if solely the first Fresnel-zone were received, the signal would be twice that from them all. Enter the real world!

PRAGMATICALLY YOURS

"Free-space" attenuation is that which might occur if the ubiquitous variables were conveniently disregarded. But alas! We must contend with the copious but "imperfect" globe dubbed "Earth," and its "defective" atmosphere, with their propagational-purity upsets (e.g., reflection, deflection, diffraction, blocking, absorption, focusing, defocusing, trapping, and scattering), each a subject in itself which we hope to treat in future issues, . . . as well as their complex mixes in dramatic temporal and spatial convolutions, spawning all manner of perturbations for which proper allowances must be made in viable radio-telecommunications circuit design.

NO SPECTRUM-SEGMENT EXEMPT

These disquieting effects embrace the spectral gamut. Beginning at VLF (Very Low Frequency—3.0 to 30.0 kHz), this range is affected adversely by *solar-proton* events on trans-polar circuits. On the opposite spectral edge, contrary to some naive notions, laser transmission is more subject to tropospherically induced outages than is radio-propagation (rain and fog mercilessly scatter light and fluctuations in tropospheric density result in corresponding refractive index variations inducing noise).

EARTH/SPACE

Earth/Space communications traversing the atmosphere suffer such deviations from "free-space" propagation as signal amplitude variations, wavefront arrival-angle shift, and polarization modification (Faraday rotation).

OUTER SPACE

Even outer "space" contains plasma and "debris" (e.g., meteor-trail fallout and solar-explosion expelled matter), not to mention cosmic-injected material and man-placed satellites, probes, and "space-junk".

Thus it may be appreciated that there is more to "free-space" than *free-space*.

Courtesy TAI Inc. From Consuletter-International Vol. 5 No. 3, entitled "Free Space."

Appendix 8
Line-of-Sight

Common application (or rather mis-application) of the term "Line-of-Sight" appears to have conferred the human attribute of vision upon microwave antennae. In an attempt to expunge the resulting confusion and to salvage a modicum of meaningful radio physics from this LOS "concept", expressions such as "Radio Line-of-Sight" and "Geometrical Line-of-Sight" seem to have come to a technical breeches birth—accomplishing little more than compounding the enigma. The LOS term has been as misapplied as it is irrelevant—merely obscuring radio propagation reality-and the fallacious phrase, "if you can see it you've got a path" (applicable only to Laser), has infelicitously become an all too persistently deluding technical old wives' tale.

SEPARATE PHENOMENA

Light ray and radio ray trajectories between two given points are not to be considered coincident but rather as separately varying phenomena, reacting differently to the prevailing meteorological ambient. That virtually no correlation exists between radio and optical k-factors is ascribable to the circumstance that optical k-factors are dependent upon only barometric pressure and temperature whereas radio k-factor is, in addition, contingent upon relative humidity. The large refractivities encountered at radio frequencies are attributable to the polar characteristics of the water molecule. Thus the wet term of the Smith-Weintraub refractivity formula accounts for the immensely greater variability of the radio k-factor as opposed to the optical.

NATURAL LIMITS

Nature imposes rather tight limits upon the possible values of the optical refractivity gradient, while the range of radio refractivity gradient is given comparative free sway. While the optical k-factor is confined essentially between true and infinite earth, radio k-factors are subject to no such constraint owing to the afore-mentioned polar water vapor influence. Observed radio k-factor spreads of $+0.1$ to -0.085 are not uncommon. Data from Columb-Bechar, Algeria show that during the summers of 1964/65 the most subrefractive optical k-factor occurring was + 0.81 while the strongest sub-refractive radio k-factor was + 0.37 producing respective midpath earth bulges of 185 and 405 feet. In Aden, Arabia, radio sub-refraction occurs for 1% to 2% of the nocturnal hours exhibiting radio k-factors smaller than + 0.33 while the corresponding optical k-factors were greater than 4/3 producing respective midpath earth bulges of greater than 450 feet and smaller than 112 feet. At locations in the Persian Gulf area, radio k-factors of the order of + 0.15 (not unlikely sustaining this effective value over the expanse on an entire microwave path), occur between 1% and 2% of the worst months. This corresponds to a midpath earth bulge of 1000 feet. That the contrast with an LOS 150 foot midpath earth bulge is astounding, is an understatement indeed.

COSTLY

It is readily seen that competent consulation, engineering, and design are indicated. Over-design (such as microwave towers excessively high or stations too close together) based in fear wrought by lack of cognizance of meteorology, would obviously prove financially embarassing. Under design (microwave towers too low or stations too far apart) equally leads to economic chagrin (due to the necessity of later rectification).

TIME FOR CHANGE

While certainly, visual flashing and similar techniques have their uses, radio path Fresnel clearance confirmation is not among them, and in view of today's ever increasing demands for telecommunications circuits of extremely high-precentage reliability, the differences between optical and radio k-factors, as well as their percentages and locations of occurrence, must be amply heeded. Perhaps the LOS term's retirement is long overdue. May we suggest WRH (Within Radio Horizon) as its successor?

Courtesy TAI Inc. From Consuletter-International, Vol. 1 No. 1, entitled "Line-Of-Sight".

Appendix 9
Structure of Our Ionosphere

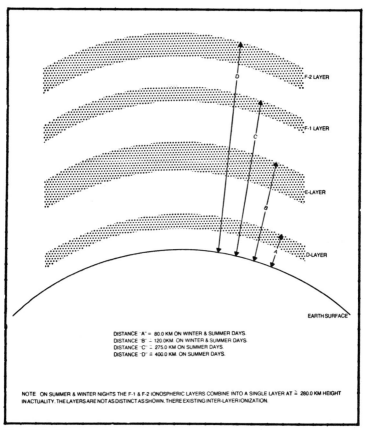

Fig. A9-1. Structure of the ionosphere.

Appendix 10
Great Circle Radio Path Calculations

A great circle, for terrestrial propagation purposes, is the shortest line between two terminal communications points on the earth's surface. Geometrically, this great circle may be defined as any circle generated upon the surface of a sphere (in our case, the Earth) by a plane passing through the center of said sphere.

While for short radio paths (e.g. to some 80.0 kilometers), a great circle may be approximated by a straight line drawn upon a map or chart, this approach is generally not recommended for longer path distances due to geometrical inaccuracies which would ensue. For these larger distances, we may calculate bearings and distances for the terminals and intermediate points. These calculations involve spherical trigonometry and require five-place (or better) tables of trigonometric functions (or their logarithms). An example spherical triangle (a triangle on a sphere instead of a plane surface) used for these calculations is shown in the accompanying figure. It is shown on the sphere (earth) as "PAB", where "A" and "B" are our radio terminals and "P" is the pole. Terminal "B" is of greater latitude than terminal "A" and "P" (the pole) is in the same earth hemisphere (herein, northern hemisphere). The spherical triangle in the illustration is as pointed out, for the northern hemisphere but might be readily inverted to apply to the southern hemisphere. "B'" is any point along the great circle path from terminal "A" to terminal "B". The triangle solved is "PAB'". The latitudes of points "A", "B", and "B'" are denoted as "ϕA", "ϕB" and "$\phi B'$". "C" and "C'" are longitude differences between "A" and "B" and "A" and "B'" respectively. "Z" and "Z'" are the corresponding path lengths. The following formulas may be employed for computation (hand computation and computer alike).

The initial bearings ("X" from terminal "A" and "Y" from terminal "B") are all measured from true North and are calculated as follows:

$$\operatorname{Tan} \frac{Y-X}{2} = \operatorname{Cot} \frac{C}{2} \left[\operatorname{Sin} \frac{\phi B - \phi A}{2} \bigg/ \operatorname{Cos} \frac{\phi B - \phi A}{2} \right]$$

$$\operatorname{Tan} \frac{Y+X}{2} = \operatorname{Cot} \frac{C}{2} \left[\operatorname{Cos} \frac{\phi B - \phi A}{2} \bigg/ \operatorname{Sin} \frac{\phi B + \phi A}{2} \right]$$

$$Y = \frac{(Y + X)}{2} + \frac{(Y - X)}{2} \text{ and } X = \frac{(Y + X)}{2} - \frac{(Y - X)}{2}$$

The great circle arc path distance, Z, is given by the following formula:

$$\operatorname{Tan} \frac{Z}{2} = \operatorname{Tan} \frac{\phi_B - \phi_A}{2} \left[\operatorname{Sin} \frac{Y + X}{2} \Big/ \operatorname{Sin} \frac{Y - X}{2} \right] \quad (1)$$

To calculate intermediate points, different formulas are used for predominantly East-West paths, than for predominantly North-South paths. In the first case, values of longitude or longitude difference are given corresponding to the map dimensions, and the solution is made for corresponding latitudes. In the second case, the opposite is true. In cases of paths close to 45° bearings, either method can be used.

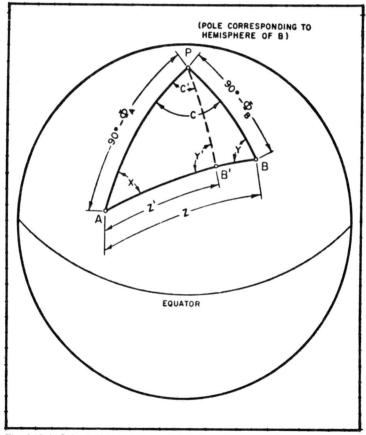

Fig. A10-1. Spherical triangle for great circle path computations.

To calculate the latitude, "$\phi B'$," for a given longitude difference "C'", use the formulas:

$$\cos Y' = \sin X \sin C' \sin \phi A - \cos X \cos C'$$

$$\cos \phi B' = \sin X \cos \phi A / \sin Y'$$

To calculate the longitude difference "C'" for given latitude "$\phi B'$", use the following formulas:

$$\sin Y' = \sin X \cos \phi A / \cos \phi B'$$

$$\cot \frac{C'}{2} = \tan \frac{Y' - X}{2} \left[\left(\cos \frac{\phi B' + \phi A}{2} \right) \bigg/ \left(\sin \frac{\phi B' - \phi A}{2} \right) \right]$$

To convert the angle, Z, obtained in degrees from formula (1), to units of length, the following constants are used (based upon the average earth radius of 6366.7 kilometers or 3956.1 statue miles):

$$d_{km} = 111.12\ Z°$$

$$d_{statute\ miles} = 69.047\ Z°$$

$$d_{nautical\ miles} = 60.0\ Z°$$

Appendix 11
Lightning

Lightning has struck and branded awe into the soul of man since the beginning, and albeit some of the mystique has been dispelled by human "enlightenment,"it still does.

LIGHTNING'S MATRIX

While lightning may be spawned in dust and snow storms or active volcanoes, we shall herein treat it in its classic arena—the thunderstorm, prototypes of which breed in steep tropospheric temperature lapse rates by convective, orographic (mountain), or front-fomented warm/moist air ascension. Creating myriad meterological "melees" for which they are dubbed **weather factories,** thunderstorms hatch groups of gigantic cumulonimbus clouds (alias convective cells, convective clouds, thunderclouds, thunderheads, or thundercells).

THUNDERCELL

A classic cell is beauteous to behold, towering and stately, reigning with anvil-top crown, against a steel blue sky—as the golden cloud with veritable silver lining recently "staged" over the Rock of Gilbratar and herein illustrated (with data overlay) from author-to-artist description. Conforming closely to a TAI electrical "tripole"/Malan modified model, the incredible shearing updraft/downdraft—the wild flurry of upper-cloud snow pellets/ice crystals and raging lower-cloud rain/hail—the dissociation, atomization, and ionization/electrification charge the upper cloud positively and mid-cloud negatively while the central cloud-base area assumes a positive "pilot" charge. The cloud-earth potential is some hundred million volts—presaging nature's *pyrotechnics.* In a flash it happens! A psyche-scorching blinding blaze and deafening "decibelladen" thunderclap!!!

DAL SEGNO

Since this drama took place in less than fifteen centiseconds, let's play back the last few bars of our "Symphony Electra" in slow motion—with editorial commentary.

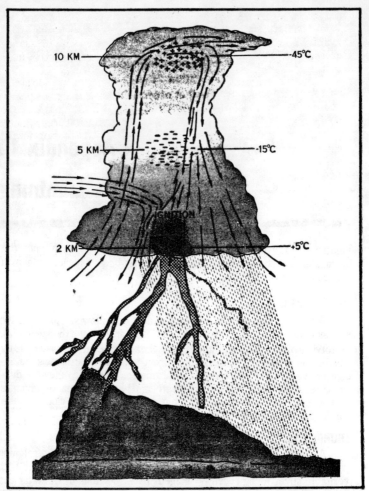

Fig. A11-1. Lightning spawned in a thunderstorm.

Upon attaining a gradient of 5000.0 volts/centimeter between negatively and positively charged cloud-regions an initial intra-cloud discharge (TAI dubbed **ignition**) occurs triggering the cloud-to-earth stroke. Despite this tremendous cloud/earth potential, note that field conditions, di-electric strength, and path breakdown impedance limit the lightning stroke's earthward progress to an incipient 50.0 meters. After a several-tens-of-microseconds "hesitation" there occurs another 50.0 meter surge and so on (thus the term **step-leader**). It characteristically propagates toward earth at 150.0 km/second and several hundred amperes in a 1.5 centimeter diameter "white-hot" current carrying core sheathed in a 5.0 meter central-chord corona, peaking at 10.0 MHz. The step-leader's 5.0 coulombs intensifies the leader-terrestrial field. Violently, an earth-discharged **return stroke** "oscillating" between a peak 200,000 and "keep alive" 500.0 amperes in a 1.0 to 10.0 kHz bandwidth, dissipating 100,000 Joules/Meter at 30,000°C(5.0 × solar temperature) with pressures in excess of 10.0 atmospheres, **crashes** into the stepped-leader 60.0 meters aloft.

At these staggering statistics it somehow seems less droll to accord a little "credulity" (albeit with scientific tongue-in-cheek) to the "pearls of wisdom" of a town twit who once burbled, "Lightning doesn't strike more than once in the same place because the same place isn't there anymore."

Hark! A second leader—this one continuous (dart)—racing down the established ionization "trail," inducing another earth-return stroke—as well as observer trepidation. Again...and another...several in rapid succession!!

The concomitant thunder is, at once, terrifying and reassuring—the former for obvious reasons and the latter because hearing it, one has escaped being struck by its lightning progenitor. Lightning's light traveling at close to 300,000,000 Meters/Second reaches the observer and passes into "history" considerably prior to the arrival of thunder's sound at 335 Meters/Second. Thunder originates along the lightning return-stroke channel by tremendously rapid heating/dissociation in a supersonic compression shock-wave. Within a meter or so this wave transmutes to sonic becoming **eminently audible.** At close range thunder is heard as a startling clapping followed by resounding/reverberating peals while at extended distances it is evidenced by more subdued rumbling rolls. Owing to a complex synergy among tropospheric temperature lapse rate, wind shear, and terrain features, thunder (peaking at 50.0 Hz) is rarely heard at distances exceeding 25.0 kilometers.

LIGHTNING TYPES

In addition to the classic lightning above described, there are several other varieties. It is interesting that in this "enlightened" age their *folk-logic* classifications have managed to survive.

Streak lightning, the most customarily observed, appears as a streak owing to a paucity or total lack of branches from the main channel flash.

Heat lightning, Characteristically reddish-orange, is a cloud-to-cloud discharge sufficiently distant that its thunder is not heard; occurring during observer-locale warm summer evenings, it "logically" acquired its name.

Ribbon lightning is a series of strokes, succeeding channels of which are displaced downwind. Retinal retention produces an **aggregate** multi-stroke "ribbon."

Forked lightning consists of principal stroke and "sprouting" branching channels. Branch earth-impact points generally lie from 0.15 to 15.0 kilometers apart. It is engaging that a forked ground-to-cloud stroke may appear "upside down."

Bead lightning results from luminosity variations (bead effect) along the flash's channel.

Ball lightning is a controversial variety, possibly resulting from retinal saturation and ensuing optical illusion. It "appears" as a small reddish ball, "carried along" at a "faltering" speed as the observer's eyes shift...and "explodes" within several seconds. Scientifically authenticated evidence is lacking.

Sheet lightning is an **intra-cloud** stroke whose diffused flash appears as a...*sheet*.

Hot lightning is so identified if ground discharges set forest fires . . .*hot*.

Cold lightning consists of flashes to trees causing explosive damage (tree splitting, etc.) but **not** fires...*cold*.

LIGHTNING HAZARDS

Our earth experiences some 44,000 thunderstorms "hurling" in excess of 8,000,000 lightning flashes **daily**. It is thus not surprising that, despite its "enchanting" mien, it is responsible for human casualties exceeding those of any other weather phenomenon. Additionally, in the U.S.A. along lightning sets 10,000 forest fires laying waste to $30,000,000 worth of valuable timber annually, and destroys capital assets of some twice this amount. The writer would be loath to contemplate the order of magnitude by which this havoc would have burgeoned in the **complete** absence of effectual lightning protection installations.

LIGHTNING PROTECTION

Although Benjamin Franklin's **basics** remain comparatively intact, much has been added to the lightning protection science since. **Effective** lightning protection saves lives, calms fears, secures capital investment and lowers insurance rates.

To realize these advantages, however, a word of caution is in order—accent the word, **effective**. **Effective** lightning protection design does **not** admit of unjustified fixed or "standard" cone and wedge dodges—these representing mere statistical values with lightning strikes having been reported within these so-called "protected" zones. Each installation is individual.

FIRST AID

As in **any** first aid, the best regarding lightning is to obviate its necessity. However, should a lightning accident occur in your presence, emergency treatment is the same as for other electrical shocks. Although not-infrequently causing death, medical evidence indicates that **prompt and prolonged** artificial respiration and cardiopulmonary resuscitation measures proffer maximal revival potential.

Courtesy TAI Inc. From Consuletter-International Vol. 5 No. 5, entitled "LIGHTENING."

Appendix 12
Diversity Reception and Combining

Diversity reception is a method of minimizing the detrimental effects of fading (e.g. worsening of voice signal-to-noise ratios or hits (errors) in record information such as teletype, data, facsimile, etc.) Two or more sources of received signal energy (e.g. one transmitter sending to two or more antenna/receiver combinations) carrying the same intelligence information, but differing in strength or S/N ratios from moment to moment, may be combined or switched automatically to accept the momentarily best signal available. This improves the communication quality or effects a diversity reception gain. While there are many forms of diversity as well as concepts, diversity reception fundamentally is based upon the fact that fading does not generally co-occur on radio circuits differing sufficiently from each other in such parameters as space (space diversity), frequency (frequency diversity), polarization (polarization diversity) or time (time diversity). Space, frequency & polarization diversity, as examples, are described as follows:

Space Diversity. Spacing of receiving antennas (of the order of ten to one hundred wavelengths) is extensively employed in receiving diversity. This allows the radio signal to arrive at the receiving location via two (or more) different radio paths and renders uncorrelated path transmission losses (fast fading). Establishment of two (or more depending upon the order of diversity) paths by spaced antennas at the receiving station allows this action.

Frequency Diversity. The employment of two (or more) different carrier frequencies, even over the same physical radio path, will also improve a signal by virtue of the fact that the instantaneous (fast or short term) transmission losses (fading) for differing frequencies will generally be uncorrelated. This procedure is generally frowned upon, though, since it requires extra on-the-air signals and besides being more costly (for the extra transmitting and receiving equipment) contributes to spectrum pollution.

Polarization Diversity. This type of diversity reception is based upon differences in fading characteristics of a given radio frequency signal as viewed simultaneously by receivers connected to antennas of differing polarization (normally vertical and horizontal).

Combining. The different antenna/receiver combinations having received their respective signals, we now require a method to combine or select the momentarily best so that as one fades, the other can take over and "fill in" to hold the signal as steady as possible, as well as maximize the signal-to-noise ratio. Three basic combiners are the "Selector" Combiner, the "Equal Gain" combiner, and the "Maximal Ratio" combiner. The selector combiner determines the larger signal or the one with the best S/N (signal-to-noise) and automatically switches this superior signal to the output for use on the circuit. The equal gain combiner (also known as linear

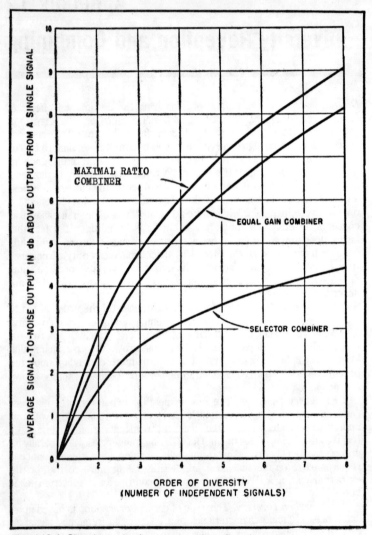

Fig. A12-1. Signal-to-noise improvement in a diversity system.

adder combiner) sums the signals from the separate receivers and applies this signal summation to the output. The maximal ratio combiner (also known as ratio squarer combiner) squares the signals prior to addition and applies the sums of these squares to the output for use. Each of the above combiner types has its advantages and also its disadvantages, a subject too voluminous to be adequately covered in this radio propagation practical handbook. The accompanying figure of this Appendix shows the advantages of some combiner types in terms of S/N (signal-to-noise ratio) average output for various orders of diversity (in this case up to eight) versus a single receiver/antenna combination.

This material is from (excerpted, abbreviated, and slightly modified) NBS Report-6767, entitled "Ground Communication Tele-communication Performance Standards—Part 5 of 6—Tropospheric Systems" by A.F. Barghausen et al, courtesy of US Department Of Commerce, National Bureau Of Standards, Boulder, Colorado.

Appendix 13
Rayleigh Fading

This is statistical distribution describing the magnitude of a phasor composed of the sum of a multitude of component phasors randomly distributed in phase and *amplitude*. Note that the definition of "Rayleigh Distribution" and "Rayleigh Distributed Fading" differ insofar as the former is theoretical with constant amplitude phasors, while the latter is propagationally pragmatic, e.g., it is not difficult to visualize that several phasors of an electromagnetic wave might traverse somewhat different paths and therefore be subjected not only to differing phase rotations but additionally to different amounts of absorption, scattering, or refraction, thereby altering their relative amplitudes. Possibly, more correctly, this term could be called "Quasi-Rayleigh Distributed Fading." Albeit this article revolves about Lord Rayleigh, it would be somewhat "isolating" and discontinuous historically to fail to include pertinent related contributions of other workers. For example, Beckon has stated the "Quasi-Rayleigh Distributed Fading" concept rather neatly, calling it a time distribution of the instantaneous field strength resulting from a combination of a large number of waves of random phases and of nearly the same amplitudes. The median signal value of such distribution is 0.83 of the intensity of the non-faded field. Thus, the "Rayleigh" faded signal is approximately 1.6 decibels less than the FSL (Free Space Loss) value, this difference being generally ignored in microwave path calculations.

It might be of additional interest that Bullington maintains the fading "rate" (speed of rise and fall of signal intensity) to be 10.0 decibels/second to 100.0 decibels/second during Rayleigh fading with the latter occurring only seldom, while Barsis and Johnson take another tack on the description of Rayleigh fading "rate" (frequency of fade occurrence). This is dependent upon radiometeorological conditions and operating wavelength, with the usual limits being from 0.1 to 10.0 Hz.

It might be interesting, in closing, to relate some CCIR theoretical Rayleigh-distributed signal levels for the percentages of time these levels are equalled or exceeded as related to the median (50% value): 1.0% (+8.22 dB), 10% (+5.21 dB), 50.0% (0.0dB), 90.0% (−8.18 dB), 99.0% (−18.39 dB), 99.99% (−28.2 dB), 99.99% (−38.5 dB).

Courtesy of TAI Inc. From Consuletter-International Vol. 4 No. 2, entitled "RAYLEIGH."

Appendix 14
White Noise

As nature abhors a vacuum, it would appear to view with equal disdain, a quiescence in the electron/hole "fortuitous-fluctuation" domain. At temperatures above absolute zero (−273.2° Centigrade, −459.7° Fahrenheit or 0° Kelvin), physical matter "simmers and seethes" yielding electromagnetic energy. Man ever struggles to achieve this quiescence—this telecommunicational utopia—this infinite S/N ratio. As of the present, this "ideal" has not been consummated—and albeit the sagacious cannot help but contemplate such philosphical fortuities that this particulate perturbation may indeed constitute the intrinsic and fundamental essence of life itself and thus should *not* be quenched, or that perhaps eliminating it might "unearth" and render audible (or visible) the plethora of worldwide signals, however weak, superposed so as to effectively re-establish the very noise so eliminated—we are, for the nonce, confronted with the pragmatical grace and grief of white noise. Albeit in its Dr. Jekyl identity it may perform the gainful function of simulating signals, its Mr. Hyde "alter-ego" may indeed have a pernicious effect upon intelligence transmission. White noise would appear to be possessed of more "aliases" than the hues and tints of its white light analogy which avers that white noise contains all frequencies as does white light all colors. The popular version of this analogue, incidentally, is somewhat in error, as Bennett points out, since spectroscopy has established an energy parity per unit wavelength—not per unit frequency. At any rate, these white noise definitional nuances, in the practical overview, appear more likely a consequence of technological "dishevelment" than scientific finesse. All manifest themselves by a hissing or rushing sound in a loudspeaker, "grass" on an oscilloscope and snow or confetti respectively in monochrome and color television. As we scan through this noise's general-usage multi-terminology, it will be engaging to regard the nomenclative/definitional "criss-cross".

THERMAL NOISE

Also termed *Johnson, resistance,* and *white* noise. *Random* motion of free electrons in a conductor gives rise to a voltage whose value is instantaneously fortuitous with time and whose spectral components uniformly embrace the electromagnetic gamut—albeit the bandwidth of many practical circuits limit both the noise spectrum, and magnitude as well. *Thermal* noise is associated with the thermodynamic interchange of energy necessary to maintain thermal equilibrium beteween a material and its thermometric ambient. *Thermal* noise is characterized by a normal distribution of levels.

GAUSSIAN NOISE

Electrical perturbation described by a probability density function following a *normal* law of statistics. Incidentally, although practically synonymous with and well nigh universally accredited to Karl Gauss, it was Abraham DeMoivre, an English mathematician of French Huguenot extraction, who discovered the normal probability distribution curve (Approximatio ad Summam Terminorum Binomii) in 1733. This *normal* distribution is the well known symmetrical "bell shaped" density function of a population which is completely defined by two independent parameters, viz: the mean and the standard deviation. The *Gaussian (normal)* amplitude distribution is of fundamental significance in statistical theory describing numerous natural phenomena. The central-limit theorem of statistics states, in essence, that the sum of a number of independent random variables approaches the *Gaussian* distribution as the number of said variables increases, regardless of the distribution of the individual variables. *Shot* noise and *thermal* noise are *Gaussian* amplitude distributed. A significant facet of *Gaussian random* noise is that it exceeds its positive root-mean-square amplitude, and attains twice that value for 16% and 2% of the time respectively.

RANDOM NOISE

Random noise comprises transient perturbations occuring at *random* with a *Gaussian* spectral distribution equivalent to *thermal* noise. *Thermal* and *shot* noise are forms of *random* noise which may be present in all portions of a telecommunications system, being more significant, of course, at points of lower S/N ratio. In a *random* noise distribution, 13.0 decibels was opted as a peak to rms value—at tangential threshold the peak of an input signal would thus equal or exceed the *random* noise peaks for 99.999% of the time. Random noise, also "pen-named" fluctuation noise, has been accorded the characteristic of a waveform the instantaneous amplitude of which is *random* and therefore unpredictable and without periodicity.

SHOT NOISE

Shot noise, in the modern idiom, is characteristic of transistors and diodes; it is directly proportional to the square root of the applied current. This noise is generated by random passage of discrete current carriers across a barrier of discontinuity, e.g., a semiconductor junction, as well as by thermal agitation in a base resistor. Actually, the *shot* noise (German "schott-effekt") concept stems from the *random* manner in which electrons in a vacuum tube collide with the plate (anode)—not unlike the "sound effect" produced by casting a handful of BB-shot against a wall with their slightly differing and "overlapping" impact times. Additionally, the electrons from a vacuum tube cathode are not emitted uniformly producing *fluctuation* noise.

PARTITION NOISE

This type of noise appears in multi-element vacuum tubes due to the *random* division of cathode current among the different electrodes; this has also been referenced as "pseudo*shot*" noise. A "distinct relative" of this variety of noise is that induced in the various vacuum tube electrodes (especially the control grid) by *random* passing electrons.

JOHNSON NOISE

Discovered by J.B. Johnson in the late twenties, this *thermal* noise may be generated by a resistor at a temperature above absolute zero. It is a *random* noise engendered by *thermal* agitation.

RESISTANCE NOISE

This *thermal, Johnson,* or *white* noise is a *thermal* agitation product, generated in a resistance (not a reactance).

FLUCTUATION NOISE

Noise resulting from undesired *fluctuations* in quantity and/or velocity or electron (or hole) flow.

BROWNIAN MOTION NOISE

Also identified as *thermal* or *Johnson* noise. Interestingly, Brownian motion noise was named after Robert Brown (1772-1858), a Scottish botanist, the discoverer of Brownian motion, a random movement of microscopic particles, in organic or inorganic fluid suspension, caused by collision with surrounding molecules.

Courtesy of TAI Inc. From Consuletter-International Vol. 4 No. 7, entitled "WHITE NOISE."

Appendix 15
VIVA HF

Notwithstanding the progressive appearance, in their turn, of more sophisticated telecommunications systems and techniques—such as microwave, tropospheric scatter and satellites—all surely ushering in greater telecommunications capacity and reliability, HF (High Frequency—3.0 to 30.0 MHz) is still very much with us and albeit parts of its physiognomy have been altered here and there over the last six decades (e.g., log periodic antennas, improved speech conditioning, time multi-diversity, data retiming, synthesized frequency control, automatic equipment turning and error detection/correction schemes). HF appears to be holding its own. One has but to dial across the HF bands to witness the tremendous amount of use of this 3.0 to 30.0 MHz portion of the radio spectrum.

HF USES

Examples of activity extant in the HF bands include aeronautical and maritime mobile, broadcasting (short-wave), point-to-point, standard frequency transmissions, space research, meteorology, military, amateur radio (many of whose members are highly placed in the professional telecommunications community), worldwide embassy connection with home country, and citizens band.

In "ye olden days" of HF simpler terms and concepts were in vogue. For example, HPF (highest probable frequency supportable by the ionospher's F_2 layer, its being defined as a frequency 15% above the MUF—the E-layer's HPF is equal to its MUF/FOT (OWF)). MUF means "Maximum Usable Frequency" and indicates a frequency supportable by the ionosphere for 50% of the time of the month of calculation while FOT comes from the French "Frequence Optimale de Travail" meaning the same as OWF (Optimum Working Frequency). Some later HF terms include "Operational MUF" (highest frequency permitting acceptable operation between geographical points at a certain time under specific ionospheric conditions) and "Classical MUF" (highest frequency capable of being propagated via a specific ionospheric transmission mode between specified terminals be ionospheric refraction alone). According to CCIR the classical MUF may be determined as the frequency at which low and high angle HF

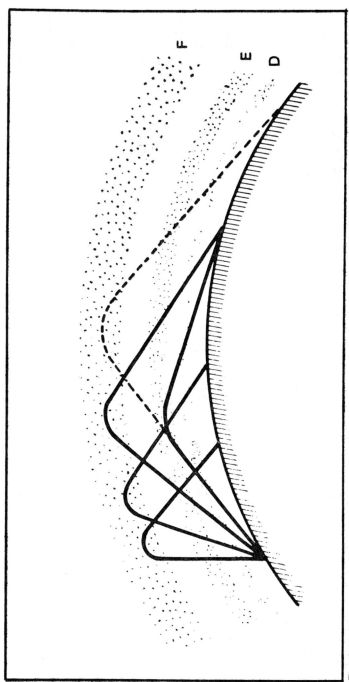

Fig. A15-1. Examples of HF ionospheric transmission.

radio rays merge into a single ray. Another relatively modern HF term is "Standard MUF" denoting approximation to the classical MUF and obtainable by application of the conventional transmission curve (W.R. Piggott and K. Rawer, 1961) together with vertical incidence ionograms and distance factor. Note, however, the URSI (International Scientific Radio Union) has recommended the term EJF (Estimated Junction Frequency) for the "Standard MUF." As a parting contribution to this paragraph, we present the term, "MOF" (Maximum Observed Frequency) ushered in by URSI; it denotes the highest frequency detectable by oblique incidence ionogram.

Other changes constitute the passing of the CRPL-D Propagation Calculation Series (USA) and its replacement by the Ionospheric Prediction Series issued monthly, three months in advance and its being superseded, in turn by the latest "permanent" ionospheric predictions by the ITS/OT (Institute For Telecommunications Sciences/Office of Telecommunications) at Boulder, Colorado (OT/TRER-13—Telecommunications Research and Engineering Report-13). The rocket and satellite age has contributed materially to our knowledge of the ionosphere. Computer methods of HF transmission path calculation have been brought to a high state of utility. While the USA has been one of the leaders in HF research, many other nations have contributed materially and have kept up in the HF telecommunications realm.

FUTURE HF USES

While it is certain that VHF, cable microwave, tropospheric scatter, and satellite have taken over some of the tasks previously performed by HF ionospheric transmission, and may continue to proceed in this direction in the foreseeable future, expansion in other areas appears likely. For example, HF broadcasting, although improvements are surely in the offing, shall continue. This if for no other reason that any significant changes might require new equipment and consequent re-education of a multi-lingual worldwide audience—not to mention the necessary time consuming international agreements. Additionally there are large investments in current plant and it does not seem unreasonable that they must be amortized. There are other fascinating facets in HF broadcasting. As an example, local broadcasting in equatorial regions has been better consummated by the employment of HF at vertical ionospheric incidence and the reflection of this wave to the service area surrounding the transmitter location. The use of the regular broadcast band (525 to 1505 kHz) or the LF broadcast band (150 to 285 kHz) is largely precluded by the copious incidence of lightning storms in these areas and the concomitant interference produced.

In addition it appears logical that the world's military sectors will continue to employ HF for over the horizon radar as well as for backup to other communications modes, to say nothing of the low frequency bands for backup, navigation, and sub-terraquatic communications.

As for amateur radio, only tremendous expansion can be forseen in this area for the future. Regarding the citizens band service, it is not difficult to prophesy that the world's populace will ultimately require and obtain more HF space. One has but to listen to the interference cacophony produced by overcrowding in this band, coupled with its popularity, to draw this conclusion.

CURRENT HF CIRCUITS AND HF PHILOSOPHY:

An example of current HF use is that Cuba's only existing international connection with Spain is by this mode. Another is that despite the fact that the country of Cameroon, Africa, possesses such sophisticated telecommunications as microwave and earth station (Intelsat), it has nevertheless seen fit to employ some 80 telegraph offices, about 60 of which employ HF, working in manual Morse. Cameroon also deemed it advantageous to increase its HF broadcasting capability. Relatively recently, in excess of a quarter million dollars was invested in equipment for HF communications by the country of Chile. As of June, 1975, almost a million dollars was allotted by a large USA firm for an overseas point-to-point and air traffic control HF system.

HF continues to "prove-in" for lighter routes particularly where flexibility and operation over hostile geography at low capital investment are requirements.

Courtesy TAI Inc. From Consuletter-International VOL. 4 NO. 5 entitled "VIVA HF."

Appendix 16
Standard Time Zone of the World and Their Relationship to UTC or GMT

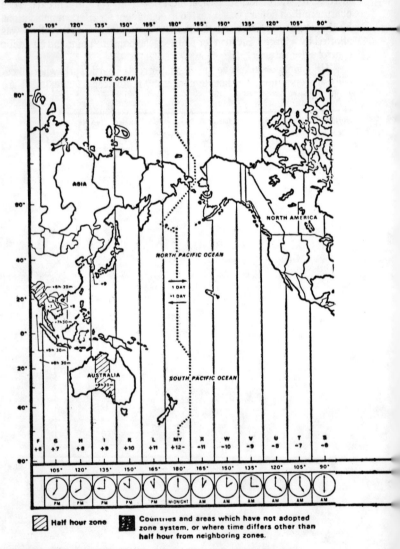

Fig. A16-1. Standard time zones and their relationship to UTC.

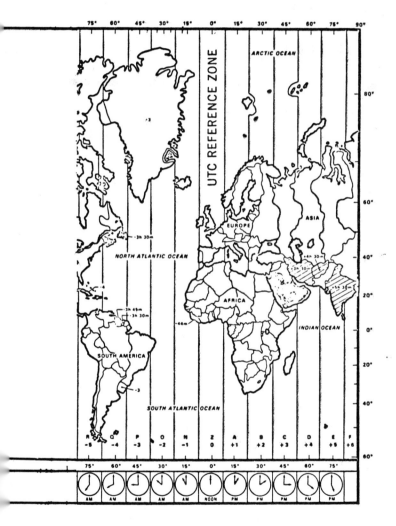

This material is taken from, "NBS Special Publication 432, Dissemination Services", entitled NBS TIME AND FREQUENCY, courtesy of the US Department Of Commerce, National Bureau Of Standards, Time and Frequency Division Institute for Basic Standards, National Bureau Of Standards, Boulder, Colorado.

Appendix 17
Secret Spectrum

Radio frequency band designation by letters was initiated immediately prior to World War II by the military forces—the whole idea was security. That it continues to confound many in the telecommunications community attests to the impression that it must have nobly exonerated itself in its original mission. The practice was gradually expanded and by 1946 became a matter of band classification convenience rather than secrecy. In the interest of clarifying these obsolete, albeit tenaciously tarrying, denominations we present the following based upon information kindly made available to us by the U.S.A. Department of Defense.

Table A16-1. Former Method of Band Classification.

P-band — 0.225 to 0.390 GHz.

L-band — L_p = 0.390 to 0.465 GHz, L_c = 0.465 to 0.510 GHz, L_1 = 0.510 to 0.725 GHz, L_y = 0.725 to 0.780 GHz, L_t = 0.780 to 0.900 GHz, L_s = 0.900 to 0.950 GHz, L_x = 0.950 to 1.150 GHz, L_k = 1.150 to 1.350 GHz, L_f = 1.350 to 1.450 GHz, L_z = 1.450 to 1.550 GHz.

S-band — S_e = 1.550 to 1.650 GHz, S_f = 1.650 to 1.850 GHz, S_t = 1.850 to 2.00 GHz, S_c = 2.00 to 2.40 GHz, S_q = 2.40 to 2.60 GHz, S_y = 2.60 to 2.70 GHz, S_g = 2.70 to 2.90 GHz, S_s = 2.90 to 3.10 GHz, S_a = 3.10 to 3.40 GHz, S_w = 3.40 to 3.70 GHz, S_h = 3.70 to 3.90 GHz, S_z = 3.90 to 4.20 GHz, S_d = 4.20 to 5.20 GHz.

X-band — X_a = 5.20 to 5.50 GHz, X_q = 5.50 to 5.75 GHz, X_y = 5.75 to 6.20 GHz, X_d = 6.20 to 6.25 GHz, X_b = 6.25 to 6.90 GHz, X_r = 6.90 to 7.00 GHz, X_c = 7.00 to 8.50 GHz, X_1 = 8.50 to 9.00 GHz, X_s = 9.00 to 9.60 GHz, X_x = 9.60 to 10.00 GHz, X_f = 10.0 to 10.25 GHz, X_k = 10.25 to 10.90 GHz.

K-band — K_p = 10.90 to 12.25 GHz, K_s = 12.25 to 13.25 GHz, K_e = 13.25 to 14.25 GHz, K_c = 14.25 to 15.35 GHz, K_u = 15.35 to 17.25 GHz, K_t = 17.25 to 20.50 GHz, K_q = 20.50 to 24.50 GHz, K_r = 24.50 to 26.50 GHz, K_m = 26.50 to 28.50 GHz, K_n = 28.50 to 30.70 GHz, K_l = 30.70 to 33.00 GHz, K_a = 33.00 to 36.00 GHz.

Q-band — Q_a = 36.0 to 38.0 GHz, Q_b = 38.0 to 40.0 GHz, Q_c = 40.0 to 42.0 GHz, Q_d = 42.0 to 44.0 GHz, Q_e = 44.0 to 46.0 GHz

V-band — V_a = 46.0 to 48.0 GHz, V_b = 48.0 to 50.0 GHz, V_c = 50.0 to 52.0 GHz, V_d = 52.0 to 54.0 GHz, V_e = 54.0 to 56.0 GHz.

C-band — This band is comprised of the frequencies between 3.90 to 6.20 GHz (bands S_z through X_y). The C-band has also been variously identified as 5.0 to 6.50 GHz.

K_1-band — This band consists of the frequency span 15.35 to 24.50 GHz (K_u through K_q).

W-band — The W-band includes frequencies between 56.0 and 100.0 GHz.

Table A16-2. Modern Method of the Up-To-Date (and more in keeping with today's metrication momentum) Frequency Band Classification System.

Band Number	Frequency Range	Wavelength Range	Metric Identification	Designation
2	30.0 to 300.0 Hz	1000.0 to 10,000.0 Km	Megametric	ELF (Extremely Low Frequency)
3	0.3 to 3.0 KHz	100.0 to 1000.0 Km	Hectokilometric	VF ("Voice Frequency")
4	3.0 to 30.0 KHz	10.0 to 100.0 KM	Myriametric	VLF (Very Low Frequency)
5	30.0 to 300.0 KHz	1000.0 to 10,000.0 Meters	Kilometric	LF (Low Frequency)
6	300.0 to 3,000.0 KHz	100.0 to 1000.0 Meters	Hectometric	MF (Medium Frequency)
7	3.0 to 30.0 MHz	10.0 to 100.0 Meters	Decametric	HF (High Frequency)
8	30.0 to 300.0 MHz	1.0 to 10.0 Meters	Metric	VHF (Very High Frequency)
9	300.0 to 3,000.0 MHz	0.1 to 1.0 Meters	Decimetric	UHF (Ultra High Frequency)
10	3,000.0 to 30,000.0 MHz	1.0 to 10.0 Centimeters	Centimetric	SHF (Super High Frequency)
11	30.0 to 300.0 GHz	1.0 to 10.0 Millimeters	Millimetric	EHF (Extremely High Frequency)
12	300.0 to 3,000.0 GHz	0.1 to 1.0 Millimeters	Decimillimetric	(None officially assigned)

The above system is simplicity itself — viz: Band n extends from (0.3×10.0^n) to (3.0×10.0^n) Hz.

Appendix 18
Parabolic World

Christopher Columbus in his persuasion that the world was round embarked on a quest westward for a shorter route to India; that he failed to achieve his aspiration, jocularly aver our aerial radar terrain profiling crews, resulted from his lack of appreciation that the earth is not a sphere but a paraboloid. That their wit and drollery are not **totally** wanting in fact, is evidenced by modern day employment of the parabola in terrain profile plotting and propagational problem resolution.

WHEREFORE PARABOLA

It all stems from the relationship between the earth's curvature and that of the tropospheric radio ray. Of two lines (geometric and radio) plotted on a spherical earth diagram, the former would be straight and the latter an arc. It is desirable then, to rectify the curvature of the arc transforming it into a tractable straight line. Having transacted this rectification, it is essential to reestablish the original ray/length relationship (consummated by converting our circular earth arc into a more suitable conformation). Enter the parabola, the best available standard mathematical curve for representing the earth while allowing radio rays to be rectilinearly negotiated.

PARABOLIC CHART

In a terrain profile chart, the smooth terrestrial surface and parallel arcs are parabolae whose curvature reflects the amount and direction of ray rectification (in turn dependent upon troposhperic refractivity gradient). Path distance is marked off along the smooth-earth arc while the intersections of the parabolae with the straight lines (for convenience constructed vertically and not radially as would be the case in a circular earth diagram) denote elevation. To clearly portray elevations versus path distance, the former's scale is magnified.

PARABOLIC CHART CHARACTERISTICS

The renunciation of circularity for parabolicity (tolerable under the conditions that path distance is small compared to earth's radius but large with respect to elevation) happily accrues with minimal penalty.

Representation of earth by partial parabolae entails the small error of 0.3%, while that accruing from scaling path distance along the smooth-earth parabola is negligible.

For terminals of different elevations, a radio ray's length is not identical with path distance as determined from the latter's scale, while additionally there accrues a Fresnel zone error (technically, zone radius is measured normal to the radio ray while in practice it is plotted along the vertical). Errors from both these sources are negligible for ray vertical angles less than 2°.

A slight inaccuracy arises from the fact that elevations are plotted along vertical lines instead of radial; however, since the angular difference is of the order of only 10.0 minutes, it may be ignored.

Contrary to extant technical old wives' tales it is not necessary to plot a terrain profile symmetrically about the parabolic apex; it can be shown mathematically and graphically that equally valid results obtain whether or not the plot is "centered."

The distended vertical scale with respect to that of path distance **does**, however, lead to the important limitation of greatly distorted vertical angle presentation making impossible direct determination without a special EBU protractor.

Terrain profile engineering makes use of various types of representations, among which are parabolic rays on rectangular coordinates (and others) as well as the herein touched upon straight line rays on parabolic grid. Each has its special advantages and preferences. The latter, for example, offers facility of graphic straight-line terrestrial reflection analysis (a "sticky" procedure using parabolic rays) and plotting Fresnol zone **ellipses** around straight line rays as well as added convenience in path-problem solutions over a given range of regional refractivity gradients employing the TAI Multi-Refractivity Gradient plotting technique.

Courtesy TAI Inc. From Consuletter-International Vol. 5 No. 7, entitled "PARABOLIC WORLD."

Appendix 19
Refractivity and K-Factor Data

The information in this Appendix is given graphically as cumulative distributions for the ground-based 100.0 meter tropospheric layer, the area in which the preponderance of VHF, UHF, and microwave signals propagate. It is a sampling of such data at locations around the world. The data were calculated from weather balloon (radiosonde or RAOB) observations of barometric pressure, temperature, and humidity versus height. While two balloon releases per day per location are made at the internationally standardized times of 0000 hours GMT (UT) and 1200 hours GMT (UT) and 1500 hours GMT (UT) and 0300 GMT (UT) prior to 1957, as part of the synoptic world-wide weather observation, balloon observation at other times are in some cases herein considered.

The graphs of cumulative distributions of refractive gradients and corresponding k-factors do not show the percentage of time of a year during which these refractivity gradients and k-factors may be expected to occur, but rather the percentage of occurrences of these values at the time of the brief daily observations in the ground-based 100.0 meter tropospheric layer.

In this Appendix, the stations are filed in alphabetical order rather than by geographical location. In tabular form beneath the graphs are given the location geographical coordinates, elevation above mean sea level in meters, period of record and climatological information. In order to provide an indication of winter/summer contrasts, the average daily minimum and average daily maximum temperatures are given. Mean monthly dewpoint temperatures for January and July are also given. Precipitation data given are in terms of annual total and averages for the wettest and driest months. A brief climatic description of the measuring station location is also included.

There may be some objection to the use of such data, as presented here, for microwave path engineering. Firstly, it is point data (i.e. data taken at one location, not along the entire radio path). Secondly, as a general rule, balloon launchings take place only twice per day and other radio-meteorological conditions could occur outside these observation times. Thirdly, there may be such objections as RAOB sensor lag in which given tropospheric data are not transmitted to the ground station at the correct altitude. Much of the data, furthermore, has been transmitted to the ground station sequentially instead of simultaneously. Some aver also that measurement at a point over the earth does not represent the entire path; while this may be in some instances true, other students of radio-meteorology hold that in "difficult" propagational areas (e.g. Persian Gulf area) a point measurement might well represent the entire path. Much more work is needed in this realm. Be that as it may, we appear to be at the state-of-the-art. Improvements will doubtless occur to the serious propagationist. For the most part, the RAOB information (a fallout of weather observation) is all that is available, and it, thus, appears that it would indeed be wasteful indeed not to "cash in" on this information as a propagational guide. It is the writer's stance that the art/science of telecommunications is a big boy now and deserves his own specially tailored radio-meteorology "set-up" rather than depending upon weather's hand-me-downs, albeit the latter does serve an interim purpose.

The real "aficionado" of propagation might do well to obtain the following works which give much more worldwide data of the type herein included:

"Refractivity Gradients In The Northern Hemisphere," OT Report 75-79, NTIA, US Dept. Of Commerce by CA Samson. NTIS Accesion Number AD-A009 503.

Monograph-1, "A World Atlas Of Atmospheric Radio Refractivity," NTIA, US Dept. of Commerce by BR Bean, BA Cahoon, CA Samson and GD Thayer.

This material is courtesy of the NTIA (National Telecommunications and Information Administration), U.S. Department of Commerce, Boulder, Colorado-80303, USA. It is from "REFRACTIVITY AND RAINFALL DATA FOR RADIO SYSTEMS ENGINEERING." OT REPORT 76-105, by CA SAMSON.

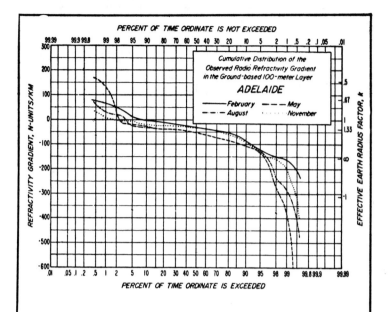

ADELAIDE, AUSTRALIA

34° 56' S, 138° 35' E   11 M M.S.L.
DATA: RADIOSONDE.   0800Z (1700 LST): 2/49 – 11/51
                                  0700Z (1600 LST): 2/52 – 8/52
                                  0400Z (1300 LST): 11/52 – 2/53
TEMPERATURE (°F):   JANUARY 86/61; JULY 59/45
MEAN DEWPOINT (°F):   JANUARY 48; JULY 44
PRECIPITATION (INCHES): ANNUAL 21.1; JUNE 3.00; FEBRUARY 0.70

NEAR THE EAST SHORT OF THE GULF OF ST. VINCENT AT THE FOOT OF THE MT. LOFTY RANGES. A MEDITERRANEAN TYPE OF CLIMATE WITH WARM, DRY SUMMERS AND COOL RAIN WINTERS.

Fig. A19-1. Cumulative distribution of refractive gradient and k-factor information for Adelaide, Australia.

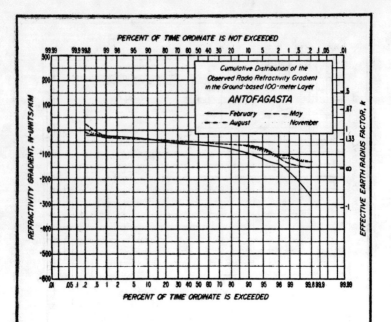

ANTOFAGASTA, CHILE

23° 28′ S, 70° 26′W
DATA: RADIOSONDE.

122 M M.S.L.
0000 and 1200Z (1900 and 0700 LST):
8/57 – 8/62

TEMPERATURE (°F):
MEAN DEWPOINT (°F):
PRECIPITATION (INCHES):

JANUARY 76/63;   JULY 63/51
JANUARY 59; JULY 48
ANNUAL : JULY 0.20; DECEMBER, JANUARY, FEBRUARY, MARCH 0.00

A PORT ON THE PACIFIC OCEAN IN NORTHERN CHILE ON THE EDGE OF A DESERT; IT HAS MODERATE TEMPERATURES AND SMALL SEASONAL VARIATIONS BECAUSE OF THE MARINE EXPOSURE.

Fig. A19-2. Cumulative distribution of refractive gradient and k-factor information for Antofagasta, Chile.

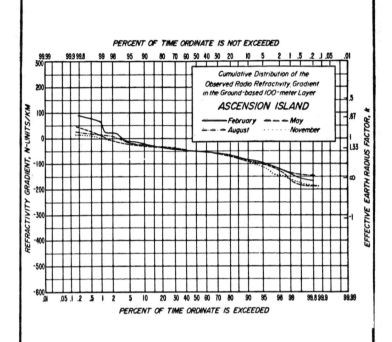

ASCENSION ISLAND

07° 58' S, 14° 24' W	79 M M.S.L.
DATA: RADISONEONDE.	0000 AND 1200Z (2300 AND 1100 LST):
	2/58 – 11/62 (0600 and 1800Z
	OBSERVATIONS INCLUDED FOR 11/59
	AND PART OF 1/8 1/8/58)
TEMPERTUURE (°F):	JANAUARY 85/73; JULY 84/72
MEAN DEWPOINT (°F):	JANUARY 68; JULY 65
PRECIPITATION (INCHES):	ANNUAL 9.0; NOVEMBER 2.10; JUNE, AUGUST 0.03

A VOLCANIC ISLAND 4KM WIDE AND 13 KM LONG; GREEN MOUNTAIN RISES TO 875 M. A MILD TROPICAL CLIMATE IN THE SOUTHEAST TRADE WIND REGION.

Fig. A19-3. Cumulative distribution of refractive gradient and k-factor information for Ascension Island.

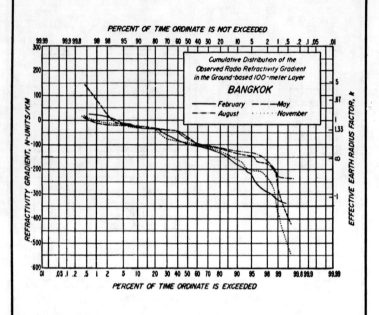

BANGKOK, THAILAND

13° 44′ N, 100° 30′ E
DATA: RADIOSCODE. 0000 AND 1200Z (0/00 and 1900 LST):
TEMPERATURE (°F): 8/57 – 11/59
MEAN DEWPOINT (°F): JANUARY 89/67; JULY 90/76
PRECIPITATION (INCHES): JANUARY 66; JULY 75
ANNUAL 57.8; SEPTEMBER 14.04;
DECEMBER 0.08

LOCATED ON THE CHAO PHRAYA RIVER NEAR THE GULF OF SIAM. A TROPICAL MONSOON TYPE OF CLIMATE WITH PRONOUNCED WET AND DRY SEASONS.

Fig. A19-4. Cumulative distribution of refractive gradient and k-factor information for Bankok, Thailand.

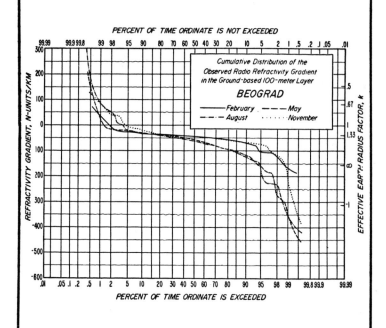

BEOGRAD, YUGOSLAVIA

44° 47' N, 20° 32' E
DATA: RADIOSONDE.

243 M M.S.L.
0200Z (0300 LST): 2/56 – 2/57
0000Z (0100 LST): 5/57 – 11/60

TEMPERATURE (°F): JANUARY 37/27; JULY 84/61

MEAN DEWPOINT (°F): JANUARY 26; JULY 57

PRECIPITATION (INCHES): ANNUAL 27.6; JUNE 3.78; FEBRUARY, MARCH 1.81

A DANUBE RIVER PORT IN NORTHEAST YUGOSLAVIA. A CONTI – NENTAL CLIMATE WITH COLD WINTERS AND WARM SUMMERS.

Fig. A19-5. Cumulative distribution of refractive gradient and k-factor information for Beograd, Yugoslavia.

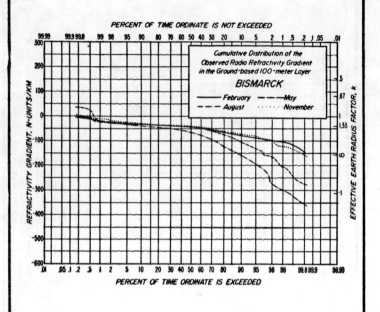

BISMARCK, N. DAKOTA

46° 46′ N, 100° 45′ W   506 M M.S.L.

DATA: RADIOSONDE.   0300 AND 1500Z (2000 and 0800 LST): 8/52 – 5/57

TEMPERATURE (°F):   JANUARY 20/0; JULY 86/58
MEAN DEWPOINT (°F):   JANUARY 1; JULY 55
PRECIPITATION (INCHES):   ANNUAL 15.2; JUNE 3.40; DECEMBER 0.03

LOCATED ON THE MISSOURI RIVER IN A SHALLOW BASIN SUR — ROUNDED BY LOW-LYING HILLS. A SEMI-ARID CONTINENTAL CLIMATE WITH COLD WINTERS AND WARM SUMMERS.

Fig. A19-6. Cumulative distribution of refractive gradient and k-factor information for Bismark, North Dakota.

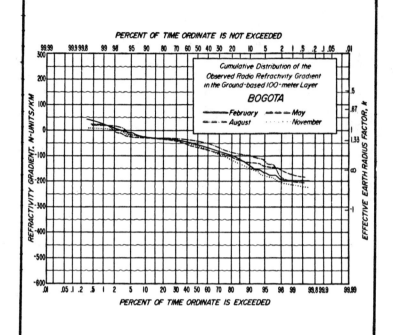

BOGOTA, COLUMBIA

4° 42′ N, 74° 09′ W   2547 M M.S.L.

DATA: RADIOSONDE.   0000 and 1200Z (1900 AND 0700 LST):
                                                  8/60 – 2/63
TEMPERATURE (°F):   JANUARY 48; JULY 48
                                                  JULY 2.01
PRECIPITATION (INCHES):   ANNUAL 41.7; OCTOBER 6.30;

LOCATED ON A HIGH ANDEAN PLATEAU IN AN AGRICULTURAL REGION. A MILD CLIMATE WITH SMALL SEASONAL CHANGES: THE EFFECTS OF THE HIGH ALTITUDE MODERATE THE PREVAILING TROPICAL AIR MASSES.

Fig. A19-7. Cumulative distribution of refractive gradient and k-factor information for Bogata, Columbia.

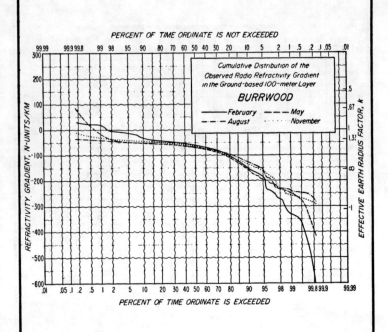

BURRWOOD, LOUISIANA

28° 58′ N, 89° 22′ W	2 M M.S.L.
DATA: RADIOSONDE.	0300 and 1500Z (2100 AND 0900 LST): 8/52 – 5/57
TEMPERATURE (°F):	JANUARY 65/60; JULY 89/76
MEAN DEWPOINT (°F):	NOT AVAILABLE
PRECIPITATION (INCHES):	ANNUAL 58.20; AUGUST 7.85; OCTOBER 3.07

LOCATED NEAR THE SOUTHWEST EXTREMITY OF THE MISSISSIPPI RIVER DELTA. A PREDOMINANTLY MARITIME CLIMATE WITH MILD WINTERS AND HOT, HUMID SUMMERS.

Fig. A19-8. Cumulative distribution of refractive gradient and k-factor information for Burrwood, Louisiana.

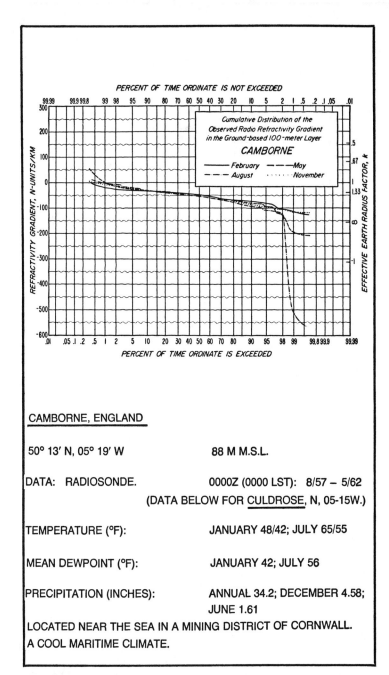

Fig. A19-9. Cumulative distribution of refractive gradient and k-factor information for Camborne, England.

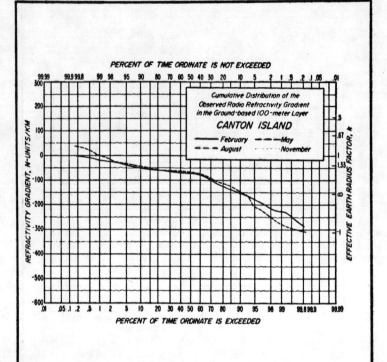

Fig. A19-10. Cumulative distribution of refractive gradient and k-factor information for Canton Island, Phoenix Islands.

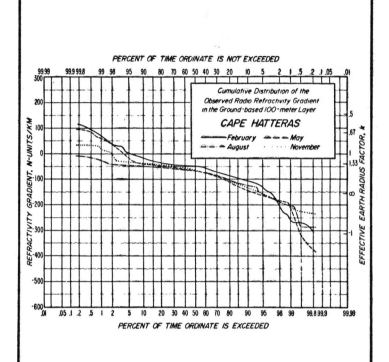

CAPE HATTERAS, N. CAROLINA

35° 16' N, 75° 33' W	3 M M.S.L.
DATA: RADIOSONDE.	0300 AND 1500Z (2200 AND 1000 LST): 8/52 – 5/57
TEMPERATURE :	JANUARY 53/40; JULY 84/72
MEAN DEWPOINT (°F):	NOT AVAILABLE
PRECIPITATION (INCHES):	ANNUAL 54.5; AUGUST 6.42; APRIL 2.99

LOCATED ON AN ELONGATED SANDY ISLAND IN THE OUTER BANKS OF NORTH CAROLINA. THE GULF STREAM OF THE ATLANTIC OCEAN PASSES 30 TO 80 KM OFFSHORE. A MARITIME CLIMATE WITH WARM SUMMERS AND COOL WINTERS. RELATIVELY WINDY.

Fig. A19-11. Cumulative distribution of refractive gradient and k-factor information for Cape Hatteras, North Carolina.

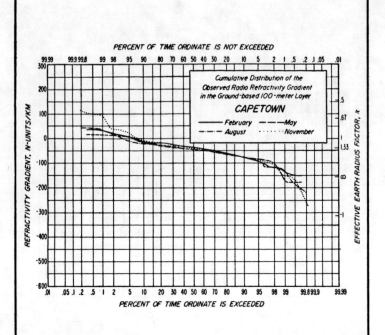

CAPETOWN, S. AFRICA

33° 58′ S, 18° 36′ E 50 M.M.S.L.
DATA: RADIOSONDE. 0400Z, 0500Z, 0600Z (0500, 0600, 0700 LST):
8/52-11/56
0600 AND 1200Z (0700 AND 1300 LST):
8/57-11/58

TEMPERATURE (°F): JANUARY 78/60; JULY 63/46

MEAN DEWPOINT (°F): JANUARY 56; JULY 47

PRECIPITATION (INCHES): ANNUAL 24.7; JUNE 4.29; FEBRUARY 0.59

LOCATED ON TABLE BAY OF THE ATLANTIC OCEAN AND AT FOOT OF 1100-M TABLE MOUNTAIN. A MEDITERRANEAN TYPE OF CLIMATE WITH WINTER RAINS AND SMALL SEASONAL TEMPERATURE CHANGES.

Fig. A19-12. Cumulative distribution of refractive gradient and k-factor information for Capetown, South Africa.

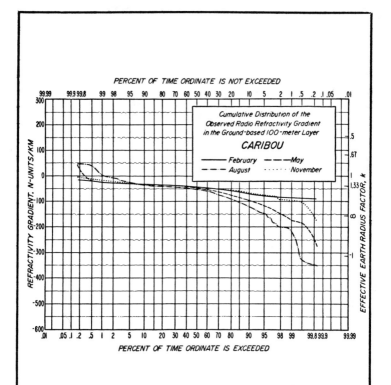

CARIBOU, MAINE

46° 52' N, 68° 01' W	191 M M.S.L.
DATA: RADIOSONDE.	0300 AND 1500Z (2200 AND 1000 LST): 8/52 - 5/57
TEMPERATURE (°F):	JANUARY 20/1; JULY 75/54
MEAN DEWPOINT (°F):	JANUARY 4; JULY 55
PRECIPITATION (INCHES):	ANNUAL 36.3; JUNE 4.07; FEBRUARY 2.02

LOCATED IN THE GENTLY ROLLING HILLS OF NORTHEASTER MIANE. ALTHOUGH ONLY 240 KM FROM THE ATLANTIC, THE CLIMATE IS PREDOMINANTLY CONTINENTAL WITH COLD, WINDY WINTERS NAD MILD SUMMERS.

Fig. A19-13. Cumulative distribution of refractive gradient and k-factor information for Caribou, Maine.

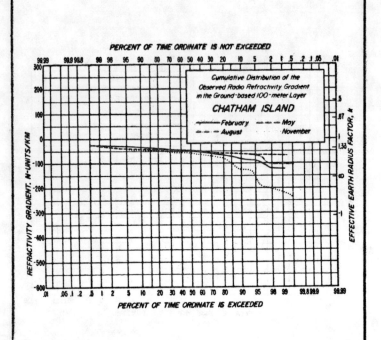

CHATHAM ISLAND, PACIFIC

43° 58' S, 176° 33' W	49 M M.S.L.
DATA: RADIOSONDE.	0000Z (1200 LST): 8/57 – 11/59
TEMPERATURE (°F):	JANUARY 63/52; JULY 50/41
MEAN DEWPOINT (°F):	NOT AVAILABLE
PRECIPITATIO (INCHES):	ANNUAL 33.6; JULY 3.8; OCTOBER 2.1

A SMALL VOLCANIC ISLAND, WITH FORESTED HILLS, GRASSY PLAINS AND A CENTRAL LAGOON. A COOL MARITIME CLIMATE.

Fig. A19-14. Cumulative distribution of refractive gradient and k-factor information for Chatham Island.

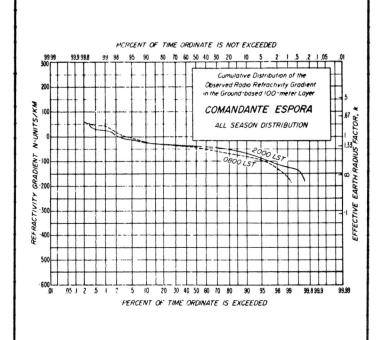

Fig. A19-15. Cumulative distribution of refractive gradient and k-factor information for Comandante Espora, Argentina.

COMMANDANTE ESPORA, ARGENTINA

38° 44′ S, 62° W	70 M M.S.L.
DATA: RADIOSONDE.	0000 AND 1200Z (2000 AND 0800 LST): 2/64 – 11/68 (FEBRUARY, MAY, AUGUST, NOVEMBER)
ANALYZED BY:	ENTEL ARGENTINA
TEMPERATURE (°F):	JANUARY 88/62; JULY 57/39
MEAN DEWPOINT (°F):	JANUARY 54; JULY 37
PRECIPITATION (INCHES):	ANNUAL 20.6; MARCH 2.50; JUNE 0.90

LOCATED ON THE ATLANTIC COAST. A MARITIME CLIMATE WITH HOT SUMMERS AND COOL WINTERS.

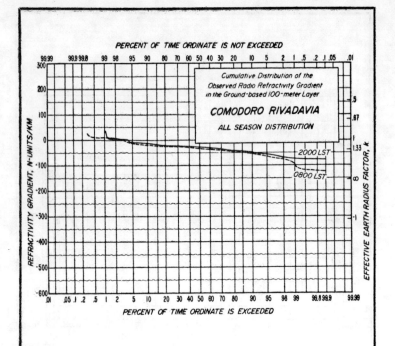

COMODORO RIVADAVIA, ARGENTINA

45° 47' S, 67° 30' W	61 M M.S.L.
DATA: RADIOSONDE.	0000 AND 1200Z (2000 AND 0800 LST):
	2/64 – 11/68 (FEBRUARY, MAY, AUGUST, NOVEMBER)
ANALYZED BY:	ENTEL ARGENTINA
TEMPERATURE (°F):	JANUARY 78/55; JULY 56/39
MEAN DEWPOINT (°F):	JANUARY 39; NOVEMBER 47 (JULY N/A)
PRECIPITATION (INCHES):	ANNUAL 7.3; AUGUST 1.10; MARCH 0.05

LOCATED ON THE GULF OF SAN JORGE IN SOUTHERN ARGENTINA. A COOL, MARITIME CLIMATE.

Fig. A19-16. Cumulative distribution of refractive gradient and k-factor information for Comodoro Rivadavia, Argentina.

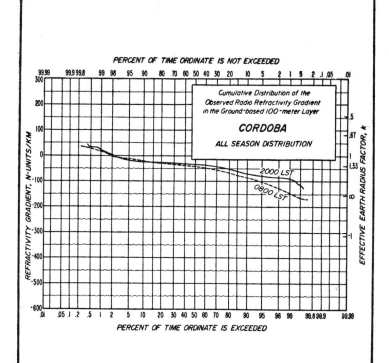

CORDOBA, ARGENTINA

31° 19' S, 64° 13' W	474 M M.S.L.
DATA: RADIOSONDE.	0000 and 1200Z (2000 AND 0800 LST): 2/64 – 11/68 (FEBRUARY, MAY, AUGUST, NOVEMBER)
ANALYZED BY:	ENTEL ARGENTINA
TEMPERATURE (°F):	JANUARY 88/61; JULY 65/38
MEAN DEWPOINT (°F):	JANUARY 61; JULY 39
PRECIPITATION (INCHES):	ANNUAL 28.3 DECEMBER 4.80; JUNE 0.30

LOCATED ON THE RIO PRIMERO ON THE EAST SLOPE OF THE SIERRA DE CORDOBA, IN AN AGRICULTURAL AREA ON THE WEST EDGE OF THE PAMPA. A MILD CONTINENTAL CLIMATE.

Fig. A19-17. Cumulative distribution of refractive gradient and k-factor information for Cordoba, Argentina.

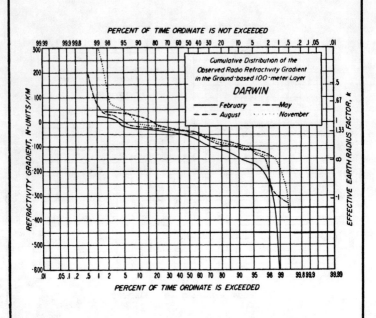

DARWIN, AUSTRALIA

12° 26' S, 130° 52' E	27 M M.S.L.
DATA: RADIOSONDE.	0700Z (1600 LST): 8/52
	0400Z (1300 LST): 11/52 – 2/53
	0000Z (0900 LST): 8/57 – 11/58
TEMPERATURE (°F):	JANUARY 90/77; JULY 87/67
MEAN DEWPOINT (°F):	JANUARY 75; JULY 60
PRECIPITATION (INCHES):	ANNUAL 58.7; JANUARY 15.20 ;
	JULY 0.03

LOCATED ON AN INLET OF THE TIMOR SEA. A TROPICAL MARITIME CLIMATE WITH MARKED WET AND DRY SEASONS.

Fig. A19-18. Cumulative distribution of refractive gradient and k-factor information for Darwin, Australia.

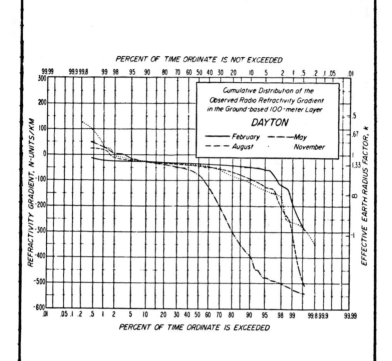

DAYTON, OHIO

39° 52' N, 84° 07' W	298 M M.S.L.
DATA: RADIOSONDE.	0300, 0900, 1500, 2100Z (2100, 0300, 0900, 1500 LST):
	9/51 – 1/53
TEMPERATURE (°F):	JANUARY 37/22; JULY 85/65
MEAN DEWPOINT (°F):	JANUARY 25; JULY 64
PRECIPITATION (INCHES):	ANNUAL 35.8; JUNE 4.10; OCTOBER 2.23

LOCATED IN THE NEARLY FLAT PLAIN OF THE MIAMI RIVER VALLEY; SURROUNDED BY ROLLING COUNTRY. A CONTINENTAL CLIMATE MODIFIED BY PROXIMITY TO THE GREAT LAKES. COLD, CLOUDY WINTERS AND HOT, HUMID SUMMERS.

Fig. A19-19. Cumulative distribution of refractive gradient and k-factor information for Dayton, Ohio.

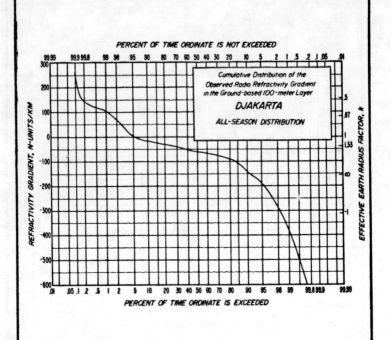

DJAKARTA, INDONESIA

06° 11' S, 106° 50' E	6 M M.S.L.
DATA: RADIOSONDE.	0000Z (0700 LST): 1/60 – 12/64 (ALL MONTHS)
ALALYZED BY:	B. SUTANTO, INSTITUTE FOR R&D OF POSTS AND TELECOMMUNICATIONS, INDONESIA
TEMPERATURE (°F):	JANUARY 84/74; JULY 86/73
MEAN DEWPOINT (°F):	JANUARY 73; JULY 71
PRECIPITATION (INCHES)	ANNUAL 71.5; JANUARY, FEBRUARY 11.80; AUGUST 1.70

LOCATED AT THE FOOT OF MT. MERAPI ON THE SOUTH SIDE OF THE ISLAND OF JAVA. A HOT AND HUMID TROPICAL MARITIME CLIMATE.

Fig. A19-20. Cumulative distribution of refractive gradient and k-factor information for Djakarta, Indonesia.

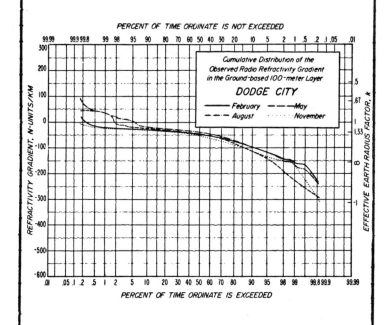

DODGE CITY, KANSAS

37° 46' N, 99° 58' W   791 M.M.S.L.

DATA: RADIOSONDE. 0300 and 1500Z (2000 and 0800 LST): 2/56-5/57
                                 000 and 1200Z (1700 and 0500 LST): 8/57-11/60

TEMPERATURE (°F): JANUARY 42/20; JULY 93/68

MEAN DEWPOINT (°F): JANUARY 19; July 60

PRECIPITATION (INCHES): ANNUAL 19.25; MAY 3.22; DECEMBER 0.47

LOCATED IN THE HIGH PLAINS REGION ON THE ARKANSAS RIVER. A SEMI-ARID CONTINENTAL CLIMATE WITH HOT SUMMERS AND LARGE TEMPERATURE FLUCTUATIONS IN WINTER. RELATIVELY WINDY; PERIODS OF CALM ARE RARE.

Fig. A19-21. Cumulative distribution of refractive gradient and k-factor information for Dodge City, Kansas.

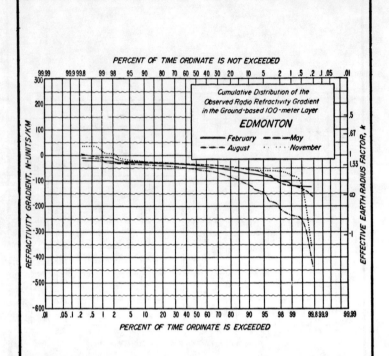

EDMONTON, ALBERTA

53° 34′ N, 113° 31′ W.	676 M M.S.L.
DATA: RADIOSONDE.	0300 AND 1500Z (1900 AND 0700 LST): 1/46 – 12/49
TEMPERATURE (°F):	JANUARY 16/–3; JULY 74/50
MEAN DEWPOINT (°F):	JANUARY 5; JULY 52
PRECIPITATION (INCHES):	ANNUAL 18.0; JULY 3.30; FEBRUARY, MARCH 0.70

LOCATED IN AN AGRICULTURLA REGION ON THE BANKS OF THE NORTH SASKATCHEWAN RIVER. A SEVERE CONTINENTAL CLIMATE WITH COLD, DRY WINTERS AND MILE SUMMERS.

Fig. A19-22. Cumulative distribution of refractive gradient and k-factor information for Edmonton, Alberta.

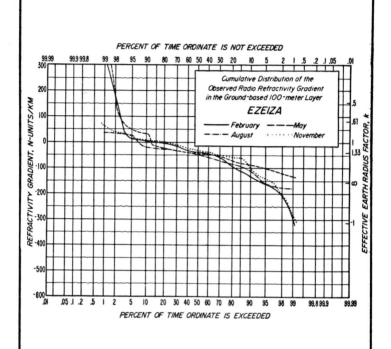

EZEIZA, ARGENTINA (BUENOS AIRES)

34° 50' S, 58° 32' W	20 M M.S.L.
DATA: RADIOSONDE.	1200Z (0800 LST): 11/57 – 11/59
TEMPERATURE (°F):	JANUARY 85/63; JULY 57/42
MEAN DEWPOINT (°F):	JANUARY 59; JULY 42
PRECIPITATION (INCHES):	ANNUAL 37.4; MARCH 4.30; JULY 2.20

LOCATED AT A LARGE APIRPORT NEAR BUENOS AIRES; NEAR A WIDE INLET FROM THE ATLANTIC OCEAN. A PREDOMINANTLY MARITIME CLIMATE.

Fig. A19-23. Cumulative distribution of refractive gradient and k-factor information for Ezeiza, Argentina.

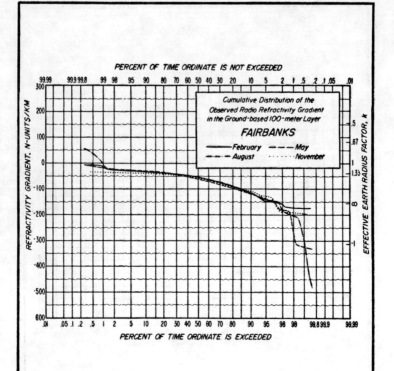

FAIRBANKS, ALASKA

64° 49′ N, 147° 52° W	138 M M.S.L.
DATA: RADIOSONDE.	0000 AND 1200Z (1400 AND 0200 LST): 2/58 – 8/60
TEMPERATURE (°F):	JANUARY −1/−21 JULY 72/48
MEAN DEWPOINT (°F):	JANUARY −19; JULY 50
PRECIPITATION (INCHES):	ANNUAL 11.3; AUGUST 2.20; PARIL 0.25

LOCATED IN THE TANANA VALLEY OF INTERIOR ALASKA: SHELTERED BY MOUNTAINS FROM MARITIME INFLUENCES. A SEVERE CONTINENTAL CLIMATE WITH VERY COLD WINTERS AND COOL SUMMERS. THE SUN IS ABOVE THE HORIZON FROM 10 TO LESS THAN 4 HRS/DAY IN THE PRIOD NOVEMBER TO MARCH, BUT IN JUNE AND JULY THERE IS SUNLIGHT FROM 18 to 21 HRS/DAY.

Fig. A19-24. Cumulative distribution of refractive gradient and k-factor information for Fairbanks, Alaska.

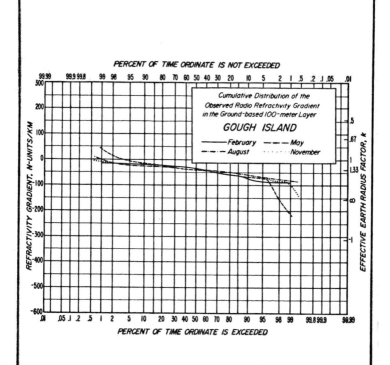

GOUGH ISLAND

40° 19′ S, 09° 54 W
DATA: RADIOSONDE.
MEAN TEMPERATURE (°F):
PRECIPITATION (INCHES):

40 M M.S.L.
0600Z (0500 LST): 8/57 – 11/59
JANUARY 58; JULY 49
ANNUAL 133.7; JUNE 16.1; FEBRUARY 7.76

A SMALL VOLCANIC ISLAND IN THE TRISTAN DE CUNHA GROUP IN THE SOUTH ATLANTIC. A RELATIVELY MILD BUT WINDY AND WET MARITIME CLIMATE.

Fig. A19-25. Cumulative distribution of refractive gradient and k-factor information for Gough Island.

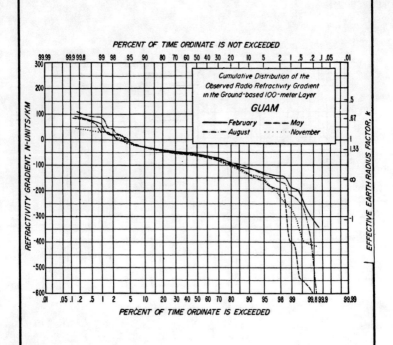

GUAM, MARIANA ISLANDS

13° 33′ N, 144° 50′ E	157 M M.S.L.
DATA: RADIOSONDE.	0300 AND 1500Z (1300 AND 0100 LST): 1/51 – 5/57 0000 AND 1200Z (100 AND 2200 LST): 6/57 – 12/57
TEMPERATURE (°F):	JANUARY 84/72; JULY 87/72
MEAN DEWPOINT (°F):	JANUARY 71; JULY 74
PRESCIPITATION (INCHES):	ANNUAL 88.5; SEPTEMBER 13.36; MARCH 2.64

AN ISLAND 45 KM LONG AND 6 TO 13 KM WIDE. A WARM, HUMID, TROPICAL MARITIME CLIMATE WITH A DRY SEASON JANUARY THROUGH APRIL AND A RAINY SEASON MID-JULY TO MID-NOVEMBER. IN THE NORTHEAST TRADE WIND BELT. OCCASIONAL TYPHOONS.

Fig. A19-26. Cumulative distribution of refractive gradient and k-factor information for Guam.

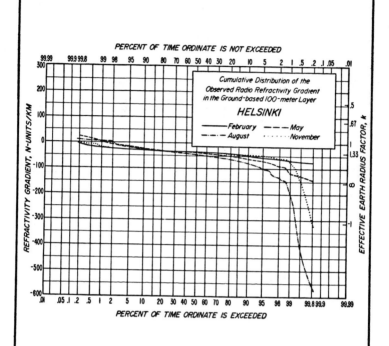

A19-27. Cumulative distribution of refractive gradient and k-factor information for Helsinki, Finland.

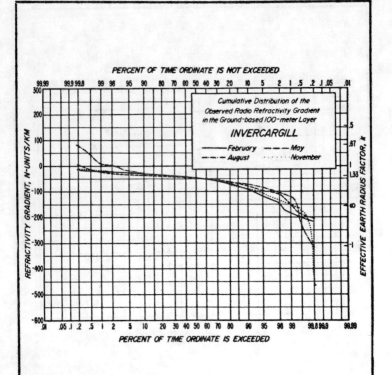

INVERCARGILL, NEW ZEALAND

46° 25' S, 168° 19' E	1 M M.S.L.
DATA: RADIOSONDE.	1100Z (2200 LST): 8/52
	0400Z (1500 LST): 11/52 – 11/53
	0300Z (1400 LST): 2/54 – 2/57
	0000Z (1100 LST): 5/57 – 11/60
TEMPERATURE (°F):	JANUARY 66/48; JULY 49/34
MEAN DEWPOINT (°F):	JANUARY 51; JULY 38
PRECIPITATION (INCHES):	ANNUAL 45.5; MAY 4.40; JULY, AUGUST, SEPTEMBER 3.20

LOCATED ON AN INLET FROM FOVEAUX STRAIT ON THE SOUTH SIDE OF SOUTH ISLAND. A COOL MARITIME CLIMATE WITH RELATIVELY FREQUENT RAINS.

Fig. A19-28. Cumulative distribution of refractive gradient and k-factor information for Invercargill, New Zealand.

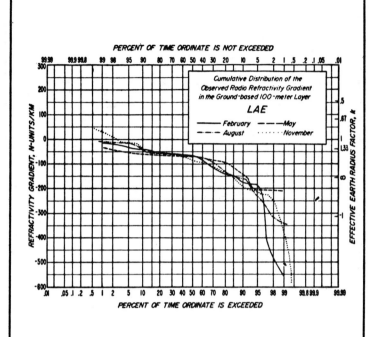

LAE, NEW GUINEA

06° 45′ S, 146° 59′ E   12 M.M.S.L.

DATA: RADIOSONDE.   0000Z (1000 LST): 8/57-11/59

TEMPERATURE (°F):   JANUARY 88/75; JULY 83/72

MEAN DEWPOINT (°F):   JANUARY 75/ JULY 73

PRECIPITATION (INCHES):   ANNUAL 130.1; AUGUST 19.80; JANUARY 3.10

LOCATED IN THE HARBOR AREA ON SOUTHEAST NEW GUINEA ISLAND. A HOT, HUMID, TROPICAL MARITIME CLIMATE.

Fig. A19-29. Cumulative distribution of refractive gradient and k-factor information for Lae, New Guinea.

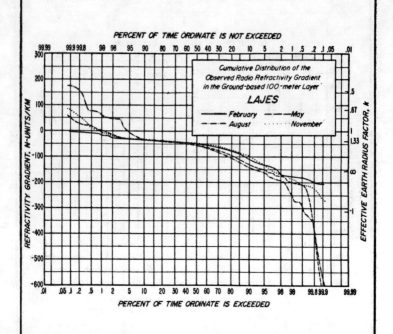

LAJES, AZORES

38° 45′ N, 27° 05′ 05′ W	112 M M.S.L.
DATA: RADIOSONDE.	0000, 0600, 1200, 1800Z (2200, 0400, 1000, 1600 LST): 11/57 - 8/62
TEMPERATURE (°F):	JANUARY 61/54; JULY 74/65
MEAN DEWPOINT (°F):	JANUARY 52; JULY 62
PRECIPITATON (INCHES):	ANNUAL 41.8; MARCH 6.13; JUNE, AUGUST 1.41

LOCATED NEAR THE NORTHEAST COAST OF TERCEIRA ISLAND IN THE AZORES GROUP: THE ISLAND HAS A STEEP COASTLINE AND A RUGGED INTERIOR. A MILD MARITIME CLIMATE WITH LITTLE SEASONAL TEMPERATURE CHANGE.

Fig. A19-30. Cumulative distribution of refractive gradient and k-factor information for Lajes, Azores.

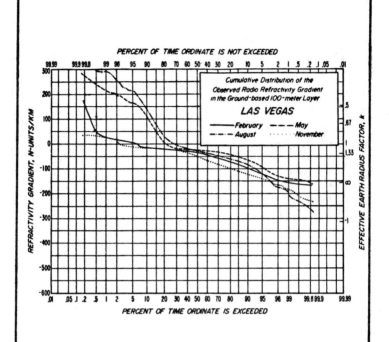

Fig. A19-31. Cumulative distribution of refractive gradient and k-factor information for Las Vegas, Nevada.

LAS VEGAS, NEVADA

36° 05' N, 115° 09' W 66 M M.S.L.
DATA: RADIOSONDE. 0300 AND 1500Z (1900 AND 0700 LST):
 2/56 – 5/57
 0000 AND 1200Z (1600 AND 0400 LST):
 8/57 – 11/60

TEMPERATURE (°F): JANUARY 54/32; JULY 104/76
MEAN DEWPOINT (°F): JANUARY 22; JULY 39
PRECIPITATION (INCHES): ANNUAL 3.9; JANUARY 0.53; JUNE 0.04

LOCATED NEAR THE CENTER OF A BROAD DESERT VALLEY SUR—
ROUNDED BY MOUNTAINS. A DESERT CLIMATE WITH LOW HUMIDITY,
VERY HOT SUMMERS, AND GENERALLY MILD WINTERS.

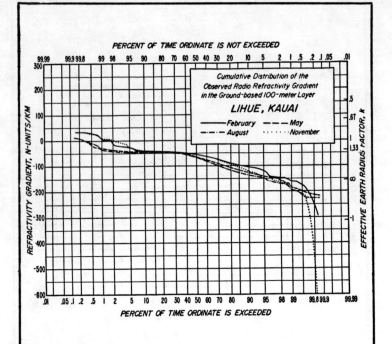

LIHUE, KAUAI, HAWAII

21° 59′ N, 159° 21′ W   36 M M.S.L.
DATA: RADIOSONDE.   0300 AND 1500Z (1600 AND 0400 LST): 1/51 – 5/57
    0000 AND 1200Z (1300 AND 0100 LST): 6/57 – 12/57

TEMPERATURE (°F):   JANUARY 78/64; JULY 83/72
MEAN DEWPOINT (°F):   JANUARY 63; JULY 68
PRECIPITATION (INCHES):   ANNUAL 43.0; JANUARY 5.51; JUNE 1.46

LOCATED NEAR THE EASTERN SHORT OF A MID-OCEAN ISLAND 53 KM LONG AND 40 KM WIDE; IN THE TRADE WIND BELT. MILD MARITIME CLIMATE WITH SMAL DIURNAL AND ANNUAL TEMPERA TURE RANGES.

Fig. A19-32. Cumulative distribution of refractive gradient and k-factor information for Lihue, Kauai, Hawaii.

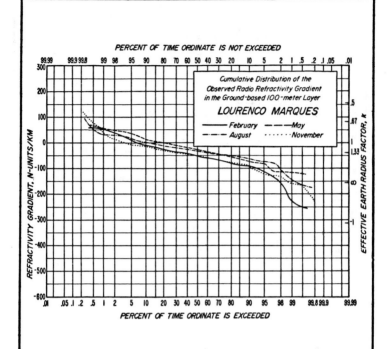

LOURENCO MARQUES, MOZAMBIQUE

25° 55' S, 32° 34' E	44 M M.S.L.
DATA: RADIOSONDE.	0300Z (0500 LST): 11/54 – 2/59
TEMPERATURE (°F):	JANUARY 86/71; JULY 76/55
MEAN DEWPOINT (°F):	JANUARY 69; JULY 55
PRECIPITATION (INCHES):	ANNUAL 29.9; JANUARY 5.10; JULY, AUGUST 0.05

LOCATED ON DELAGOA BAY OFF THE INDIAN OCEAN. A WARM MARITIME CLIMATE WITH WET AND DRY SEASONS.

Fig. A19-33. Cumulative distribution of refractive gradient and k-factor information for Lourenco Marques, Mozambique.

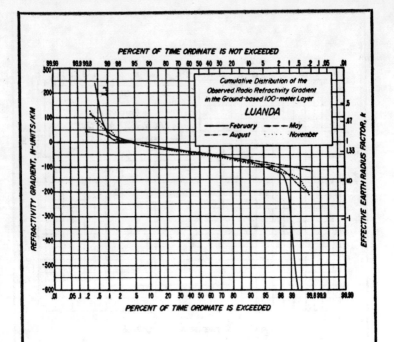

LUANDA, ANGOLA

08° 49' S, 13° 13' E	70 M M.S.L.
DATA: RADIOSOSNDE.	0700Z (0800 LST): 2/53 – 5/53
	0500Z (0600 LST): 8/53 – 5/54
	0300Z (0400 LST): 8/54 – 11/54
	0200Z (0300 LST): 2/55 – 11/55
	0600 AND 1200Z (0700 AND 1300 LST):
	8/57 – 11/57
	0000 AND 1200Z (0100 AND 1300 LST);
	2/58 – 11/58
	0000Z (0100 LST): 2/59 – 11/59
TEMPERATURE (°F):	JANUARY 83/74; JULY 74/65
MEAN DEWPOINT (°F):	JANUARY 70; JULY 63
PRECIPITATION (INCHES):	ANNUAL 12.8; APRIL 4.60; JULY 0.00

A PORT ON THE ATLANTIC OCEAN. A WARM, HUMID MARITIME CLIMATE.

Fig. A19-34. Cumulative distribution of refractive gradient and k-factor information for Luanda, Angola.

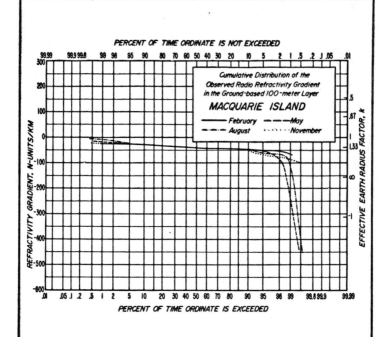

MACQUARIE ISLAND, PACIFIC

54° 30′ S, 158° 57′ E   6 M M.S.L.
DATA: RADIOSONDE.   0000Z (1100 LST): 8/57 – 11/59 AND
                                          5/62 – 11/62
TEMPERATURE (°F):   JANUARY 47/40; JULY 41/34
MEAN DEWPOINT (°F):   JANUARY 41; JULY 35
PRECIPITATIO (INCHES): ANNUAL 35.5; JUNE 4.11; DECEMBER 1.67

A VOLCANIC ISLAND IN THE SOUTH PACIFIC: 34 KM LONG AND 5 KM WIDE. ROCKY TERRAIN WITH SMALL GLACIAL LAKES. A COOL MARITIME CLIMATE WITH SMALL SEASONAL CHANGES.

Fig. A19-35. Cumulative distribution of refractive gradient and k-factor information for Macquarie Island.

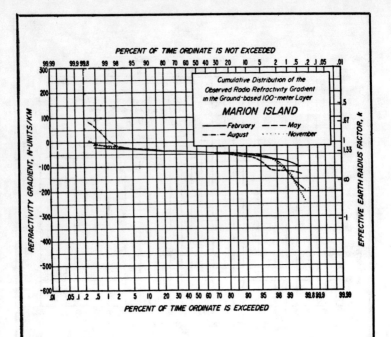

MARION ISLAND

46° 53′ S, 37° 52′ E 26 M M.S.L.
DATA: RADIOSONDE. 0500Z (0800 LST): 8/52 – 5/54
　　　　　　　　　　　0600Z (0900 LST): 8/54 – 11/54
　　　　　　　　　　　0400Z (0700 LST): 2/55 – 11/56
　　　　　　　　　　　1200Z (1500 LST): 8/57 – 11/58

MEAN TEMPERATURE (°F): JANUARY 44; JULY 39
PRECIPITATION (INCHES): ANNUAL 96.5; MAY 9.25; OCTOBER 6.57

AN ISLAND 21 KM LONG BY 13 KM WIDE IN THE SOUTHERN INDIAN OCEAN. A COOL MARITIME CLIMATE WITH HEAVY RAINFALL THROUGHOUT THE YEAR.

Fig. A19-36. Cumulative distribution of refractive gradient and k-factor information for Marion Island.

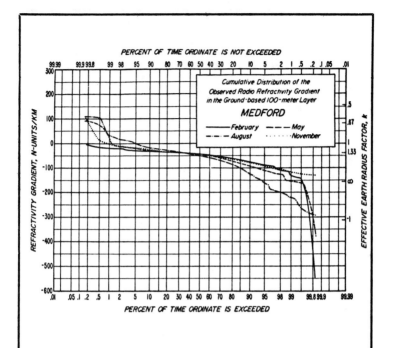

MEDFORD, OREGON

42° 23' N, 122° 52' W — 405 M M M.S.L.

DATA: RADIOSONDE. 0000 AND 1200Z (1600 AND 0400 LST): 11/57 - 8/62

TEMPERATURE (°F): JANUARY ½¼/28; JULY 88/56

MEAN DEWPOINT (°F): JANUARY 33; JULY 49

PRECIPITATION (INCHES): ANNUAL 19.8; DECEMBER 3.38; AUGUST 0.18

LOCATED IN A MOUNTAIN VALLEY. A MODERATE CONTINENTAL CLIMATE MODIFIED BY THE MARITIME AIR MASSES DURING THE CLOUDY, DAMP, AND COOL WINTERS; SUMMERS ARE WARM AND DRY..

Fig. A19-37. Cumulative distribution of refractive gradient and k-factor information for Medford, Oregon.

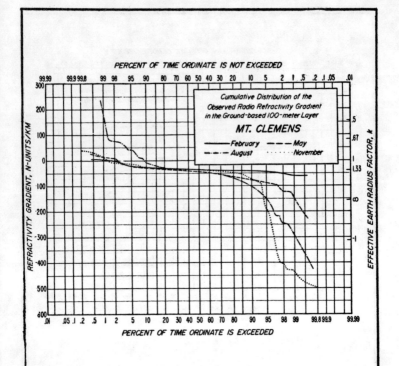

MT. CLEMENS, MICHIGAN

42° 36′ N, 82° 49′ W	178 M M.S.L.
DATA: RADIOSONDE.	0300, 0900, 1500, 2100Z (2100, 0300, 0900, 1500 LST): 7/51–1/53
TEMPERATURE (°F):	JANUARY 32/17; JULY 83/61
MEAN DEWPOINT (°F):	JANUARY 19; JULY 61
PRECIPITATION (INCHES):	ANNUAL 28.2; MAY 3.05; JANUARY 1.74

LOCATED ON FLAT LAND ON THE WEST SIDE OF LAKE ST. CLAIR JUST NORTHEAST OF DETROIT. A CONTINENTAL CLIMATE MODIFIED BY THE INFLUENCE OF THE GREAT LAKES. COLD, CLOUDY WINTERS AND WARM, SUNNY SUMMERS.

Fig. A19-38. Cumulative distribution of refractive gradient and k-factor information for Mt. Clemens, Michigan.

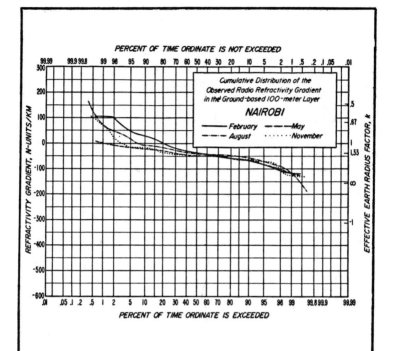

NAIROBI, KENYA

01° 18′ S, 36° 45′ E   1798 M M.S.L.

DATA: RADIOSONDE.
0400Z (0600 LST): 8/52 – 2/53
1200Z (1400 LST): 5/53 – 2/54
0600Z (0800 LST): 5/54 – 8/54
0000Z (0200 LST): 8/57 – 2/59
MISSING DATA: 2/55 – 5/57

TEMPERATURE (°F): JANUARY 77/54; JULY 69/51

MEAN DEWPOINT (°F): JANUARY 55; JULY 54

PRECIPITATION (INCHES): ANNUAL 34.3; APRIL 8.30; JULY 0.54

LOCATED IN THE EAST AFRICAN HIGHLANDS BETWEEN LAKE VIC — TORIA AND THE INDIAN OCEAN. A MILD UPLAND CLIMATE, ALTHOUGH IT IS NEAR THE EQUATOR.

Fig. A19-39. Cumulative distribution of refractive gradient and k-factor information for Nairobi, Kenya.

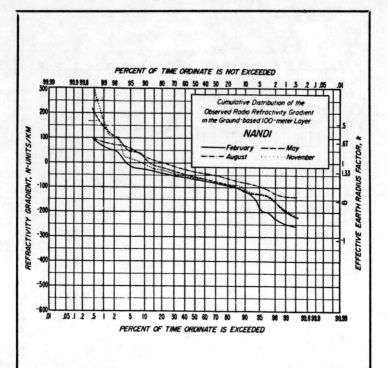

NANDI, FIJI ISLANDS

17° 45′ S, 177° 27′ E   16 M M.S.L.
DATA: RADIOSONDE.   1100Z (2300 LST): 8/49 – 2/51
                               1000Z (2200 LST): 5/51 – 8/52
                               0400Z (1600 LST): 11/52 – 8/53
TEMPERATURE (°F):   JANUARY 89/72; JULY 83/64
MEAN DEWPOINT (°F):   JANUARY 72; JULY 64
PRECIPITATION (INCHES) ANNUAL 72.7; MARCH 15.08; JULY 0.46

LOCATED AT THE MOUTH OF THE NANDI RIVER ON VITI LEVU ISLAND OF THE FIJI GROUP IN THE SOUTHWEST PACIFIC OCEAN. A HUMID TROPICAL MARITIME CLIMATE.

Fig. A19-40. Cumulative distribution of refractive gradient and k-factor information for Nandi, Fiji Islands.

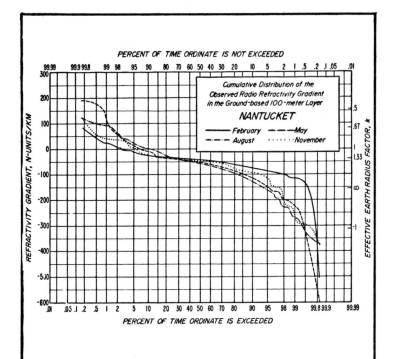

NANTUCKET, MASSACHUSETTS

41° 15′ N, 70° 04′ W   14 M M.S.L.
DATA: RADIOSONDE.   0300 AND 1500Z (2200 AND 1000 LST): 8/52 – 5/57
TEMPERATURE (°F):   JANUARY 39/27; JULY 74/62
MEAN DEWPOINT (°F):   JANUARY 26; JULY 63
PRECIPITATION (INCHES):   ANNUAL 43.7; MARCH 4.54; JULY 2.71

LOCATED ON NANTUCKET ISLAND SOUTH OF CAPE COD. A PREDOMINANTLY MARITIME CLIMATE BUT WITH SIGNIFICANT CONTINENTAL INFLUENCE IN WINTER. COOL, HUMID SUMMERS AND RELATIVELY MILD BUT WINDY WINTERS. FREQUENT FOG.

Fig. A19-41. Cumulative distribution of refractive gradient and k-factor information for Nantucket, Massachusetts.

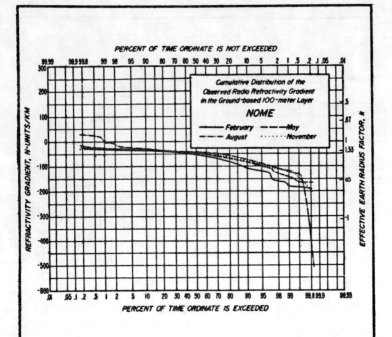

NOME, ALASKA

64° 30′ N, 165° 26′ W	14 M M.S.L.
DATA: RADIOSONDE.	0300 AND 1500Z (1600 AND 0400 LST): 2/56 – 5/57 0000 AND 1200Z (1300 AND 0100 LST): 8/57 – 11/60
TEMPERATURE (°F):	JANUARY 12/–3; JULY 55/44
MEAN DEWPOINT (°F):	JANUARY 12/–3; JULY 55/44
MEAN DEWPOINT (°F):	JANUARY –1; JULY 46
PRECIPITATION (INCHES):	ANNUAL 17.9; AUGUST 3.80; MAY 0.69

LOCATED ON FLAT GROUND NEAR NORTON SOUND OF THE BERING SEA. A VERY COLD, CONTINENTAL CLIMATE EXCEPT FOR THE MARITIME INFLUENCE OF THE OPEN WATER OF THE SOUND FROM JUNE TO MID-NOVEMBER. RELATIVELY WINDY.

Fig. A19-42. Cumulative distribution of refractive gradient and k-factor information for Nome, Alaska.

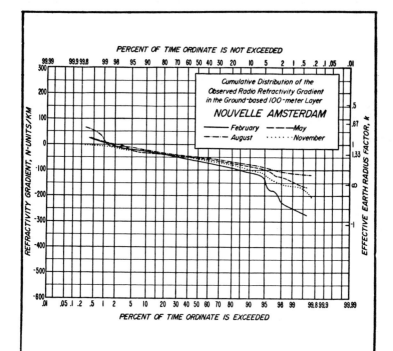

Fig. A19-43. Cumulative distribution of refractive gradient and k-factor information for Nouvelle Amsterdam Island.

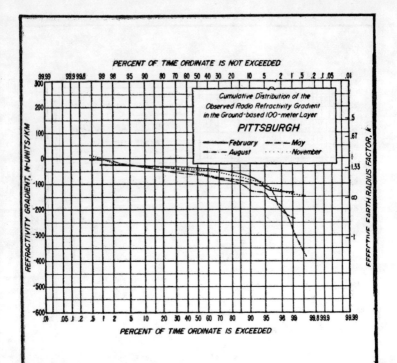

PITTSBURGH, PENNSYLVANIA

40° 30' N, 80° 13' W	351 M M.S.L.
DATA: RADIOSONDE.	0300 AND 1500Z (2200 AND 1000 LST): 7/51 – 1/53
TEMPERATURE (°F):	JANUARY 37/21; JULY 83/61
MEAN DEWPOINT (°F):	JANUARY 19; JULY 60
PRECIPITATION (INCHES):	ANNUAL 36.1; MAY 3.91; FEBRUARY 2.19

LOCATED AT THE FOOTHILLS OF THE ALLEGHENY MOUNTAINS AT THE CONFLUENCE OF THE ALLEGHENY AND MONGAHELA RIVERS. IT HAS A CONTINENTAL CLIMATE MODIFIED TO SOME EXTENT BY THE GREAT LAKES. WINTERS ARE COLD AND CLOUDY AND SUMMERS WARM AND HUMID.

Fig. A19-44. Cumulative distribution of refractive gradient and k-factor information for Pittsburgh, Pennsylvania.

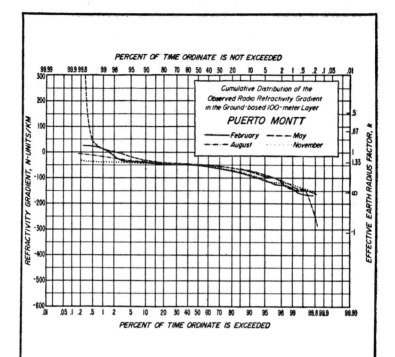

PUERTO MONTT, CHILE

41° 29 S, 72° 51' W	6 M M.S.L.
DATA: RADIOSONDE.	0000 AND 1200Z (1900 AND 0700 LST): 8/57 – 8/61
TEMPERATURE (°F):	JANUARY 68/52; JULY 51/41
MEAN DEWPOINT (°F):	JANUARY 54; JULY 43
PRECIPITATION (INCHES):	ANNUAL 86.0; MAY 10.88; FEBRUARY 4.30

A PORT ON RELONCAVI SOUND; THERE ARE FORESTED HILLS AND A LAKE DISTRICT NEARBY AND THE ANDES RANGE TO THE EAST. A COOL MARITIME CLIMATE.

Fig. A19-45. Cumulative distribution of refractive gradient and k-factor information for Puerto Montt, Chile.

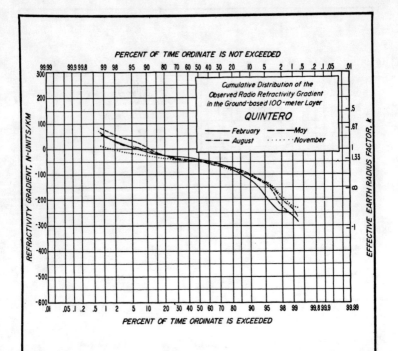

QUINTERO, CHILE

32° 47′ S, 71° 32′ W 7 M M.S.L.
DATA: RADIOSONDE. 0000 AND 1200Z (1900 AND 0700 LST);
3 YEARS (IN PIEROD 1957-67)
ANALYZED BY: DEPT. OF GEOPHYSICS AND GEODESY,
UNIVERSITY OF CHILE, SANTIAGO
TEMPERATURE (°F): JANUARY 72/56; JULY 60/47
MEAN DEWPOINT (°F): JANUARY 58; JULY 49
PRECIPITATION (INCHES): ANNUAL 19.9: JUNE 5.90; FEBRUARY 0.05

LOCATED ON THE PACIFIC COAST NEAR VALPARAISO IN CENTRAL CHILE. A COOL MARITIME CLIMATE.

Fig. A19-46. Cumulative distribution of refractive gradient and k-factor information for Quintero, Chile.

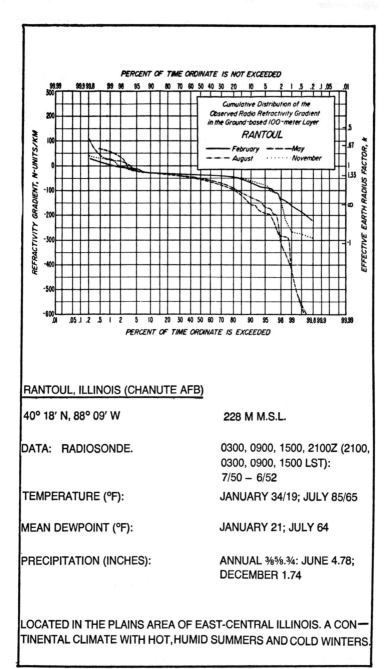

Fig. A19-47. Cumulative distribution of refractive gradient and k-factor information for Rantoul, Illionis (Chanute AFB).

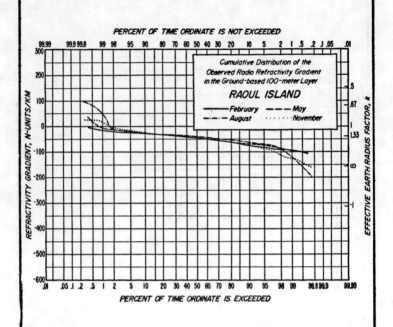

RAOUL ISLANDS, KERMADEC ISLANDS, PACIFIC

29° 15' S, 177° 55' W 49 M M.S.L.

DATA: RADIOSONDE.
1000Z (220 LST): 8/52
0400Z (1600 LST): 11/52 – 2/57
0000Z (1200 LST): 5/57 – 11/57

TEMPERATURE (°F): JANUARY 75/65; JULY 65/56

MEAN DEWPOINT (°F): NOT AVAILABLE

PRECIPITATION (INCHES): ANNUAL 57.2; MAY 6.9; NOVEMBER 2.0

A FORESTED, MOUNTAINOUS, VOLCANIC ISLAND OF ABOUT 18 SQ. KM AREA. A MILD MARITIME CLIMATE.

Fig. A19-48. Cumulative distribution of refractive gradient and k-factor information for Raoul Island, Kermadec Islands.

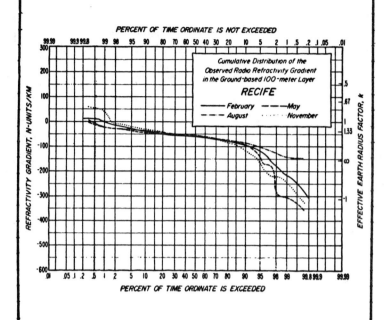

Fig. A19-49. Cumulative distribution of refractive gradient and k-factor information for Recife, Brazil.

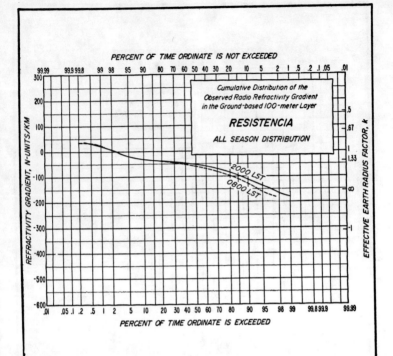

RESISTENCIA, ARGENTINA

27° 28' S, 58° 59' W	51 M M.S.L.
DATA: RADIOSONDE.	0000 AND 1200Z (2000 AND 0800 LST): 2/64 – 11/68 (FEBRUARY, MAY AUGUST, NOVEMBER)
ANALYZED BY:	ENTEN ARGENTINA
TEMPERATURE (°F):	JANUARY 93/71; JULY 71/53
MEAN DEWPOINT (°F):	JANUARY 70; JULY 52
PRECIPITATON (INCHES):	ANNUAL 51.0; JANUARY 7.14; JULY 2.12

LOCATED NEAR THE PARANA RIVER IN NORTHEASTERN ARGENTINA; IN AN AGRICULTURAL AND LUMBERING AREA. A RELATIVELY HUMID SUBTROPICAL CLIMATE WITH HOT, RAINY SUMMERS AND MILD WINTERS.

Fig. A19-50. Cumulative distribution of refractive gradient and k-factor information for Resistencia, Argentina.

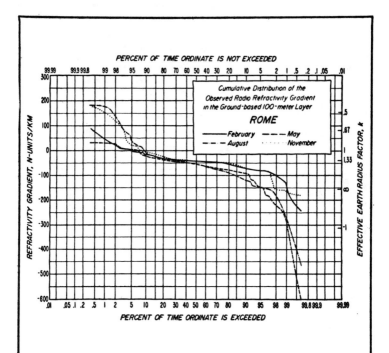

ROME, ITALY

41° 48′ N, 12° 36′ E   131 M M.S.L.
DATA: RADIOSONDE.   0200Z (0300 LST): 2/56 – 2/57
                                0000Z (0100 LST): 5/57 – 11/60
TEMPERATURE (°F):   JANUARY 52/37; JULY 88/65
MEAN DEWPOINT (°F):   JANUARY 36; JULY 60
PRECIPITATION (INCHES): ANNUAL 32.7; OCTOBER 5.04; JULY 0.67

LOCATED ON THE BANKS OF THE TIBER A SHORT DISTANCE INLAND FROM THE TYRRHENIAN SEA AND NEAR THE APPENINES MOUNTAINS. A MEDITERRANEAN TYPE CLIMATE WITH MILD, RAIN WINTERS AND HOT, HUMID SUMMERS.

Fig. A19-51. Cumulative distribution of refractive gradient and k-factor information for Rome, Italy.

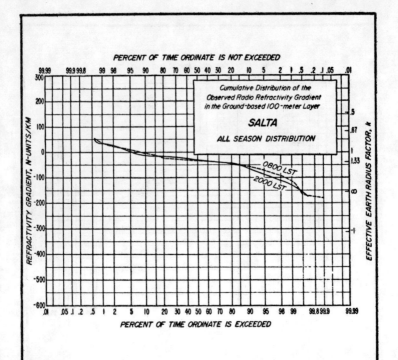

SALTA, ARGENTINA

24° 51′ S, 65° 29′ W	1226 M M.S.L.
DATA: RADIOSONDE.	0000 AND 1200Z (2000 AND 0800 LST): 2/64 – 11/68 (FEBRUARY, MAY, AUGUST, NOVEMBER)
ANALYZED BY:	ENTENL ARGENTINA
TEMPERATURE (°F):	JANUARY 83/59; JULY 70/39
MEAN DEWPOINT (°F):	JANUARY 62; JULY 41
PRECIPITATION (INCHES):	ANNUAL 27.8; JANUARY 6.50; JULY 0.00

LOCATED ON THE RIO SALADO IN AN IRRIGATED VALLEY. A SUB TROPICAL CLIMATE WITH WARM, RAINY SUMMERS AND MILD, DRY WINTERS.

Fig. A19-52. Cumulative distribution of refractive gradient and k-factor information for Salta, Argentina.

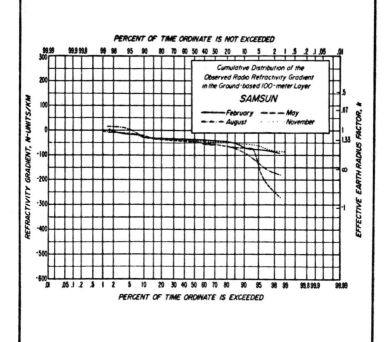

SAMSUN, TURKEY

41° 17' N, 36° 20' E
DATA: RADIOSONDE.

44 M M.S.L.
0300Z (0500 LST): 5/53 – 11/15/53
0200Z (0400 LST): 11/16/53 – 2/57
0000Z (0200 LST): 5/57 – 11/60

TEMPERATURE (°F): JANUARY 50/38; JULY 79/65
MEAN DEWPOINT (°F): JANUARY 33; JULY 63
PRECIPITATION (INCHES): ANNUAL 29.1; NOVEMBER 3.50; AUGUST 1.30

LOCATED IN AN AGRICULTURAL AND FORESTED AREA ON THE SOUTH SHORE OF THE BLACK SEA; CANIK MOUNTAINS TO THE SOUTH. A RELATIVELY MILD PREDOMINANTLY MARITIME CLIMATE.

Fig. A19-53. Cumulative distribution of refractive gradient and k-factor information for Samsun, Turkey.

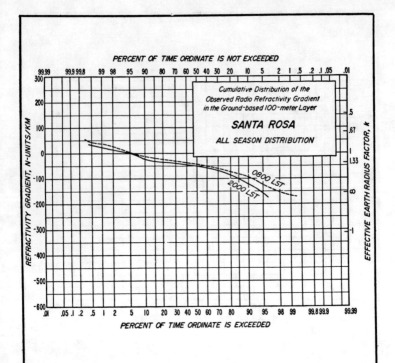

SANTA ROSA, ARGENTINA

36° 35′ S, 64° 16′ W 189 M M.S.L.
DATA: RADIOSONDE. 0000 AND 1200Z (2000 AND 0800 LST):
2/64 – 11/68 (FEBRUARY, MAY, AUGUST,
NOVEMBER)

ANALYZED BY: ENTEL ARGENTINA
(CLIMATOLOGICAL DATA NOT AVAILABLE.)

LOCATED IN THE PLAINS REGION NORTHWEST OF BAHIA BLANCA.
A CONTINENTAL CLIMATE.

Fig. A19-54. Cumulative distribution of refractive gradient and k-factor information for Santa Rosa, Argentina.

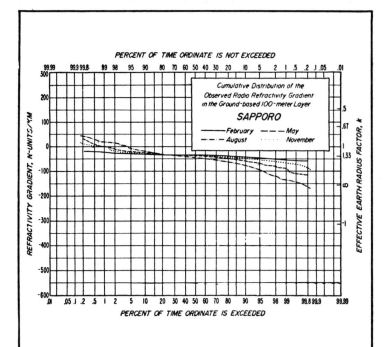

SAPPORO, JAPAN

43° 03' N, 141° 20' E 18 M M.S.L.
DATA: RADIOSONDE. 0300 AND 1500Z (1200 AND 000 LST): 8/52 – 11/56
TEMPERATURE (°F): JANUARY 29/11; JULY 75/58
MEAN DEWPOINT (°F): JANUARY 16; JULY 61
PRECIPITATION (INCHES): ANNUAL 41.1; SEPTEMBER 5.00; APRIL 2.20

LOCATED ABOUT 25 KM FROM COAST ON HOKKAIDO ISLAND OF JAPAN. A COOL MARITIME CLIMATE WITH CONTINENTAL CHARACTERISTICS IN WINTER BECAUSE OF THE SLIGHT MODIFICATION OF POLAR AIR MASSES MOVING OFF THE ASIAN MAINLAND.

Fig. A19-55. Cumulative distribution of refractive gradient and k-factor information for Sapporo, Japan.

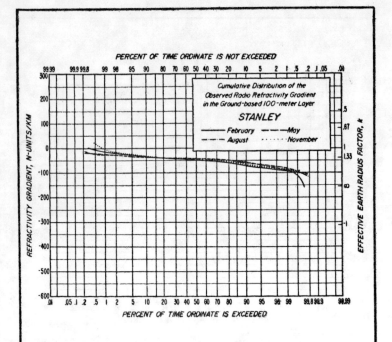

STANLEY, FALKLAND ISLANDS

51° 42' S, 57° 52' W	53 M M.S.L.
DATA: RADIOSONDE.	1400Z (100 LST): 8/52 – 2/57
	1100Z (0700 LST): 5/52 – 11/59
TEMPERATURE (°F):	JANUARY 56/42; JULY 40/31
MEAN DEWPOINT (°F):	NOT AVAILABLE
PRECIPITATION (INCHES):	ANNUAL 26.8; DECEMBER, JANUARY 2.8; SEPTEMBER 1.5

LOCATED ON PORT WILLIAM INLET ON THE NORTHEAST COAST OF EAST FALKLAND ISLAND. A COOL MARITIME CLIMATE.

Fig. A19-56. Cumulative distribution of refractive gradient and k-factor information for Stanley, Falkland Islands.

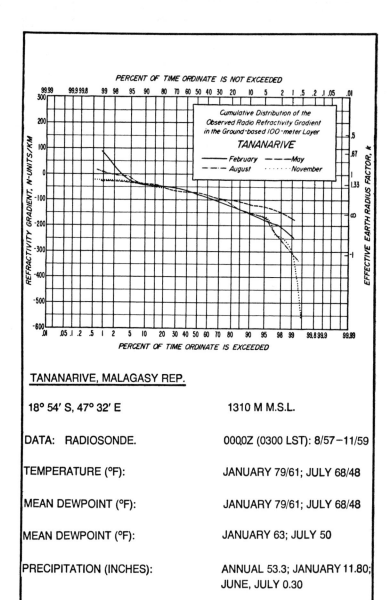

Fig. A19-57. Cumulative distribution of refractive gradient and k-factor information for Tananarive, Malagasy Republic.

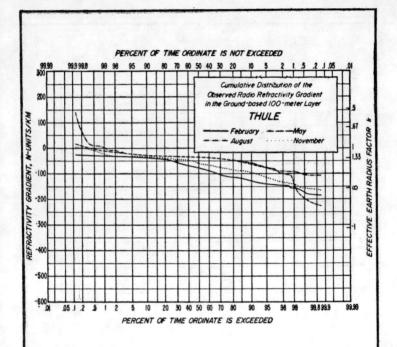

THULE, GREENLAND

76° 31′ N, 68° 44′ W   76 M M.S.L.

DATA: RADIOSONDE.   0300 AND 1500Z (2200 AND 100 LST): 1/51 – 12/53 0200 AND 1400Z (2100 AND 0900 LST): 1/54 – 5/57

TEMPERATURE (°F):   JANUARY –4/–17; JULY 46/38

MEAN DEWPOINT (°F):   JANUARY –20; JULY 34

PRECIPITATION (INCHES):   ANNUAL 4.8; JULY 0.67; APRIL 0.16

LOCATED ON AN INLET OF NORTH BAFFIN BAY IN NORTHWEST GREENLAND. A DRY ARTIC CLIMATE WITH VERY COLD WINTERS AND COOL SUMMERS.

Fig. A19-58. Cumulative distribution of refractive gradient and k-factor information for Thule, Greenland.

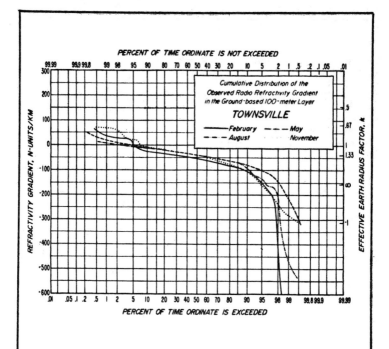

TOWNSVILLE, AUSTRALIA

19° 16′ S, 146° 45′ E
DATA: RADIOSONDE.

4 M M.S.L.
0800Z (1800 LST): 11/49 – 11/51
0700Z (1700 LST): 2/52 – 8/52
0400Z (1400 LST): 11/52 – 5/54

TEMPERATURE (°F): JANUARY 87/76; JULY 75/59
MEAN DEWPOINT (°F): JANUARY 73; JULY 57
PRECIPITATION (INCHES): ANNUAL 45.7; FEBRUARY 11.2; AUGUST 0.5

LOCATED ON CLEVELAND BAY IN NORTHEAST AUSTRALIA. A MARITIME SUBTROPICAL CLIMATE WITH HOT, HUMID, RAINY SUMMERS AND MILD, DRY WINTERS.

Fig. A19-59. Cumulative distribution of refractive gradient and k-factor information for Townsville, Australia.

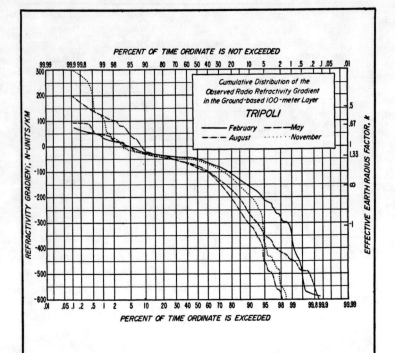

TRIPOLI, LIBYA

32° 54′ N, 13° 17′ E   10 M M.SL.L.

DATA: RADIOSONDE. 0300 AND 1500Z (0400 AND 1600 LST):
                              1/51 – 10/55
                     0200 AND 1400Z (0300 AND 1500 LST):
                     11/55 – 4/57
                     0000 AND 1200Z (0100 AND 1300 LST):
                     5/57 – 12/57

TEMPERATURE (°F):   JANUARY 61/47; JULY 85/71
MEAN DEWPOINT (°F): ANNUAL 14.6; DECEMBER 3.76 JULY 0.02

A PORT CITY ON THE MEDITERRANEAN SEA. A MARITIME CLIMATE WITH HOT, DRY SUMMERS AND COOL WINTERS.

Fig. A19-60. Cumulative distribution of refractive gradient and k-factor information for Tripoli, Libya.

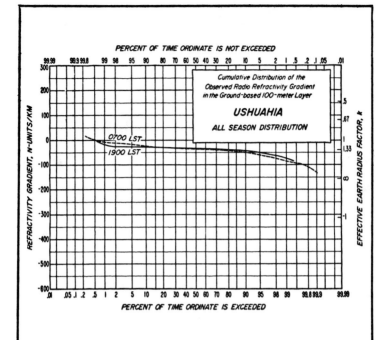

USHUAHIA, ARGENTINA

54° 48′ S, 68° 19′ W   6 M M.S.L.
DATA: RADIOSONDE.   0000 AND 1200Z (1900 AND 0700 LST): 2/64 – 11/68 (FEBRUARY, MAY, AUGUST NOVEMBER)
ANALYZED BY: ENTEL ARGENTINA
TEMPERATURE (°F): JANUARY 57/41; JULY 39/25
MEAN DEWPOINT (°F): JANUARY 38; JULY 31
PRECIPITATION (INCHES): ANNUAL 19.9; FEBRUARY 2.60; AUGUST 1.10

A PORT ON BEAGLE CHANNEL OF THE TIERRA DEL FUEGO JUST NORTHWEST OF CAPE HORN. A COOL MARITIME CLIMATE

Fig. A19-61. Cumulative distribution of refractive gradient and k-factor information for Ushuahia, Argentina.

Appendix 20
Radio Fresnel Zone Radii

It is an important problem for the designing of VHF or micro-wave communication circuits within the radio horizon to keep its propagation path sufficiently clear from such an obstacle such as a mountain ridge.

For the estimation of the path clearance, the diagrams of Fresnel zone are displayed here.

The n-th Fresnel zone is given by following equation

$$l_n = \sqrt{\frac{n \lambda d_1 d_2}{d_1 + d_2}}$$

and for first Fresnol zone this equation is written

$$l_1 = \sqrt{\frac{\lambda d_1 d_2}{d_1 + d_2}}$$

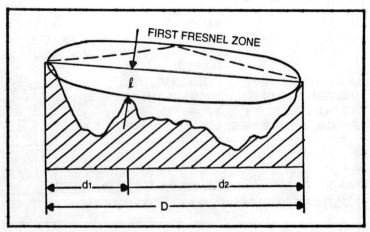

Fig. A20-1. Geometry of First Fresnel Zone.

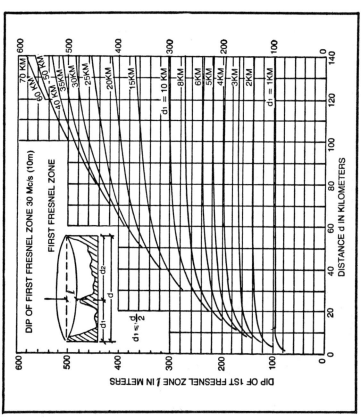

Fig. A20-2. Diagram for first Fresnel zone for 30 Mc/s.

The notations are illustrated as follows and are prepared to find the radius of the first Fresnel zone for given frequencies according to distance D at any point on the propagation path.

(Example)
Find the length of first Fresnel Zone at a distance 30 km from the transmitter in the following conditions.
Distance between transmitting antenna and receiving antenna, $D = 100$ km. and frequency $f = 30$ Mc/s

(Solution)
$l = 460$ m.

This material is courtesy of Radio Research Laboratories, Ministry of Postal Services, Tokyo, JAPAN, from the work "ATLAS OF RADIO WAVE PROPAGATION CURVES FOR FREQUENCIES BETWEEN 30.0 and 10,000.0 MHz."

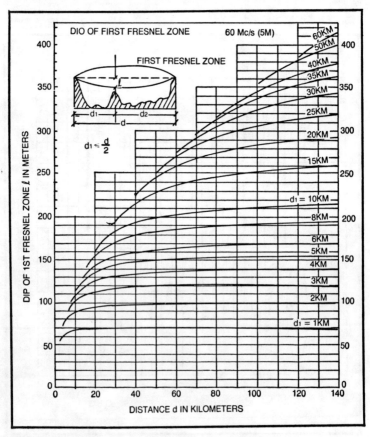

Fig. A20-3. Diagram for first Fresnel zone for 60 Mc/s.

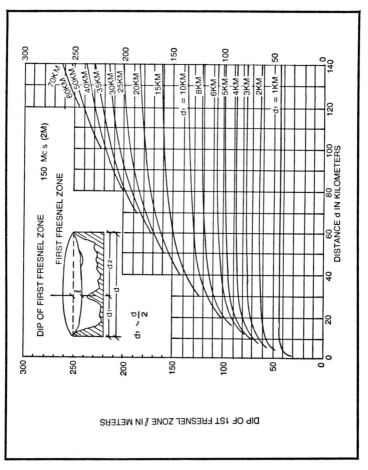

Fig. A20-4. Diagram for first Fresnel zone for 150 Mc/s.

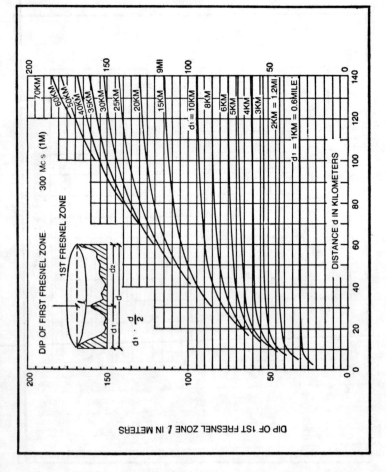

Fig. A20-5. Diagram for first Fresnel zone for 300 Mc/s.

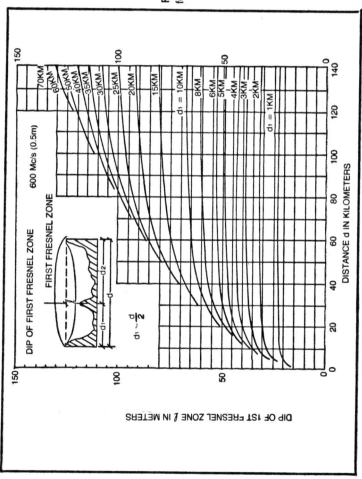

Fig. A20-6. Diagram for first Fresnel zone for 600 Mc/s.

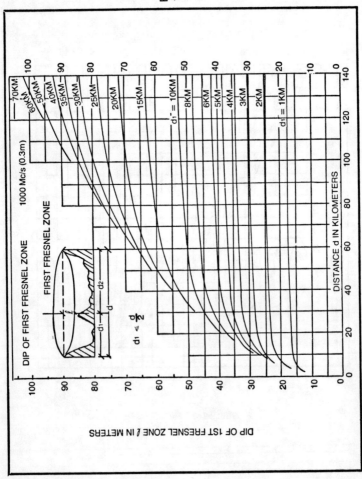

Fig. A20-7. Diagram for first Fresnel zone for 1000 Mc/s.

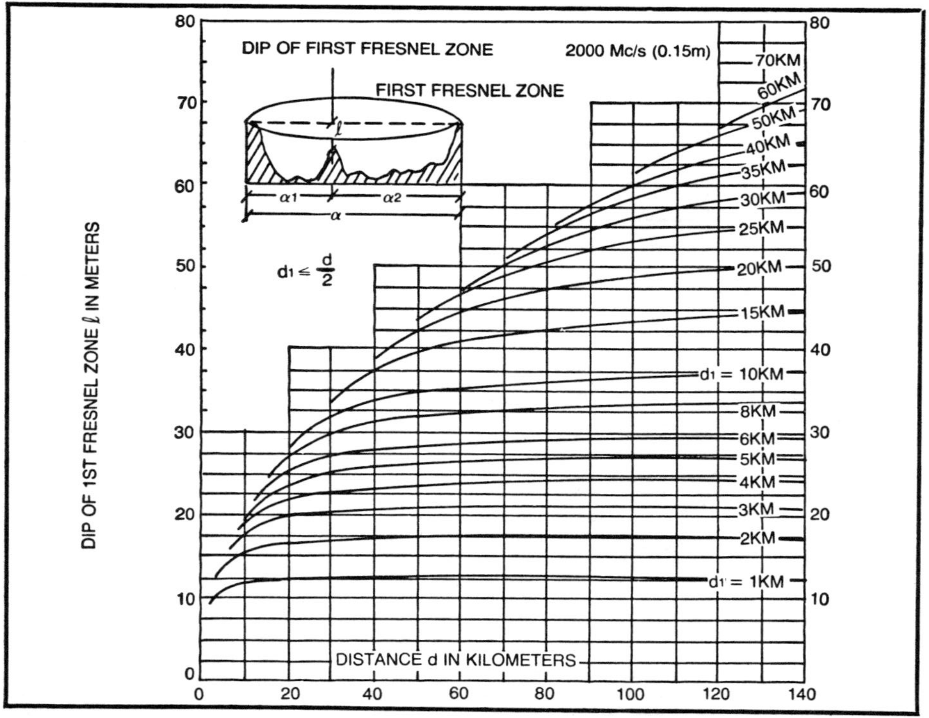

Fig. A20-8. Diagram for first Fresnel zone for 2000 Mc/s.

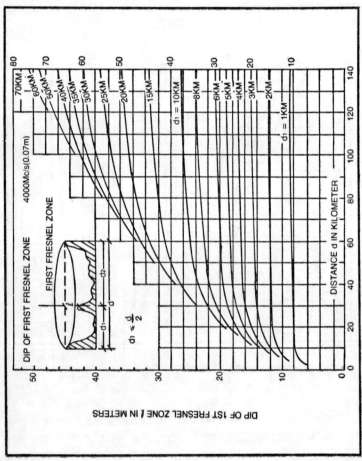

Fig. A20-9. Diagram for first Fresnel zone for 4000 Mc/s.

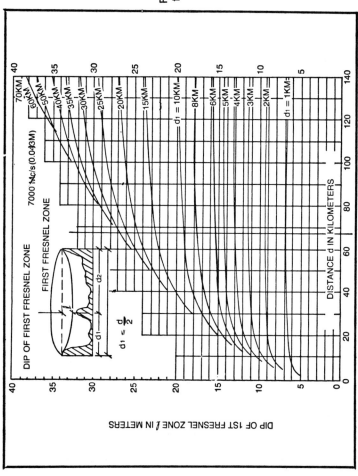

Fig. A20-10. Diagram for first Fresnel zone for 7000 Mc/s.

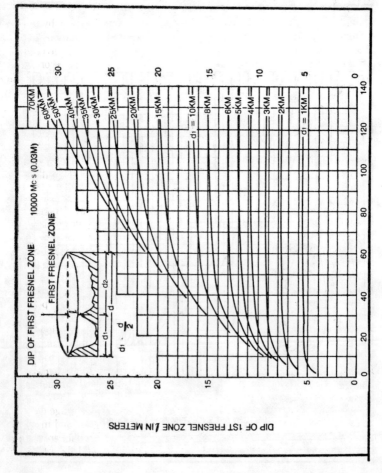

Fig. A20-11. Diagram for first Fresnel zone for 10000 Mc/s.

Appendix 21
Rain Attenuation

The attenuation versus frequency tables have been put together by the author of this book from myriad sources, experience, and consultations and represent a going rate for rain attenuation. This attenuation data is given in terms of dB/Mile versus rain in Inches/Hour. Should you prefer to work in the metric system, simply perform the following algebraic conversions.

1. Rainfall Rate in Millimeters/Hour = 25.4 × R
   Where R = Rainfall Rate in Inches/Hour
2. Rain Attenuation in dB/Kilometer = 0.62 × Rain Attenuation In dB/Mile

(Note—It is important to underscore that the Kilometers and Miles above are not necessarily the length of the path, but the portion of the path in Kilometers or Miles in which rainfall exists).

Although attenuation of electromagnetic energy is actually a complex function of frequency of operation, temperature, drop size, drop shape, drop velocity, canting angle (rain slant-of-fall), and total water volume, it has per force (there is a general lack of needed data available) usually been related to point and instantaneous rainfall rates. Further complicating the issue is that updrafts can cause U-Turns of down-coming rainfall, causing ground measurements of rainfall rates and volumes to be non-representative of the actual rain in the radio path. Be this as it may, point rainfall measurements are the most plentiful, and in many cases the only data on rainfall available, and this (or nothing) must be used in calculations of rain attenuation. The graphical distributions in the Rain Rate samples herein are based upon point measurement rates averaged over one to five minutes. The rainfall rate data herein additionally includes geographical coordinates of locations of measurement, elevation above msl (Mean Sea Level), source of rainfall rate analysis and, climatological data. Providing summer/winter comparisons, the January and July values of average daily maximum temperature, average daily minimum temperature, and mean monthly dew point temperature are presented. Precipitation data are the annual total, average for the wettest and driest months, and the yearly average for the test period. If the test period covered two years with great variability in annual totals, the total for each year is given instead of the two year average. Where available, the average number of days per year with precipitation totaling 0.01 inches or more and the average number of days per year having thunderstorms are also listed. Since complete climatological data were not available at all rain-gauge rainfall measuring sites, in some cases the data are given for a nearby station which is identified by parenthesis after the climatological element to which it applies.

It should be kept in mind that while rain attenuation has been traditionally given in dB/Kilometer (or dB/Mile), the entire radio path is not generally in a rain area in any given rainstorm. This is worth repeating! Rain cells (or thunderheads) possibly cover from about one kilometer to several kilometers of a path during a given rainfall period. Consequently, we must prorate the per unit distance rainfall attenuations given. To employ the data herein most effectively, an investigation of rain characteristics should be made via the nearest available weather organization. A thorough assessment by the microwave path engineer should be made regarding direction and movements of rainstorms in the area of interest, terrain effects, and time of day (as well as other pertinent data) in which rainfall occurs (e.g. should rainfall occur in the wee hours of the morning when traffic is generally minimal, perhaps less weight might be assigned to the gravity of rainfall attenuation).

The graphic and tabular information under the graphs is extracted from *Refractivity & Rainfall Data For Radio Systems Engineering* by Mr. C.A. Samson courtesy of the NTIA (National Telecommunications and Information Agency), Boulder, Colorado 80303, USA.

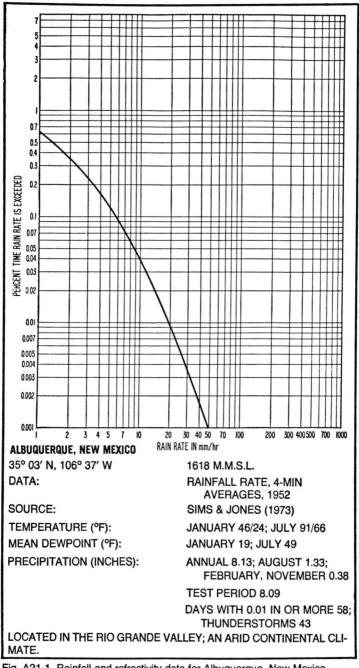

Fig. A21-1. Rainfall and refractivity data for Albuquerque, New Mexico.

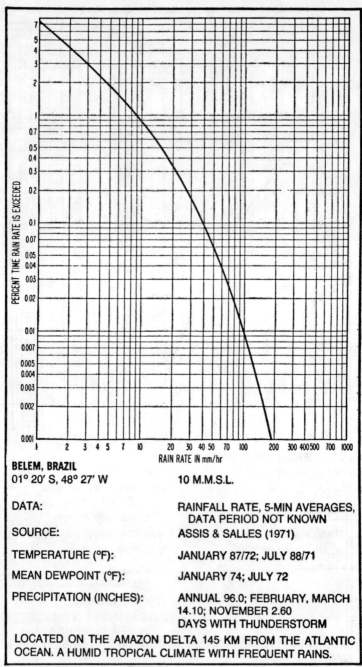

Fig. A21-2. Rainfall and refractivity data for Belem, Brazil.

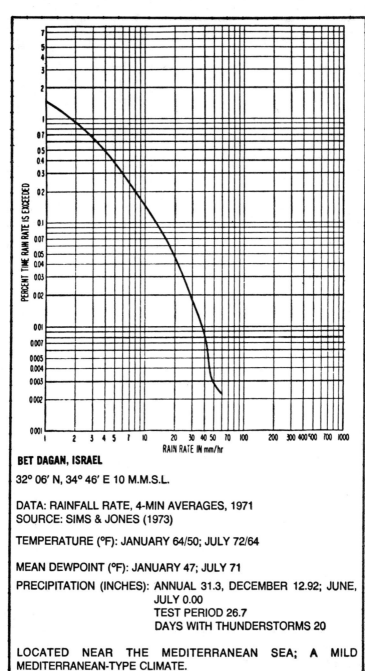

BET DAGAN, ISRAEL

32° 06′ N, 34° 46′ E 10 M.M.S.L.

DATA: RAINFALL RATE, 4-MIN AVERAGES, 1971
SOURCE: SIMS & JONES (1973)

TEMPERATURE (°F): JANUARY 64/50; JULY 72/64

MEAN DEWPOINT (°F): JANUARY 47; JULY 71

PRECIPITATION (INCHES): ANNUAL 31.3, DECEMBER 12.92; JUNE, JULY 0.00
TEST PERIOD 26.7
DAYS WITH THUNDERSTORMS 20

LOCATED NEAR THE MEDITERRANEAN SEA; A MILD MEDITERRANEAN-TYPE CLIMATE.

Fig. A21-3. Rainfall and refractivity data for Bet Dagan, Israel.

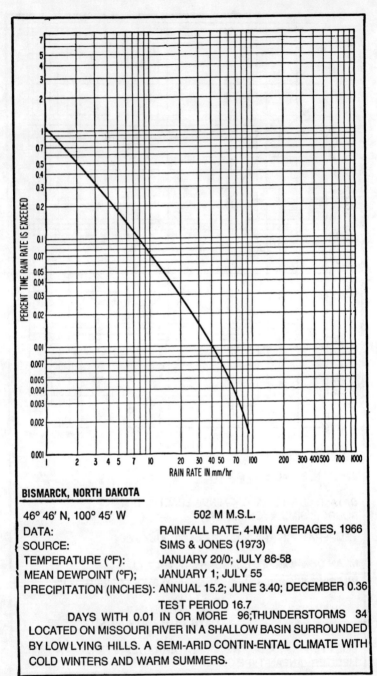

Fig. A21-4. Rainfall and refractivity data for Bismarck, North Dakota.

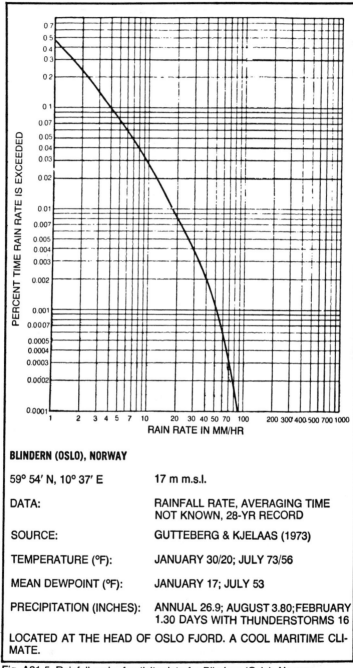

Fig. A21-5. Rainfall and refractivity data for Blindern (Oslo), Norway.

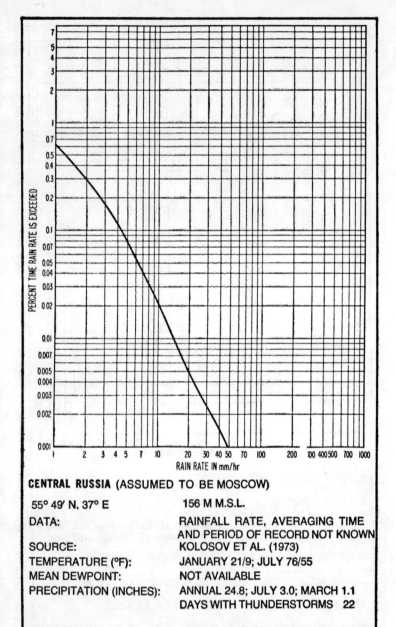

Fig. A21-6. Rainfall and refractivity data for Central Russia (assumed to be Moscow).

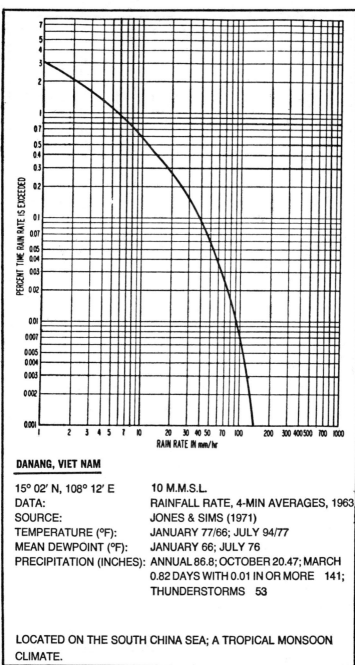

DANANG, VIET NAM

15° 02′ N, 108° 12′ E	10 M.M.S.L.
DATA:	RAINFALL RATE, 4-MIN AVERAGES, 1963
SOURCE:	JONES & SIMS (1971)
TEMPERATURE (°F):	JANUARY 77/66; JULY 94/77
MEAN DEWPOINT (°F):	JANUARY 66; JULY 76
PRECIPITATION (INCHES):	ANNUAL 86.8; OCTOBER 20.47; MARCH 0.82 DAYS WITH 0.01 IN OR MORE 141; THUNDERSTORMS 53

LOCATED ON THE SOUTH CHINA SEA; A TROPICAL MONSOON CLIMATE.

Fig. A21-7. Rainfall and refractivity data for Danang, Viet Nam.

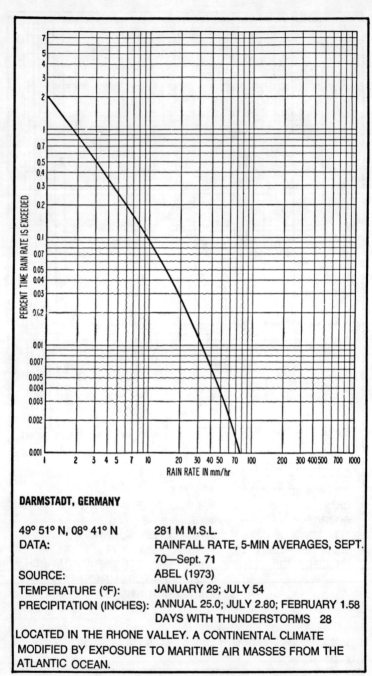

Fig. A21-8. Rainfall and refractivity data for Darmstadt, Germany.

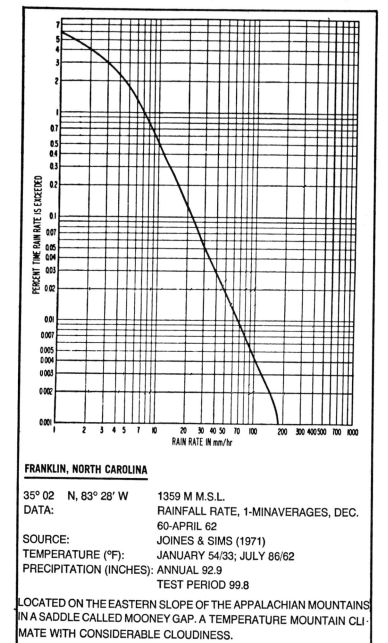

FRANKLIN, NORTH CAROLINA

35° 02 N, 83° 28' W	1359 M M.S.L.
DATA:	RAINFALL RATE, 1-MIN AVERAGES, DEC. 60-APRIL 62
SOURCE:	JOINES & SIMS (1971)
TEMPERATURE (°F):	JANUARY 54/33; JULY 86/62
PRECIPITATION (INCHES):	ANNUAL 92.9
	TEST PERIOD 99.8

LOCATED ON THE EASTERN SLOPE OF THE APPALACHIAN MOUNTAINS IN A SADDLE CALLED MOONEY GAP. A TEMPERATURE MOUNTAIN CLIMATE WITH CONSIDERABLE CLOUDINESS.

Fig. A21-9. Rainfall and refractivity data for Franklin, North Carolina.

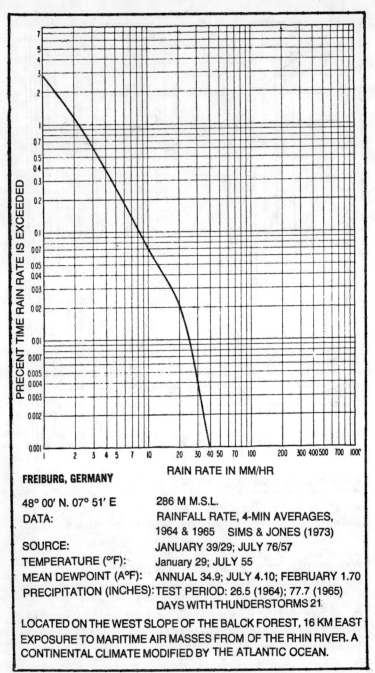

FREIBURG, GERMANY

48° 00' N. 07° 51' E	286 M M.S.L.
DATA:	RAINFALL RATE, 4-MIN AVERAGES, 1964 & 1965 SIMS & JONES (1973)
SOURCE:	JANUARY 39/29; JULY 76/57
TEMPERATURE (°F):	January 29; JULY 55
MEAN DEWPOINT (A°F):	ANNUAL 34.9; JULY 4.10; FEBRUARY 1.70
PRECIPITATION (INCHES):	TEST PERIOD: 26.5 (1964); 77.7 (1965) DAYS WITH THUNDERSTORMS 21

LOCATED ON THE WEST SLOPE OF THE BALCK FOREST, 16 KM EAST EXPOSURE TO MARITIME AIR MASSES FROM OF THE RHIN RIVER. A CONTINENTAL CLIMATE MODIFIED BY THE ATLANTIC OCEAN.

Fig. A21-10. Rainfall and refractiviry data for Freiburg, Germany.

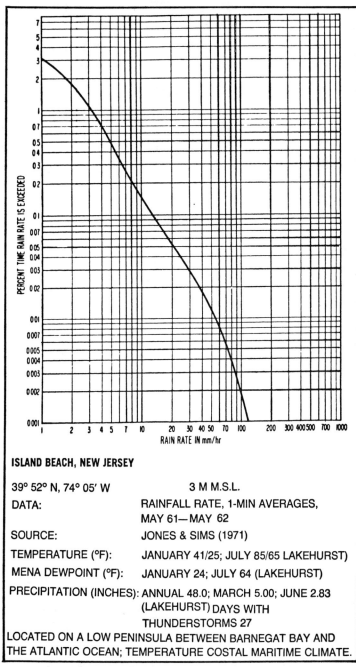

Fig. A21-11. Rainfall and refactiviry data for Island Beach, New Jersey.

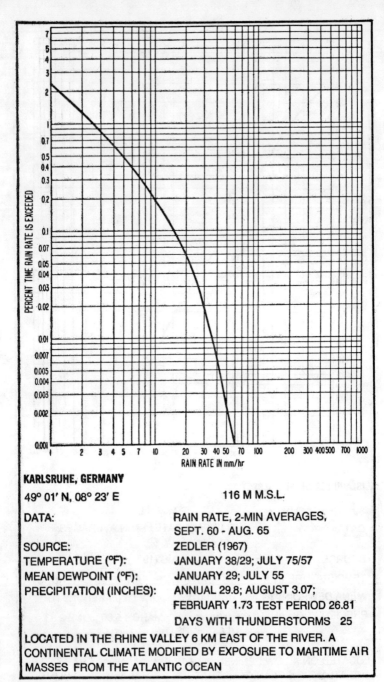

KARLSRUHE, GERMANY

49° 01′ N, 08° 23′ E   116 M M.S.L.

DATA:	RAIN RATE, 2-MIN AVERAGES, SEPT. 60 - AUG. 65
SOURCE:	ZEDLER (1967)
TEMPERATURE (°F):	JANUARY 38/29; JULY 75/57
MEAN DEWPOINT (°F):	JANUARY 29; JULY 55
PRECIPITATION (INCHES):	ANNUAL 29.8; AUGUST 3.07; FEBRUARY 1.73 TEST PERIOD 26.81
	DAYS WITH THUNDERSTORMS 25

LOCATED IN THE RHINE VALLEY 6 KM EAST OF THE RIVER. A CONTINENTAL CLIMATE MODIFIED BY EXPOSURE TO MARITIME AIR MASSES FROM THE ATLANTIC OCEAN

Fig. A21-12. Rainfall and refractiviry data for Karlsruhe, Germany.

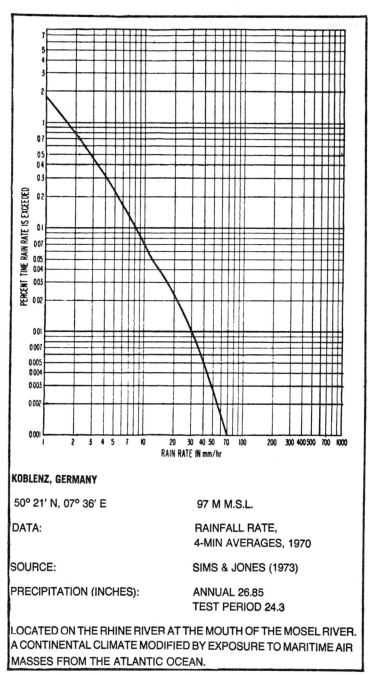

Fig. A21-13. Rainfall and refractivity data for Koblenz, Germany.

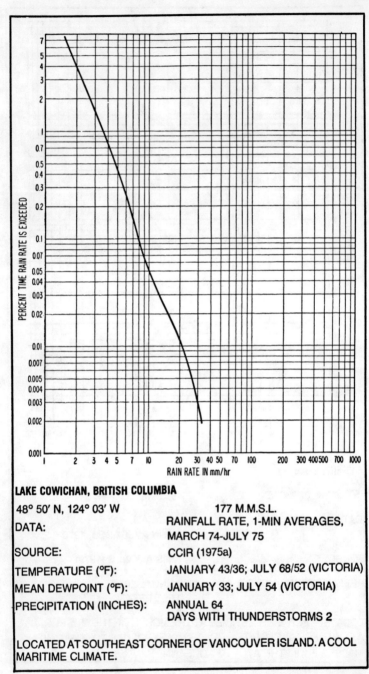

LAKE COWICHAN, BRITISH COLUMBIA

48° 50′ N, 124° 03′ W	177 M.M.S.L.
DATA:	RAINFALL RATE, 1-MIN AVERAGES, MARCH 74-JULY 75
SOURCE:	CCIR (1975a)
TEMPERATURE (°F):	JANUARY 43/36; JULY 68/52 (VICTORIA)
MEAN DEWPOINT (°F):	JANUARY 33; JULY 54 (VICTORIA)
PRECIPITATION (INCHES):	ANNUAL 64 DAYS WITH THUNDERSTORMS 2

LOCATED AT SOUTHEAST CORNER OF VANCOUVER ISLAND. A COOL MARITIME CLIMATE.

Fig. A21-14. Rainfall and refractivity data for Lake Cowichan, British Columbia.

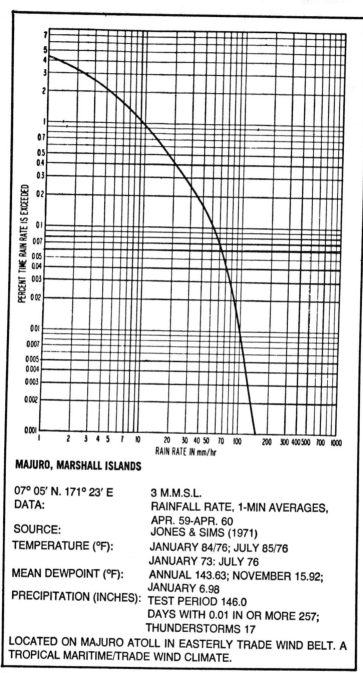

Fig. A21-15. Rainfall and refractivity data for Majuro, Marshall Islands.

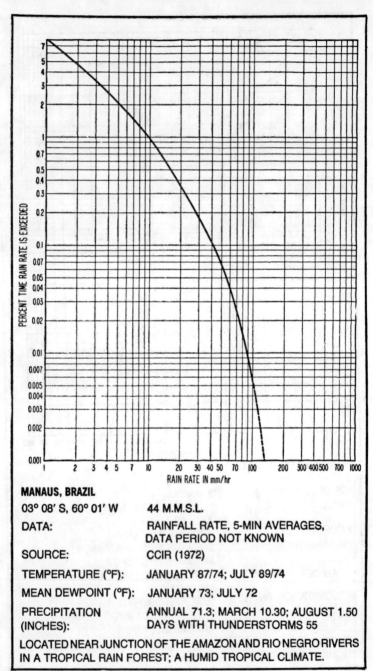

Fig. A21-16. Rainfall and refractivity data for Manaus, Brazil.

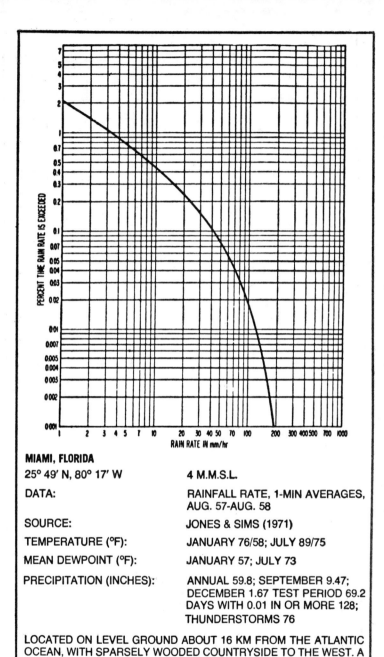

MIAMI, FLORIDA

25° 49′ N, 80° 17′ W   4 M.M.S.L.

DATA:	RAINFALL RATE, 1-MIN AVERAGES, AUG. 57-AUG. 58
SOURCE:	JONES & SIMS (1971)
TEMPERATURE (°F):	JANUARY 76/58; JULY 89/75
MEAN DEWPOINT (°F):	JANUARY 57; JULY 73
PRECIPITATION (INCHES):	ANNUAL 59.8; SEPTEMBER 9.47; DECEMBER 1.67 TEST PERIOD 69.2 DAYS WITH 0.01 IN OR MORE 128; THUNDERSTORMS 76

LOCATED ON LEVEL GROUND ABOUT 16 KM FROM THE ATLANTIC OCEAN, WITH SPARSELY WOODED COUNTRYSIDE TO THE WEST. A SUBTROPICAL MARITIME CLIMATE, WITH WARM, SHOWERY SUMMERS AND MILD, DRY WINTERS. PREVAILING EASTERLY WINDS.

Fig. A21-17. Rainfall and refractivity data for Miami, Florida.

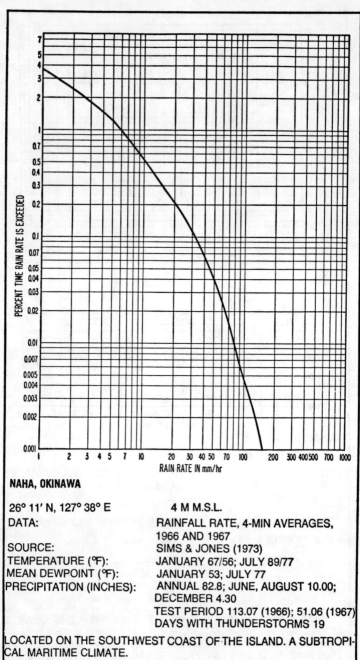

Fig. A21-18. Rainfall and refractivity data for Naha, Okinawa.

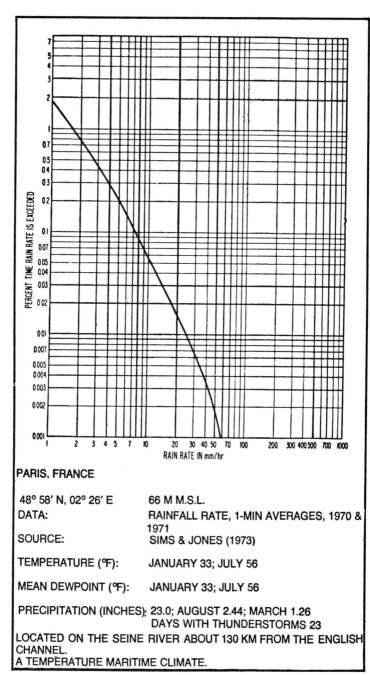

Fig. A21-19. Rainfall and refractiviry data for Paris, France.

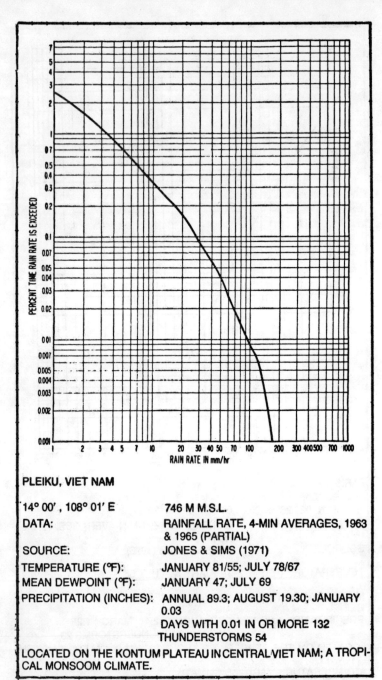

Fig. A21-20. Rainfall and refractivity data for Pleiku, Viet Nam.

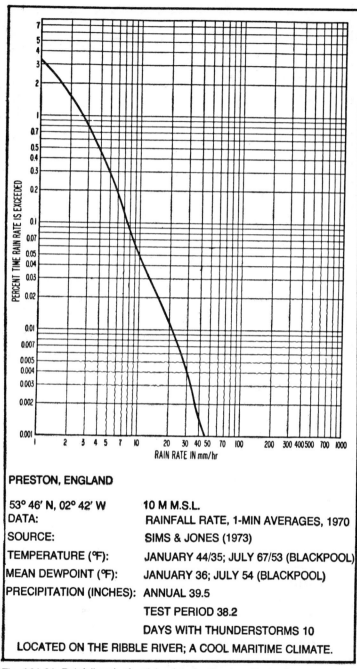

Fig. A21-21. Rainfall and refractivity data for Preston, England.

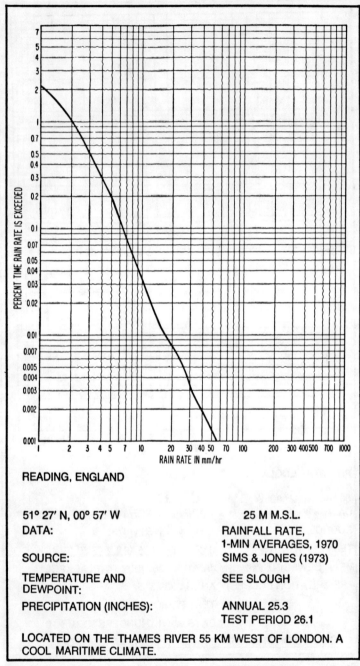

Fig. A21-22. Rainfall and refractivity data for Reading, England.

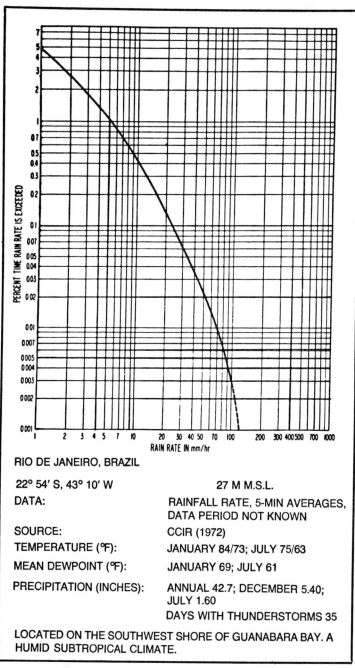

Fig. A21-23. Rainfall and refractivity data for Rio de Janeiro, Brazil.

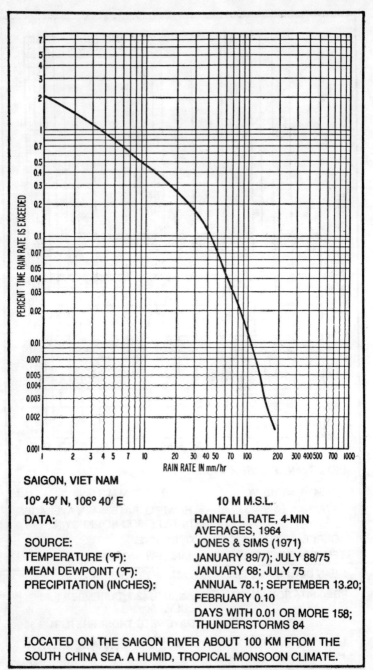

Fig. A21-24. Rainfall and refractivity data for Saigon, Viet Nam.

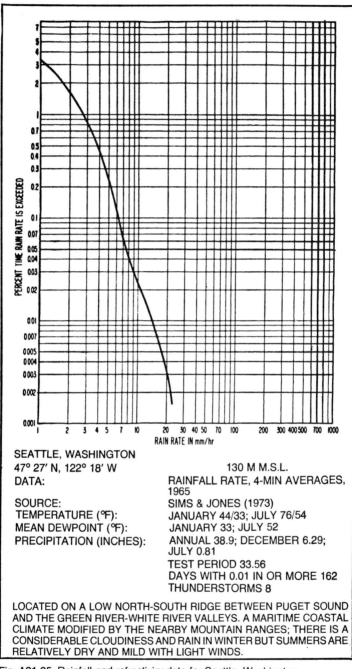

SEATTLE, WASHINGTON
47° 27' N, 122° 18' W   130 M M.S.L.
DATA: RAINFALL RATE, 4-MIN AVERAGES, 1965
SOURCE: SIMS & JONES (1973)
TEMPERATURE (°F):   JANUARY 44/33; JULY 76/54
MEAN DEWPOINT (°F):   JANUARY 33; JULY 52
PRECIPITATION (INCHES):   ANNUAL 38.9; DECEMBER 6.29; JULY 0.81
                                   TEST PERIOD 33.56
                                   DAYS WITH 0.01 IN OR MORE 162
                                   THUNDERSTORMS 8

LOCATED ON A LOW NORTH-SOUTH RIDGE BETWEEN PUGET SOUND AND THE GREEN RIVER-WHITE RIVER VALLEYS. A MARITIME COASTAL CLIMATE MODIFIED BY THE NEARBY MOUNTAIN RANGES; THERE IS A CONSIDERABLE CLOUDINESS AND RAIN IN WINTER BUT SUMMERS ARE RELATIVELY DRY AND MILD WITH LIGHT WINDS.

Fig. A21-25. Rainfall and refractiviry data for Seattle, Washington.

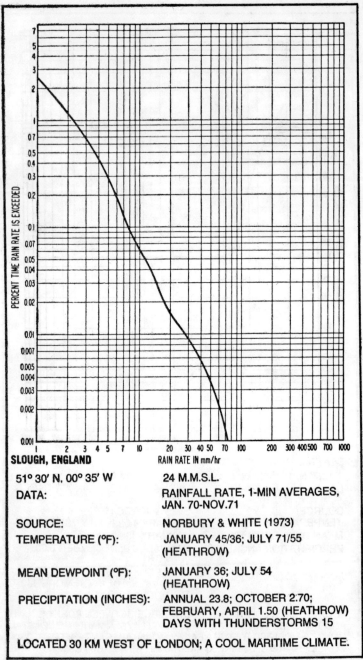

SLOUGH, ENGLAND

51° 30′ N, 00° 35′ W	24 M.M.S.L.
DATA:	RAINFALL RATE, 1-MIN AVERAGES, JAN. 70-NOV. 71
SOURCE:	NORBURY & WHITE (1973)
TEMPERATURE (°F):	JANUARY 45/36; JULY 71/55 (HEATHROW)
MEAN DEWPOINT (°F):	JANUARY 36; JULY 54 (HEATHROW)
PRECIPITATION (INCHES):	ANNUAL 23.8; OCTOBER 2.70; FEBRUARY, APRIL 1.50 (HEATHROW) DAYS WITH THUNDERSTORMS 15

LOCATED 30 KM WEST OF LONDON; A COOL MARITIME CLIMATE.

Fig. A21-26. Rainfall and refractivity data for Slough, England.

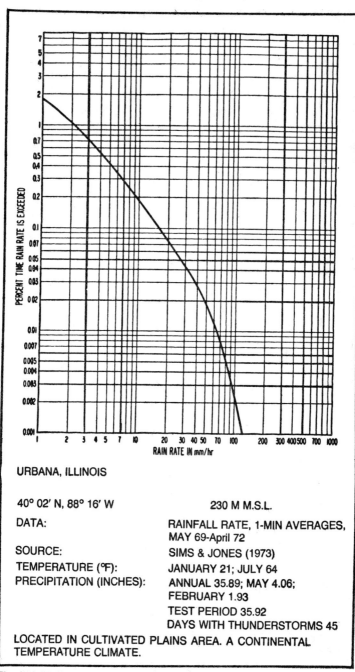

Fig. A21-27. Rainfall and refractiviry data for Urbana, Illinois.

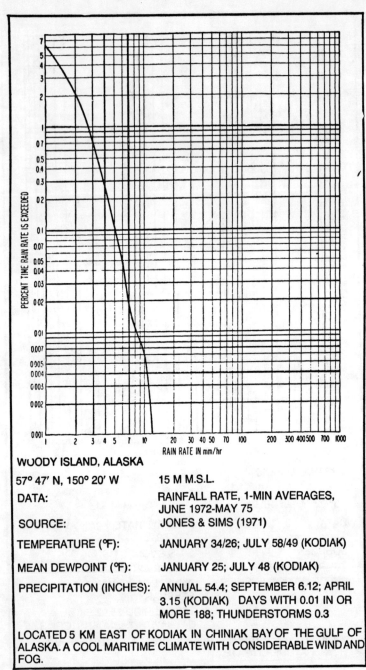

WOODY ISLAND, ALASKA

57° 47′ N, 150° 20′ W	15 M M.S.L.
DATA:	RAINFALL RATE, 1-MIN AVERAGES, JUNE 1972-MAY 75
SOURCE:	JONES & SIMS (1971)
TEMPERATURE (°F):	JANUARY 34/26; JULY 58/49 (KODIAK)
MEAN DEWPOINT (°F):	JANUARY 25; JULY 48 (KODIAK)
PRECIPITATION (INCHES):	ANNUAL 54.4; SEPTEMBER 6.12; APRIL 3.15 (KODIAK) DAYS WITH 0.01 IN OR MORE 188; THUNDERSTORMS 0.3

LOCATED 5 KM EAST OF KODIAK IN CHINIAK BAY OF THE GULF OF ALASKA. A COOL MARITIME CLIMATE WITH CONSIDERABLE WIND AND FOG.

Fig. A21-28. Rainfall and refractivity data for Woody Island, Alaska.

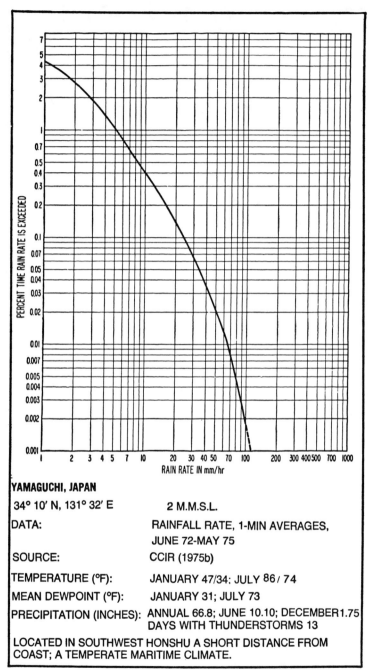

Fig. A21-29. Rainfall and refractivity data for Yamaguchi, Japan.

DRIZZLE (0.05 Inches/Hour)		HEAVY (0.6 Inches/Hour)	
FREQUENCY (Gc)	ATTENUATION (dB/Mile)	FREQUENCY (Gc)	ATTENUATION (dB/Mile)
3.0	0.001	3.0	0.016
4.0	0.0035	4.0	0.045
5.0	0.005	5.0	0.08
6.0	0.01	6.0	0.17
7.0	0.018	7.0	0.30
8.0	0.03	8.0	0.41
9.0	0.04	9.0	0.50
10.0	0.05	10.0	0.70
11.0	0.07	11.0	0.85
12.0	0.085	12.0	1.2
MODERATE (0.2 Inches/Hour)		TORRENTIAL (4 Inches/Hour)	
FREQUENCY (Gc)	ATTENUATION (dB/Mile)	FREQUENCY (Gc)	ATTENUATION (dB/Mile)
3.0	0.004	3.0	0.1
4.0	0.012	4.0	0.3
5.0	0.02	5.0	0.5
6.0	0.04	6.0	1.0
7.0	0.07	7.0	1.65
8.0	0.12	8.0	3.0
9.0	0.16	9.0	4.0
10.0	0.19	10.0	4.9
11.0	0.22	11.0	6.9
12.0	0.35	12.0	8.0

Fig. A21-30. Attenuation values for different frequencies and rainfall rates.

Appendix 22
Fading Distributions

Worldwide experience with tropospheric propagation has indicated that **universal** application of the "worst-case" Rayleigh distribution to short term (within the hour) fading is imprudent. Measurements show a substantial amount of such fading to have distributions **between** those of Rayleigh and Dirac.

STATE OF THE ART

Barnett of Bell Laboratories has taken a significant stride forward in this area producing some excellent theoretical/empirical results. The following formula is based upon his work—"% Time Ordinate Exceeded = $100 - (a \times b \times F \times d^3 \times 6.0 \times 10^{-5} \times 10^{-f/10})$", where "a" = 4.0 for very smooth terrain including over water, 1.0 for average terrain with some roughness and 0.25 for mountainous, very rough or very dry while "b" = 0.5 for Gulf Coast or similar hot humid areas, 0.25 for normal interior temperate or northern and 0.125 for mountainous or very dry. "F" = frequency (GHz), "d" = path distance (Km) and "f" = fade depth (dB).

CRITIQUE

Figure A22-1 depicts the Rayleigh distribution (interestingly referred to "free-space" 1.6 dB different from that of self reference), the distribution of Dirac and typical measured WRH (Within Radio Horizon) path summer/winter values with corresponding **calculated** worst-month/year-round-best-month distributions.

While Barnett's approach is meritorious indeed, it would appear that certain refinements are indicated. The terrain and climate factors seem a bit indelicate and their distinction somewhat blurred. There certainly exist terrains between those of *very smooth* and *average* (the latter an inelegant classification open to interpretative error) as well as between the latter and *very rough* or *mountainous*. Water bodies being subject to varying degrees of surface perturbation do not always qualify as smooth. **Earth return,** the **real** characteristic in question, is inadequately described by smoothness or roughness (sub-dependent upon frequency and grazing angle) alone; it is additionally subject to **absorption** and polarization as well as such

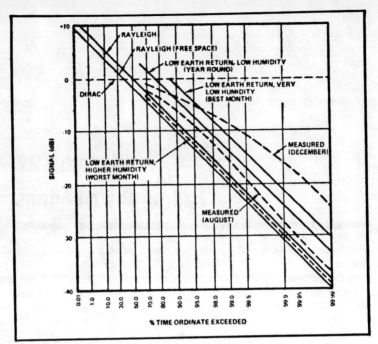

Fig. A22-1. Rayleigh distribution.

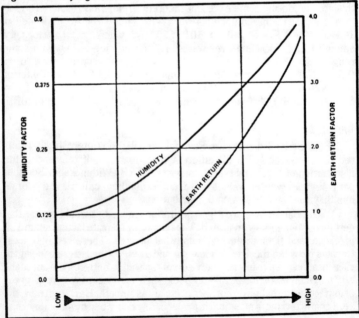

Fig. A22-2. Terrain and climatic factors.

mechanisms as shallow, highly sub-refractive ground-based layers negating terrain diffusion/absorption qualities. Mountainous/rough terrain is not invariably dry (orographic rains are not rare). Hot areas are not inevitably humid (e.g., desert) as are not humid areas always hot (e.g., cold, damp swamplands) while the climatological "geography" is over-generalized ignoring important local-scale meteorological domains existing **within** larger weather systems.

Figure A22-2 proposes to separate the terrain and climate factors conferring parametric independence upon each as well as to connect Barnett's points permitting finer honed factor assignments corresponding more closely to locale-of-interest conditions. It is noteworthy that in computing the Figure 1 curve labelled "Low Earth Return—Very Low Humidity" we employed a factor outside Barnett's factorial range. It would also appear beneficial to introduce into Barnett's equation, factor(s) providing for curve slopes and shapes reflecting closer conformance with those of measured data (in lieu of "Rayleigh-parallel" lines).

Within the confines of manageable mathematics, while heeding that problem oversimplification is not tantamount to utilitarian solution tractability, TAI is continuing investigation of the improvements herein touched upon as well as exploration of possible additional **separate** factors for other effects (e.g., temperature and turbulence).

Decibels represent equipment and energy-consumption dollars—note that at the higher reliability percentages the measured curves reveal some 20.0 dB December/August difference, suggesting (in addition to the already pointed out less-than-Rayleigh fading allowances) savings from seasonal energy-consumption adjustment—not to mention spectrum pollution reduction.

Courtesy of TAI Inc.

Appendix 23
Outage

Although it may at first glance appear that, "an outage is an outage and that's that," the very pragmatic questions which may arise when this subject is considered in some depth might prove astonishing.

TIME

Time has ever appeared to be the paramount dimension in outage definition and measurement. But should time be the sole valid outage criterion? Let us consider a voice signal (characteristically voice embodies considerable redundancy—some have claimed up to 75%). Such a signal is able, with comparative ease, to survive a large number of successive "punctures" (short-impulse noise) of a given total duration and still be completely intelligible (this might be looked upon as a natural "time/redundancy" diversity mechanism). This same impulse noise, however, might totally destroy the intelligence in a data stream. On the other hand, this total outage, taken as occurring in a continuous time period could certainly be responsible for considerable intelligence loss in the voice transmission (as well as data). Is, therefore, time alone a convenient oversimplification? Should time outages be weighted by, for example, circuit importance? Outages of a few seconds through a few minutes might result in anything from inconsequence through inconvenience and nuisance to gravity and major disaster; as an example, two people engaged in light conversation could most certainly tolerate a minute outage with nonchalance while a pilot flying in the fog with only several minutes fuel remaining in his tanks, emergency landing his aircraft by GCA (Ground Controlled Approach), might view such an outage with alarm to say the least.

OTHER OUTAGE FACETS

Let us examine another outage facet. A microwave system built for daytime use exclusively, would hardly reflect its realistic reliability should it be located in an area notorious for nocturnal ducting and tested on a 24-hour basis.

What if an outage is one-way? Should this be considered a "half-outage," the latter on the possible grounds that an entire conversation, per se, was interrupted? And how would a two-way simultaneous outage be classified in this contest?

Another interesting case would be one in which microwave links are employed by ARSR (Air Route Surveillance Radar). These microwave links provide air traffic ontrol centers with communications channels for the purpose of remoting the radar displays. Now then, should a short microwave link failure occur, a secondary effect causing synchronization loss at the controller's console might incapacitate the system for several minutes since this amount of time might be required for synchronization restoration. What is the real time duration of the outage in this case?

Let us consider outage from another "time-alone" viewpoint. Certainly a 10-second outage would cause more data loss at 9600 bits per second than at 50 bits per second. Ergo, should a data outage time be weighted in accordance with speed of data transfer?

Outage considerations, it appears, cannot ignore the type of traffic and end use of the message. For example, the transmission of record information (data and telegraphic information which is recorded—such as facsimile, graphics, and teletype) dictates different criteria of acceptable error rates for these differing modes. A social message conveying a birthday greeting in plain language typically contains sufficient redundancy in the wording to allow a character error rate of approximately one in six without losing the message intent. On the other hand, a business message directing security sales or fund transfer might require a maximal error rate of three characters in a hundred thousand and at least a collation of repetition of the message's key alphanumeric characters (repetition of key symbols) in order that the transaction's accuracy might be ensured.

DEFINITION

It must be kept in mind that outage is largely a matter of definition. In FDM/FM short haul voice systems, an outage might be said to occur when the TT/N (Test Tone to Noise) ratio is below 30.0 decibels. On a CCIR microwave system, however, the outage point is 43.0 decibels S/N ratio. In digital microwave systems, the general trend toward outage delineation has been a BER (Bit Error Rate) of one bit per million—this being a function of the modulation schemes employed—e.g., QPSK (Quaternary Phase Shift Keying), or "quadriphase" and Dr. Lender's improved duo-binary technique—among other things.

Courtesy TAI Inc.

Appendix 24
Digital Microwave

Like the greased piggy at the county fair, the subject of **digital microwave** seems at times to defy a solid grasp. There are manifold reasons therefore, not the least of which are the melange of incompatible digital microwave systems (in turn incongruous with those of FDM-FM) and the exacerbating exaggerations of advocates and adversaries. We shall, herein, attempt to lend some perspective to the situation.

WHEREFORE DIGITAL MICROWAVE

The world is going digital by leaps and bounds; the majority of today's business machines (e.g., computers, data files, terminals and facsimile) utilize digital formatting. The development of more desirable means for digital information **transport,** therefore, is logical. Enter **digital microwave.**

DIGITAL MODULATION MODES

Digital modulation is simply the impression of digital information upon a radio carrier. Example methods follow:
Qam (Quadrature amplitude modulation) produces four digital data streams (two in quadrature on each sideband).
Qpsk (Quaternary phase shift keying) consists of phase modulation by orthogonal binary signals producing a four level phase shift.
Opsk (Octal phase shift keying) is a phase keyed eight level signal.
Okflpsk (Offset keyed four level phase shift keying) comprises a partial response phase keyed system.

DIGITAL MICROWAVE ADVANTAGES

Digital microwave is largely immune to fading noise. Fades short of threshold (not-atypically -80.0 to -90.0 dBm for 10^{-6} BER or -65.0 to -73.0 dBm for a BER of 10^{-7}) produce little degradation enabling concomitant economy in system gain (e.g., lower transmitter powers or reduced antenna sizes) while repeater regeneration makes noise buildup virtually independent of system length.

Interference insensitivity is characteristic of digital microwave, allowing an increased number of closely frequency-spaced stations in our overcrowded spectrum as well as cross-polarization/co-frequency capacity doubling. In digital microwave/FDM-FM microwave mixes, the former suffers minimally in an interference ambient of the latter while the reverse is not true.

Digital microwave betters buried cable and FDM-FM microwave cost-effectivity on many short-haul routes.

Digital microwave is directly compatible with PCM carrier. At traffic centers, through-groups may be routed via intermediate offices as T-lines. Digital microwave multiplexing is less expensive.

For D-bps data rate at 10^{-6} BER, two, four and eight level digital microwave data requires respective powers of only x, x, and x + 3.3 dBm as compared with x + 2.9, x + 5.4 and x + 9.6 dBm for FDM-FM.

DIGITAL MICROWAVE DISADVANTAGES

For D-bps data rate at 10^{-6} BER four and eight level digital microwave data require bandwidths of D and 0.6D respectively as compared with 0.6D and 0.4D for FDM-FM, while rule-of-thumb bandwidth comparison for digitized and FDM-FM voice information is respectively in the approximate ratio of 2.5:1.0.

Higher capacity digital radios are expensive—this being true of any progressive equipment in the pre-standardization stages of evolution and shakedown.

At distances exceeding short-haul, the economical advantages of digital microwave tend to break down (due in no small part to digital radio price excess over that of FDM-FM).

With respect to FDM-FM microwave, a digital system carries fewer channels per radio and requires more d.c. power.

DIGITAL MICROWAVE'S FUTURE

Upon equipment and methodology maturation and standardization, digital microwave will take its rightful place in the telecommunications field. As its utilization and ensuing competition grow, cost-effectiveness may be expected to improve.

Paradoxically, however, while economic pressures lend impetus to digital microwave advancement, they conspire toward retention of FDM-FM microwave plant. This dichotomy (decisions, decisions, always decisions) is far from a simplistic, *"digital is in, FDM-FM is out"* situation. While digital microwave shows many advantages, surely with promise of more to come, FDM-FM microwave will continue to be with us for a time—if only for investment amortization coupled with its proclivity and potential for survival as evidenced by existing "digital-to-analog" schemes and devices for enabling digital transmission over FDM-FM microwave systems.

Courtesy TAI Inc.

Appendix 25
Fade Margin Curves

This Appendix shows two sets of fade margin curves for WRH (Within Radio Horizon) microwave paths (also VHF/UHF WRH paths). While in general fast fading is less for the lower frequencies under consideration herein, these curves are excellent for planning purposes. The literature is virtually full of various fading formulas and curves and this can be confusing. However, meteorological location has alot to do with fading patterns and no single neat rule can cover *all* possible cases. These curves were developed by the author over the years as Telecommunications Engineer in various parts of the world.

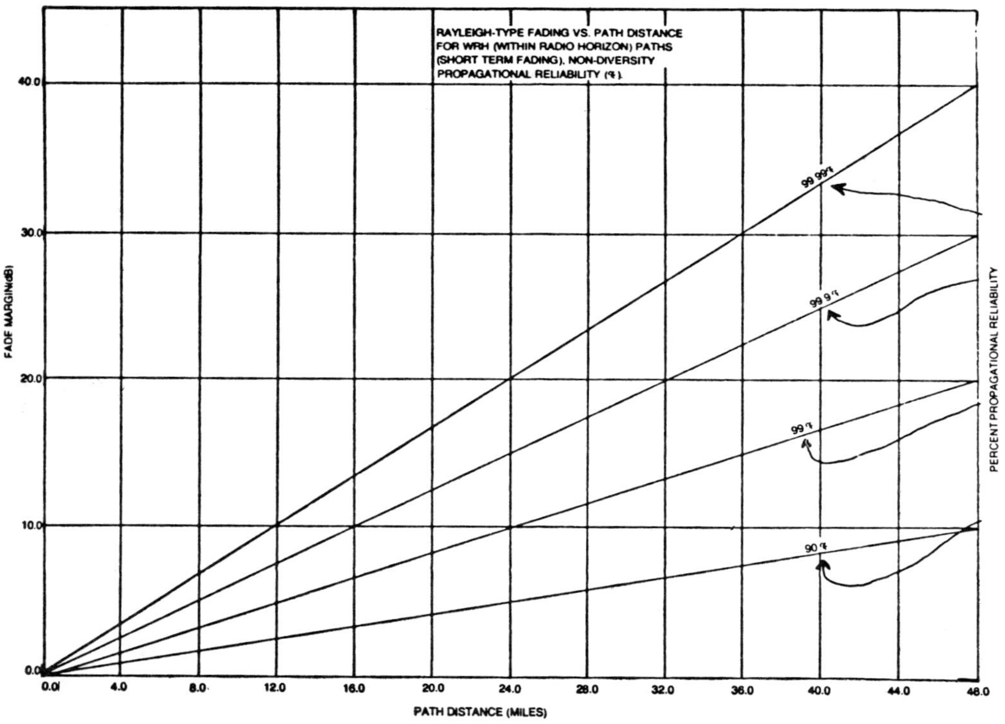

Fig. A25-1. Curve of Rayleigh-type fading vs. path distance for short-term fading.

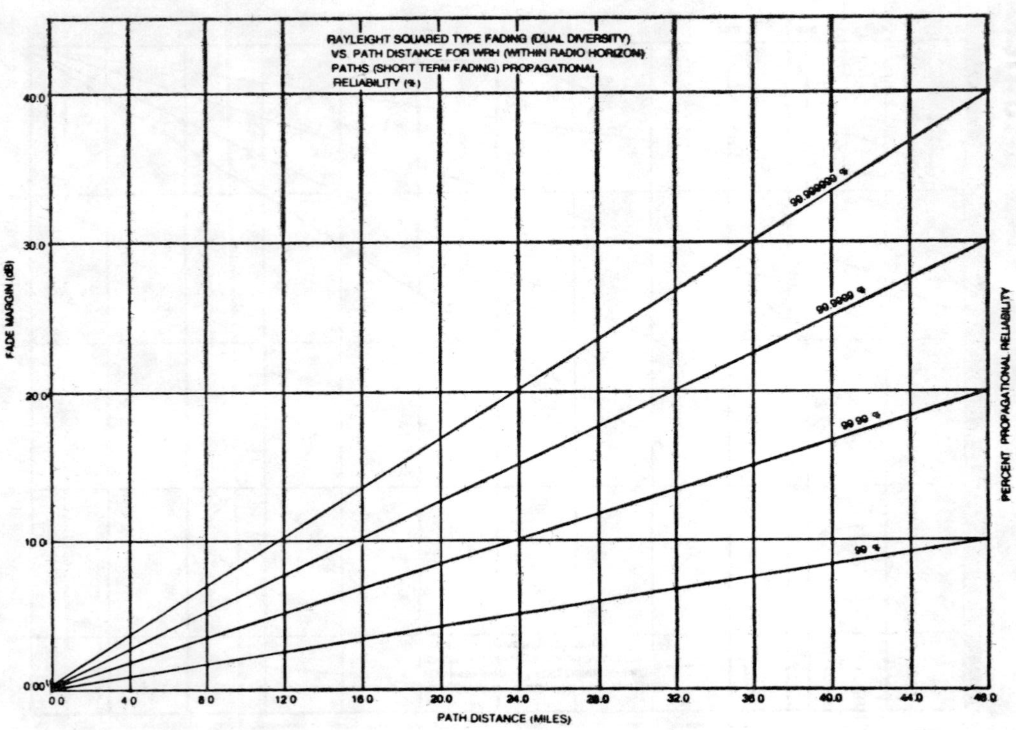

Fig. A25-2. Curve of Rayleigh squared type of fading vs. path distance for short-term fading.

Appendix 26
Special Protractor to Measure Angles on K-Factor Profiles

A nice little handy-dandy protractor which can be used to measure angles directly from k-paper is described as follows. To make one, since the horizontal and vertical scales are different, we must have a scale factor. If, for example, 1.0 centimeter on the path distance scale = 1.0 Kilometer

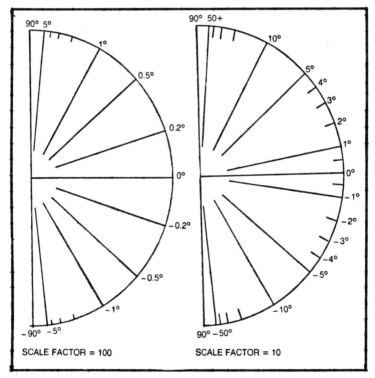

Fig. A26-1. Diagram of two versions of a special protractor for reading angles directly from the k-factor paper.

473

and 1.0 centimeter on the height scale = 100.0 meters, we have a scale factor of 1.0 kilometers (or 1,000.0 meters) over 100.0 meters or 1,000.0/100.0 = 10.0. Now, angles on the profile, can be quite simply related to their real-life counterparts. Take an angle, Θ, on a real propagational path and its "mate" on a profile chart, ϕ. The relationship between these two angles is as follows:

$$n \, \mathrm{Tan} \, \Theta = \mathrm{Tan} \, \phi$$

To construct a special protractor we simply calculate, by the above formula, various values of ϕ for corresponding values of Θ on a standard regular protractor. It is convenient to graduate the quadrants as shown in Fig. A26-1. In use, do not rotate this special protractor, but keep it always aligned with the vertical and horizontal coordinate axes of the k-paper. To determine, for example, the angle between the direct ray and the reflection ray at the transmitting antenna as an "orgin," perform the following. If the rays are not parallel to either horizontal or vertical axes, simply read off the angle each of these rays makes with the horizontal and deduce the desired angle from these values. Angles corresponding to rays below the horizon are taken as negative. The accompanying illustration shows two examples, one for a scale factor of 10.0 and the other for a scale factor of 100.0. These specially constructed protractors might also be marked off in radians or fractions thereof if desired.

Courtesy of European Broadcasting Union (E.B.U.).

Appendix 27
System Performance

Figure A27-1 is a generic curve reflecting the performance (Signal-to-Noise Ratio) of a multi-channel FDM/FM (voice) UHF hop as a function of receiver signal input. It may be used to determine the path propagational reliability (circuit availability). If, for example, our receiver's unfaded signal input (measured or estimated by computation as appropriate) is −60.0 dBm, we see from the graph that our voice-channel Signal-to-Noise ratio is 60.0 dB. If we consider that a UHF hop's service is unacceptable at a voice channel Signal-to-Noise ratio of, say 35.0 dB, we can immediately deduce our available fade margin as 60.0 dB − 35.0 dB = 25.0 dB. With this 25.0 dB fade margin, we enter Appendix No. 25. Assuming that our path length is 32.0 statute miles, we see that for our 25.0 dB fade margin we have a non-diversity propagational reliability of approximately 99.97% and 99.999979% for dual diversity. This means that these percentages represent the portions of the total time of a given year that the considered circuit will meet requirements.

For multiple-hop systems, the propagational reliability for each hop may be determined as above and the outages simply summed and subtracted from 100.0% as in the following example:

Hop Number	Propagation Reliability (%)	Hop Outage (%)
1	99.9999%	0.0001%
2	99.9987%	0.0013%
3	99.998%	0.0020%
4	99.9976%	0.0024%
5	99.9960%	0.0040%
	Total Outage =	0.0098%

Accordingly, the multiple hop system propagational reliability is equal to:

$$100.0\% - 0.0098\% = 99.9902\%$$

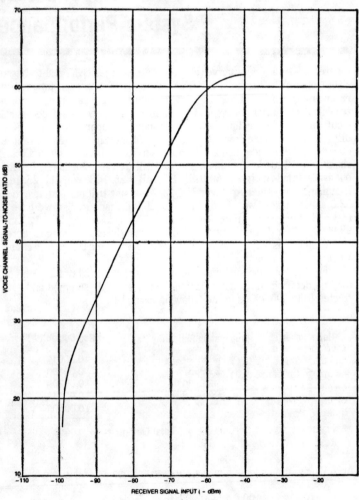

Fig. A27-1. Generic curve depicting the performance of a voice channel vs. the receiver's input signal.

The above method of calculating multi-hop system propagational reliability for FDM/FM microwave systems assumes that fading on more than one hop does not occur simultaneously, but temporally (in the time domain) independently. In actual practice there may be some occurrence of simultaneous fading on various combinations of the hops of a multi-hop radio system, thus actually rendering a better system propagational reliability. With this latter concept the individual path propagational reliabilities are all multiplied for a final propagational reliability figure. As an example, if we have a system of two tandem hops, each of 90.0% propagational reliability, we might obtain two different answers depending upon our method of calculation. Using the first described method of adding up the individual path outages, we obtain:

Hop Number	Propagational Reliability (%)	Hop Outage (%)
1	90.0%	100% − 90% = 10%
2	90.0%	100% − 90% = 10%
	Total Outage =	20.0%

The propagational reliability for the above two-hop system is, by this addition method, $100.0\% - 20.0\% = 80.0\%$

Using the second above described method (outage multiplication), we obtain a two-hop system propagational reliability of:

$$0.90 \times 0.90 = 0.81 = 81.0\%$$

As to which method to use, this, again, is a matter of judgement, available money, system importance, etc.; it involves many considerations outside the realm of the topic of radio propagation.

There are still other methods of computing a system's propagational reliability, such as that of the C.C.I.R. (International Consultative Committee of Radio) and those used by the various worldwide telephone companies and PTTs (Organizations of Post, Telephone and Telegraph), as well as military agencies, but these might be considered fundamentally similar.

It might be mentioned here that for digital radio systems, interestingly, the digital information is usually regenerated at each radio hop and, therefore, the signal at every such hop is "cleaned up," that is "brand new" and undistorted bit streams are generated from the old noisy ones. Noise buildup in the sense of FDM/FM microwave, is thus impossible. The noise on a system, then, is virtually independent of the system length or number of hops. From a noise point of view, fading on a given path is virtually independent from that on all the other paths of the system. As does everything, though, all this "bounty" has its price. The BER (Bit Error Rate) is cumulative on the system.

Appendix 28
Radiogeology

This subject treats of electromagnetic-energy/earth material interaction.

MODUS OPERANDI

"Catalyzing" this electro-tellural synergy are the terraquatic constants **conductivity** and **permittivity**, which "activated" by incident radio energy, respectively give rise to frequency-responsive/quadrature, **conduction** and **displacement** earth currents.

TERRAQUATIC-CONSTANT FUNDAMENTALS

The geologic factors determining terraquatic constants are copious and complex indeed! Soil characteristics vary in accordance with physiochemical composition and moisture absorbing/retaining capacity, as well as thermal ambient. Regarding the latter, measurements have revealed ground conductivity and permittivity coefficients of small percentages at "normal" temperature ranges while at freezing point both constants manifest dramatic variations. Additionally these constants vary with operating frequency and polarization.

GROUND GROUPING

Following are three cardinal terraquatic delineations:

Excellent ground (e.g., Sea Water—lower frequency propagation via which constitutes the closest real-world approximation to a perfectly conducting earth). Conductivity = 4.0 to 5.0 Mhos/Meter and permittivity = 73.0 to 81.0. At 300.0 MHz and below, conductivity current governs while at 3000.0 MHz and above, displacement current rules—crossover at 900.0 MHz.

Good ground (e.g., Rich agricultural moist loamy lowlands—typical of the soils found in Illinois or Ohio, U.S.A.)—underlying composition may be exemplified by that of Jurassic (Mesozoic era including that between the Permian and Tertiary). Conductivity = 0.01 to 0.02 Mhos/Meter whereas permittivity = 15.0 to 25.0. At 6.0 MHz and below, conductivity current is paramount while at 55.0 MHz and above displacement current reigns—crossover at 18.0 MHz.

Poor ground (e.g., Industrial/Urban areas or dry rocky land)—a prototype of underlying geology might be Archaen and Igneous (molten magmatic material producing igneous rock upon cooling). Conductivity = 0.001 Mhos/Meter and permittivity = 4.0. At 1.5 MHz and below conductivity controls while at 11.0 MHz and above permittivity is supreme—crossover at 4.0 MHz.

EARTHY PRAGMATICS

Three earth reactions to impinging radio energy are **transmission, reflection,** and **absorbing** (plus "Murphy's mixes" thereof). Following are some real-world across-the-spectrum responses.

SHF through VHF—While antennas at nominal structural heights exceeding several meters are, per se, less influenced by terraquatic constants, reflection coefficients along a path are of considerable moment. Reflected and direct signals might combine to render resultants ranging from the classic 6.0 dB gain over "free-space" loss to virtual cancellation.

HF—Earth constants profoundly affect an antenna's effective height, gain, and pattern/lobing. Additionally, multihop transmission in this frequency range depends, among other things such as reflection (grazing) angle, upon earth reflection efficiency; typical per-reflection loss values might be <1.0 to 40.0 decibels at conductivity = 0.001 Mhos/Meter—permittivity = 4.0, and <0.1 to 40.0 decibels at conductivity = 5.0 Mhos/Meter—permittivity = 80.0. Intermediate values depend upon a gamut of reflection area integrated terraquatic characteristics.

MF through VLF—Surface communications field strengths are calculable using area geological data. *Submarine signals* (e.g., Omega Navigation) at 10.0 kHz might encounter attenuation of some 4.5 dB/Meter while those at 100.0 kHz (e.g., Loran-C) suffer a per-meter loss 10.5 dB. *Subterranean* attenuation, important for example, in **mining** or **survival** operations where surface-oriented radio-communications are ineffectual or neutralized," might yield attenuations of 0.01 dB/Meter at 10.0 kHz to 0.775 dB/Meter at 100.0 kHz through metamorphosed low-ore rock.

Courtesy of TAI Inc.

Appendix 29
Antenna Height

Antenna heights may not be what they seem. The antenna radiating center's effective height is frequently at striking variance with its structural, or physical, counterpart. Antennas, ranging from a small interval above the earth's surface to satellite mounted, lend themselves to computational divestation of the dilemma born of this elector/physical "illusion" per their amenability to assignment of effective heights (the ones propagationally consequential) constrained between Bullington's minimum effective height on the one hand, and a disquietingly desultory (albeit alluded) maximum effective height on the other. Antennas of the surface juxtaposed, surface, and subsurface variety comprise a special species.

MINIMUM EFFECTIVE HEIGHT

Frequently dominant in broadcasting and land or maritime (as well as low flying aircraft) vehicular communications is the ground wave. Classically (ignoring classicism "dilutants" typified by trapping, scatter, diffraction, and feuillets), the ground wave is comprised of three subcomponents which are the direct wave, the terrestrially reflected wave (these two constituting the space wave), and the surface wave—the latter not infrequently representing the signal's paramount propagational mode. Generated by telluric currents (induced by incident primary waves), this surface wave, in "flood-like" diffusion, propagates along the earth/atmosphere interface in a "stratum" of one wavelength over land and several wavelengths over sea water. Compensating for signal deficits in the direct and/or earth reflected wave "deadspot" areas, it conduces to a phenomenon establishing an antenna height below which field strength tends to be essentially constant—this minimum effective height accruing as a result of a complex interaction among ground constants (terrestrial conductivity and permittivity), frequency, and polarization. A case in point reveals that a 35-MHz vertical antenna one meter above sea water (conductivity 4.5 mhos-per-meter/permittivity 74, at 10 degrees centigrade) manifests a minimum effective height of 60 meters, whereas oriented horizontally over poor soil (dry uncompacted sand or non-metallic rocks—typified by nominal conductivity and permittivity of 0.001 mhos-per-meter and 6.4), it exhibits a 0.83 meter minimum effective height—a rather dramatic electrical dichotomy.

MAXIMUM EFFECTIVE HEIGHT

Maximum effective height is similarly a product of a plurality of factors, e.g., CCIR interdecile terrain height range (or ESSA's asymptote), antenna siting, meteorological effects, and the path's basic classification (within or beyond radio horizon). The elementary smooth-earth/within-radio-horizon, terrestrial-tangent model is well known. Various administrations consider the equivalent antenna height that of its phase center above the average of terrain elevations along a particular segment of a given service radial. Military tactical operations (not unlike the requirements of the forestry radio service) may demand reckoning with area features while antenna height computations in practically all walks of radio (and TV) are plagued by ecologic clutter.

Beyond-the-horizon situations are comparably negotiated in that arcs are fitted to the terrain by statistical methods for some significant portion of the great circle distances between antennas and their horizons.

VERY HIGH AND VERY LOW ANTENNAS

Determining effective heights of satellite (or high flying aircraft) antennas requires unique technical tacks. One such case, an antenna at 200 kilometers true altitude, exhibited a maximum effective height of 130 kilometers at a 400 N surface refractivity (Ns 400 "corresponding" to a long-term effective earth radius of 12,090 kilometers or k-factor 1.9).

"Height"-wise a maverick genus, the very low antenna category embraces the range from intimate surface proximity, through flush, to subsurface (e.g., as employed in the various subterraquatic transmission techniques), and notwithstanding the rather bewhiskered bon mot that once an underground antenna was "hood-winked" into electrical exoneration by a counterpoise 15 meters in the air, this antenna is and will increasingly be, in various neo-forms, very much with us. The subject of very low antennas is intimately involved with radio-geology (and selenognosy) disciplines and we propose to dedicate a future issue to this absorbing topic.

CURIOUS CASES

Albeit quasi-normal that an antenna may manifest a multiplicity of effective heights (owing to the aforementioned terrestrial (or lunar) factors along various radials), this technical "curio" seems ever reluctant to completely shed its cloak of mystery and astonishment. A not-atypical calculation disclosed that an antenna reflected a maximum effective height of 150 meters along the east vector and neg-ten to the southwest.

Another "engaging" occasionally occurring perplexer is the anomaly whereby the calculated minimum effective height exceeds the computed effective maximum height, requiring a bit more than a "folk-wisdom" acquaintance with antenna/propagation principles for arrival at the correct solution. An elementary example is one in which static terrain features (hills, dales, etc.), or complicating semistatic ones (created by snow drifts, rock slides, dune movements, inundations, and deciduous forest leaf-on/leaf-off seasonal changes), are substituted by dynamic pelagic perturbations (sea waves) in maritime mobile and petroleum platform communications.

A provocative and at times chagrining effective height circumstance is the chameleon-like "pageant" in which an antenna, exemplified by those employed in the railroad service, may gyrate over an entire gamut of effective height "hues"—cutting helter-skelter obliquely across hard-won technical guidelines, insisting, moment to moment, upon the rectitudes of radio physics. The operating ambient of such an antenna may range from underground (tunnels)—with or without Shanklin extension, through city clutter (tall buildings and other structures), over causeways crossing salt (or fresh) water (or swamps), over bridges, through valleys, up sinuous mountain roads and to the mountain top—all the while nature unrolling the "red carpet" of an infinitude of ground constant admixtures and topography before its technical "highness." One such case occurred on a TAI contract in South America whereby a mobile antenna began its effective height trek at sea level and culminated at an altitude of 17,020 feet in the Andes Mountains.

A tantalizing factor requiring engineering address (not to mention slant-installed antennas in which intermediate polarization brings about transitional effective heights), is the vacillating effective height resulting from the erratic lashing of flexible spring-mounted mobile antennas.

Another interesting effect, called at TAI the "Nimbus Nemesis," is one in which the effective minimum height of an antenna operating in a sandy alkaline area, may vary between approximately 1 meter and some 15 meters, depending upon whether the ground is dry or drenched.

A puzzler is the apparent one in which a circularly polarized antenna "boasts" a variable (oscillatory) minimum effective height (a "tidal" swelling and ebbing of the "flux stratum" depth at radio frequency rate) depending upon the instantaneous direction (effective polarization) of the field's electric vector. At 12 MHz, for example, the effective minimum height varies an astounding 305 meters over sea water, 30 meters over good soil, and 7 meters over poor soil. This phenomenon gives rise to myriad practical applications such as radio-geological prospecting, ground constant metrology, and the TAI-dubbed, "Tide Diversity." Concomitant oscillatory radio energy flux intensity variation in the case of elliptical polarization (depending upon the polarization ellipse's spatial orientation) compounds this electromagnetic conundrum.

Courtesy of TAI Inc.

Appendix 30
Nature's Combiner

In borderline propagation cases, where both propagation mechanisms (diffraction and troposcat) contribute to the resultant received signal field, it is necessary to perform both the diffraction and troposcat (see Chapters 6 and 7 respectively) solutions and combine them. To aid in deciding whether a given path is diffractive or troposcat, the following rough guides, extant in the telecommunications field, may be used, remembering nonetheless that in case of any doubt, both the diffractive and tropospheric scatter solutions should be combined as per attached the curve and instructions herein.

Guide-1: If the scattering angle (or diffraction angle, or path angular distance), Θ, is equal to or less than $1.0°$ (0.02 radians, or 20.0 milliradians), calculate both the diffraction and tropospheric scatter path attenuations in accordance with Chapters 6 and 7 respectively. If the angle, Θ, is larger than $1.0°$. The tropospheric scatter propagation mode has been said to "usually" rule.

Guide-2: When the actual (great circle) path distance, d, or path angular distance, Θ, is large, the forward scatter path attenuation may be less than the diffraction attenuation. Therefore, when the product of the actual (great circle) path distance, d, in kilometers and path angular distance, Θ, in radians exceeds the value of 0.5, compute both the diffraction path loss and the troposcatter path loss and combine them via the curve and directions herein.

Guide-3: This is the BEST rule. In case of any doubt, compute both the diffraction and the troposcat path attenuation solutions and combine as per the curve and directions herein.

Let's take a couple of simple examples.

Example-1: The 99.99% propagational reliability path attenuation, as determined by the diffraction solution, is 176.0 dB, while the 99.99% propagation reliability path attenuation as determined by the tropospheric scatter solution is 178.0 dB. Note that these path attenuation values are taken as positive numbers. Simply subtract the troposcat path attenuation of 178.0 dB from the diffraction path attenuation of 176.0 dB, and this, of course, equals -2.0 dB. With this -2.0 dB difference value, enter the abscissa of the graph of Fig. A30-1. On the graph's ordinate, read correction factor of 1.8 dB. Now, all we do is subtract this 1.8 dB correction

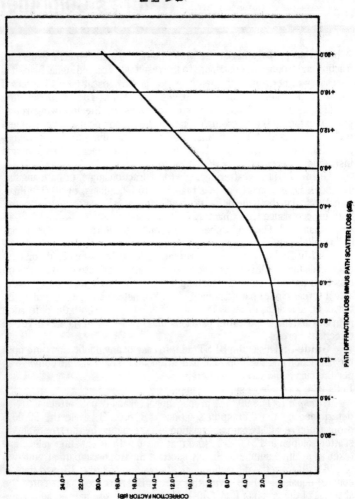

Fig. A30-1. Curve for determining in borderline propagational cases whether the mechanism is diffraction or troposcatter.

factor from the diffraction attenuation and we have our answer. The combined diffraction and tropospheric scatter path attenuation, not exceeded for 99.99% of the time, is (176.0 − 1.8) = 174.2 dB. (Be certain to always combine only *equal* diffraction and troposcat path reliability values, e.g. 90.0%, 99.0%, 99.9%, etc.).

Example-2: The 99.9% propagation reliability path attenuation, as determined by the diffraction solution, is 190.0 dB, while the 99.9% propagational reliability path attenuation, as determined by the tropospheric scatter solution is 170.0 dB. Note that these path attenuation values are taken as positive numbers. Simply subtract the troposcat path attenuation of 170.0 dB from the diffraction path loss of 190.0 dB. This equals (190.0 − 170.0) = + 20.0 dB. With this + 20.0 dB difference value, enter the abscissa of the graph of accompanying Fig. A30-1. On the graph's ordinate, read the correction factor, which just happens to be + 20.0 dB also. Now all we do is subtract this + 20.0 dB correction factor from the diffraction attenuation and we have our answer. The combined diffraction and tropospheric scatter path attenuation, not exceeded for 99.9% of the time, is then (190.0 dB − 20.0 dB) = 170.0 dB, which happens to be the troposcat attenuation value. (Be certain to always combine only equal diffraction and troposcat path reliability values, e.g. 90.0%, 99.0%, 99.9%, etc.).

Appendix 31
Aperture-to-Medium Coupling Loss

The role of an antenna is that of efficiently coupling a transmitter or receiver to the atmosphere (transmission medium). This coupling may be considered to occur at the "interface" of the effective antenna aperture and the medium.

EFFECTIVE APERTURE

In an elementary analogy, the antenna's effective aperture may be considered equal to that portion of a frontal area of a "passing wave" from which it could extract energy. This effective surface has, among other terms, been variously referenced as "capture area." Effective antenna aperture differs from physical aperture. It may be larger or smaller. Antennas such as high-Q parasitic arrays exhibit effective apertures greater than their corresponding physical apertures, while antennas exemplified by the paraboloid manifest effective apertures smaller than their physical counterparts.

ESSENCE OF APERTURE/MEDIUM COUPLING LOSS

The gain of a receiving antenna may be defined as the ratio of power absorbed by its aperture from a plane wave at an angle of incidence corresponding to its maximum response, to that power absorbed by an electrical isotrophy immersed in the identical electrical environment. Accordingly an antenna is subjected to a diminution in gain for plane wave components arriving from directions other than the one germane to its maximal directivity. Elaboration of this circumstance yields the concept that propagational modes, exemplified by tropospheric scatter, engender at a receiving antenna, in effect, an angular spectrum or distribution of equivalent planar components of random time-varying phase relations. Should the angular width of this dispersal exceed that of the antenna's pattern, the power collected ceases to be a linear function of the antenna's design aperture. This loss of gain is termed Aperture-To-Medium Coupling Loss, and it becomes evident that it is due to multipath propagation resulting in phase differences in various portions of the wave front incident upon a large aperture—the product being a time-dependent antenna gain-reduction.

APERTURE/MEDIUM COUPLING LOSS DECIBEL BUDGETING

This gain-loss phenomenon (a function of scatter angle, normalized horizon ray crossover height, and antenna beamwidth) does not readily succumb to a universally effortless quantitative solution. Levatich has stated that, except for VHF long paths and SHF short paths, it may be taken into account by a frequency/aperture product divided by distance. Other investigators have been somewhat less optimistic. Nevertheless, aperture coupling loss should be taken into proper propagational account. Exhaustive theoretical investigation and measurement have yielded some tractable mathematical tools for appraising this propagational parameter.

SOME ASTONISHING FACETS

It is to be stressed that, although scatter signals do not fade as markedly with smaller antennas and large antennas render greater signal peaks, aperture/medium coupling loss is considered a propagational effect, since even a very large antenna receiving iso-phased energy in the plane of its effective aperture does not exhibit this phenomenon. Another fascinating effect is the importance of including aperture/medium coupling loss in ionospheric E_s (Sporadic-E) paths. And, curiously enough, aperture/medium coupling loss occurs in WRH (Within Radio Horizon) as well as diffraction paths albeit to a lesser degree. Measurements of angle of arrival on WRH paths revealed departures of circa 0.75 degrees from the optimum direction. This variation is comparable to the beamwidth of a 40.0-dB-gain antenna and accounts for the limitation to this general value of antenna gains on WRH microwave paths.

RECAPITULATION

Scattered signals arrive at the receiving antenna's aperture with a phase incoherence in the aperture's plane, in which case narrow beam (large aperture) antennas do not deliver outputs commensurate with their theoretical gains. This gain-reduction is Aperture-To-Medium Coupling Loss.

Courtesy TAI Inc.

Appendix 32
Effective Distance

As an antenna's *structural* and *effective* heights may differ, there may exist an equally significant variance between a tropospheric radio path's *actual* and *effective* distances. Historically, three principal effective path distance concepts have, in turn, emerged as paramount propagational computation criteria. The first, *great circle path distance (d)*, is simply the interval (normally the shorter) between two points on the earth's surface. The second, *path angular distance* (θ_1 or θ_2), is the angle formed by the intersection of lines drawn from each antenna phase center through their respective horizons. The path *effective distance*, newest of the triad, being the least amenable to qualitative illustration, in addition to being further complicated by esoteric—sometimes contradicting—terminology and definition, is the one to which we shall herein attempt to lend some clarification. With this, then, let us venture into the didactics of various individuals and organizations regarding this "slippery" subject.

TECHNICAL NOTE NBS-101

This propagational "encyclopedia" published by the then Telecommunications Arm of the U.S. Bureau of Standards—now OT/ITS (Office of Telecommunications/Institute for Telecommunications Sciences)—states that *empirical* estimates of long-term power (attenuation) fading depends upon an *effective distance*, which has been thus far found a parameter superior to the aforementioned *great circle path length (d)*, as well as distance between *actual* horizons (d_1), that between *theoretical* horizons over a smooth earth (d_2), or the corresponding path *angular distances* (θ_1 or θ_2). This *effective distance* concept takes into consideration *great circle distance (d)*, *effective* antenna heights (h_{te} and h_{re}), and the operating frequency. This *effective distance* may be determined in the following manner. When the *great circle path (d)* is equal to or less than the qunatity, $3(\sqrt{2h_{te}} + \sqrt{2h_{re}}) + 65(100/f)^{1/3}$, the *effective distance* equals $130d/3(\sqrt{2h_{te}} + \sqrt{2h_{re}}) + 65(100/f)^{1/3}$.

When, however, the *great circle path distance (d) exceeds* the quantity $3(\sqrt{2h_{te}} + \sqrt{2h_{re}}) + 65(100/f)^{1/3}$, then the *effective distance* equals $130 + d - \{3(\sqrt{2h_{te}} + \sqrt{2h_{re}}) + 65(100/f)^{1/3}$. In the above, f is the operating

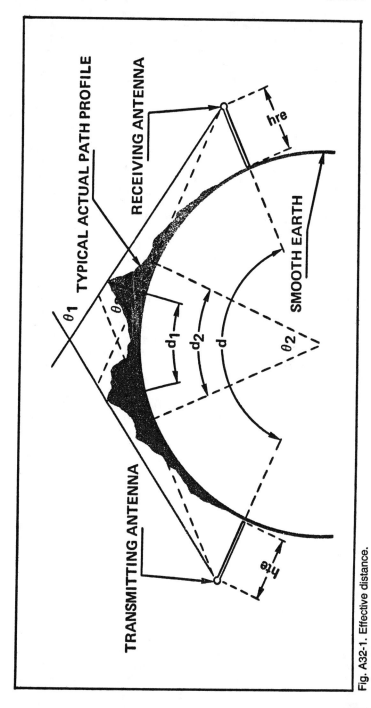

Fig. A32-1. Effective distance.

frequency in MHz, hte and hre are respectively the *effective* transmitting and receiving antenna heights in meters, and all distances are in kilometers. Incidentally, it was felt justifiable to use, in the illustration, the most elementary example of *effective* antenna heights *(structural* and *effective* heights equal) since this antenna subject, in addition to the probability of compounding confusion, is not central to the thrust of this paper. (Please see Consuletter-International Vol. 2 No. 1 June 1973 entitled "Antenna Height".)

OTHER VERSIONS

The CCIR, XII Plenary Assembly, New Delhi 1970, Vol. 2 Part 1 (Propagation in Non-Ionized Media), q.v., reiterates the above Technical Note NBS-101 information (page 91). However, page 92 mentions an *equivalent* distance defined as the product of the path angular distance and the effective earth radius of 8500 kilometers—this for the purpose of determining by means of graphs on pages 110 and 111, the loss not exceeded for 99% of the worst month and 1% of the year plus the standard deviation for various radioclimatologies. By a printing error (unfortunately further obscuring an already disarrayed situation), these graphs were entitled "effective distance." It will be appreciated (as graciously ceded by the then CCIR Director, Mr. Jack Herbstreit, in personal correspondence with the writer), *effective* and *equivalent* distances are not the same.

DuCastel of France speaks of an *equivalent distance* and defines it as the *path angular distance* multiplied by the effective earth radius.

As if matters weren't already inexorable, NBS Technical Note 98 (Synoptic Radio Meteorology) gives Gray's definition of *effective distance* (not equivalent distance) as the product of the *path angular distance* and the earth's radius modified for "normal" refraction.

Siemens of Germany makes use of a term called, *"Characteristic Scatter Distance"*, which is, as they put it, closely related to *path angular distance*, also known as *scatter angle* and is germane to long-term fading, propagational variation as a radio climatology function, and aperture-to-medium coupling loss. (Please see Consuletter-International Col. 2 No. 4 1974, entitled "Aperture-To-Medium Coupling Loss.") In short, this *"Characteristic Scatter Distance"* is essentially the *great circle path distance* minus the sum of the smooth-earth distances from each antenna to its radio-horizon. This would be (d₂) in the illustration and its geometric correspondence with angle (θ_2) will be apparent.

An important symposium on electromagnetic propagation (International Conference Sponsored By The Post And Telecommunications Group Of The Brussels Universal Exhibition) defined *effective distance* as was delimited *characteristic scatter distance* above.

IMPORTANCE

The moment of this distance matter may be appreciated by two dramatic ratios between *actual* and *effective* path distances occurring between an aircraft at 6400 meters altitude and ground—236.9/82.5 and 318.6/101.9 kilometers, while in another case the *actual path* to *characteristic scatter* proportion was 228.1/200.0 kilometers. That the net angle of tilt of the path's antennas influences *effective distance* lends at least some qualitative scope to the concept as well as another dramatic example—in this case, at 0° net antenna tilt angle, at true earth, the *effective* and *actual* path lengths were equal. At net tilt angles of minus and plus ninety minutes, however, the *effective distances* respectively were shorter and longer than the path actual distance, $(d - 160)$ and $(d + 240)$ kilometers.

While the *effective distance* phenomenon exists in other portions of the spectrum (e.g., ionospherically propagated HF band—3.0 to 30.0 MHz) we must succumb to the limitations of space and defer its treatment to a future issue.

Courtesy TAI Inc.

Appendix 33
Neither Snow nor Rain . . .

Radio propagation, like the postman, is beset by a formidable phylum of vexing atmospheric particles—rain, hail, snow, sleet, fog, sand, dust, smoke . . . even molecules and atoms. This vexation (attenuation), a function of particulate magnitude, concentration, refractive index and temperature, accrues from **scattering** and **absorption**. Although interrelated in a complex manner, it is instructive to view them as individual entities.

SCATTERING

This mechanism may be considered a **dispersal** brought about by **deflection** of electromagnetic energy from its intended target (the receiving antenna).

ABSORPTION

Consisting in two distinct modes (true and pseudo), this may be regarded as an energy dissipation.

True absorption occurs upon incident radio energy traversal of differing dielectric properties (e.g., air-to-rain). A portion of this radiation is absorbed **heating** the absorbing (lossy dielectric) particles, the process being enhanced as particle dimension and wave-length approach equivalence.

Pseudo-absorption is **dipole-moment** related, the **dipole** being an electrically or magnetically polarized particle. A passing electromagnetic wave "pumps" these atmospheric corpuscles into end-over-end rotation (or other "oscillatory" modes) by delivering (losing) to them discrete units of energy which are, in turn, re-radiated in a **random** reverse transition.

With this then, let us look at the various propagation plaguing "elements."

RAIN

Attenuation owing to this source, generally considered inconsequential in the 7.5 MHz microwave band and below, limits hops at 12.0 GHz to 15.0-25.0 kilometers in areas of enhanced rainfall incidence. Assessment of this loss by rate-of-fall methods, albiet a debatable practice (updraft rain U-turns, crosswind slant rains and the like exacerbating the anti-correlation between ground measured quantities and rain-in-the-path), takes advantage of this most available (and in a not-inconsiderable number of instances **only**) parameter. CCIR figures reflect attenuations of 0.01 to > 10.0 dB/Km across the 3.0 to 100.0 GHz spectrum for 0.25 to 100.0 Millimeter/Hour rainfall rates. Falls of 2.5 Millimeters/Hour might exert diminished effects upon millimetric waves while rates of 5.0 to 10.0 Millimeters/Hour could reduce their range to some 15.0 kilometers.

HAIL

Except for the millimeter wave region, attenuation by dry hailstones is generally small compared with that from raindrops of equivalent water content or precipitation rates. At 10.0 to 30.0 GHz hailstones sized $\cong$ 0.1λ attain cross-sections corresponding to rain when less than 10% of the stone is melted. At 10% to 20% melted mass the attenuation approaches twice that of the rain particle.

SNOW

This precipitation produces little dielectric attenuation compared with rain, even at fall rates of 15.0 Centimeters/Hour—this amount melting down to 1.5 Centimeters of water.

State-of-the-art measurements/calculations place snow attenuations for this rate of fall at $\cong$ 0.01 to 0.4 dB/Km over the 18.0 to 100.0 GHz frequency ranges.

FOG

Fog droplets act as small scatters/absorbers. At small drop size absorption is proportional to per-volume water content. It is considered convenient to estimate fog attenuation by visibility. Based upon Ryde's work Saxon and Hopkins have produced some dB/Km attenuation evaluation criteria. At 1° C typical visibility/frequency versus loss figures are as follows: **30 Meters**—3.0 GHz/0.02 dB, 9.4 GHz/0.2 dB, 24.0 GHz/1.25 dB; **90 Meters**—3.0 GHz/0.004 dB, 9.4 GHz/0.04 dB, 24.0 GHz/0.25 dB; **300 Meters**—3.0 GHz/0.001 dB, 9.4 GHz/0.007 dB, 24.0 GHz/0.045 dB. Fog attenuation assumes increasing importance at millimeter through laser wavelengths progressively limiting their ranges.

CLOUDS

Clouds may be divided into two categories—**water** and **ice**. **Water clouds** generally render attenuations similar to fog. **Ice clouds**, owing to dielectric property differences, yield attenuations approximately two orders of magnitude smaller than liquid clouds of the same water content.

ATMOSPHERIC GASES

The major energy debilitating "elements" in the 100.0 to 50,000 MHz span are water-vapor and oxygen with lines at 22,235 MHz and a series of frequencies centered about 60,000 MHz respectively (Van Vleck). The

combined water-vapor/oxygen attenuation at a temperature of 20° C and 10.0 Grams/Cubic Meter ≅ 0.2 to 5.0 dB from 100.0 MHz to 10.0 GHz at 600 kilometers range.

SAND

While sand particle diameters might be comparable to some raindrops, Ryde has pointed out that in **true** sandstorms sand particles rarely exist more than about two meters above the ground and it is therefore logical that in well designed radio links their effect should be minimal.

DUST

These particles, being finer than those of sand (10.0 microns or less), are able to remain suspended in the air for appreciable periods. Unless the dust particles are polarized, however, the effect is generally considered negligible except for such as laser links.

SMOKE

The average cross-section of smoke particles is in the neighborhood of 1.0 micron with a density less than 1.0 Gram/Cubic Meter. At these figures, coupled with relatively rare incidence, the effect should "normally" be minimal short of possible particle polarization.

Courtesy TAI Inc.

Index

Index

A

Absorbing	479
Absorption	463, 492
true	492
Antennas, very high	481
very low	481
Antenna parameters	120
Aperture, effective	486
Aperture/medium coupling loss,	
decibel budgeting	487
essence	486
Atmospheric gases	493
noise	64

B

Ball lightning	337
Bead lightning	337
Brownian motion noise	345

C

Circuit parameters	116
Clouds	493
Cold lightning	338
Combining	340
Cosmic noise	64
Critique	463

D

Dal segno	335
dB power addition	320
dB power, subtraction	320
Diffraction problems	216-222
propagation	190
Diffraction	
propagation, general rules	190
Diffraction propagation problems,	
operations	192-216
Digital microwave,	
advantages	468
disadvantages	469
it's future	469
wherefore	468
Digital modulation modes	468
Diversity	258
frequency	339
millimeter wave	258
polarization	339
space	339
Diversity reception	339
Ducting, millimeter wave	258
Dust	494

E

Earth return	463
Earth/space	328
Earthy pragmatics	479
Effective height, maximum	481
minimum	480
Excellent ground	478

F

Fading, millimeter wave	254
Field strength, examples	22-24
First aid	338

Fluctuation noise	345
Fog	257, 493
Forked lightning	337
Forward scatter, examples	146
problem	146
Forward scatter	
path calculations	224
procedure	224
Free space	327
Frequency compliment	116
diversity	339
Fresnel clearance	253

G

Gaussian noise	344
Good ground	478
Ground, excellent	478
grouping	478
wave problems	12-20

H

Hail	493
millimeter wave	258
Heat lightning	337
HF	479
future uses	348
uses	346
HF circuits, current	349
HF philosophy, current	349
HF propagation,	
general information	84
program control information	87
prediction by computer	82
HF calculation procedures	26
HF long path procedure	74
HF short path manual procedure	26
Hot lightning	338

I

Ionosphere	21
characteristics	21

J

Johnson noise	345

K

K-factor	252

L

Lightning, first aid	338
hazards	338
protection	338
types	337
Lightning's matrix	335

M

Man-made noise	64
Maximum effective height	481
MF through VLF	479
Microwave path selection,	
field phase	172
office phase	148
Millimeter wave, diversity	258
ducting	258
hail	258
snow	258
fading	254
fog	257
polarization	256
rain	255
windows	252
Millimeter wave path, design	258
Minimum effective height	480
Modus operandi	478

N

Natural limits	329
Noise, thermal	343

O

Okflpsk	478
Opsk	468
Outage, definition	467
facets	466
time	466
Outer space	328

P

Parabola, wherefore	355
Parabolic chart	355
characteristics	355
Partition noise	345
Phenomena, separate	329
Polarization	256
diversity	339
Poor ground	479
Pseudo-absorption	492

Q

Qam	468
Qpsk	468

R

Rain	492
millimeter wave	255
Random noise	344
Rayleigh fading	144
Recapitulation	487
Receiving antennas	142
Reflection	479
Refractivity	252
Resistance noise	345
Ribbon lightning	337

S

Sand	494
Scattering	492
Sheet lightning	338
SHF through VHF	479
Shot noise	344
Smoke	494
Snow	493
millimeter wave	258
Space diversity	339
Spectrum-segment exempt	327
State of the art	463
Streak lightning	337
Surface wave	9
System parameters	116

T

Terraquatic-constant, fundamentals	478
Terrestrial reflections	254
Thermal noise	343
Thundercell	335
Transmission	479
Transmitting antennas	142

W

Wherefore digital microwave parabola	468 355
Windows, millimeter wave	252